Inductive Powering

ANALOG CIRCUITS AND SIGNAL PROCESSING SERIES

*Consulting Editor*: **Mohammed Ismail.** ***Ohio State University***

For other titles published in this series, go to
www.springer.com/series/7381

Koenraad Van Schuylenbergh • Robert Puers

# Inductive Powering

## Basic Theory and Application to Biomedical Systems

Springer

Koenraad Van Schuylenbergh
3WIN nv
Ko Pa Electronics Consulting
ko.pa@telenet.be

Robert Puers
Department of Electrical Engineering
(ESAT-MICAS)
Katholieke Universiteit Leuven
Kasteelpark Arenberg 10
3001 Leuven
Belgium
robert.puers@esat.kuleuven.be

ISBN: 978-90-481-2411-4 e-ISBN: 978-90-481-2412-1

Library of Congress Control Number: 2009926322



Printed on acid-free paper

9 8 7 6 5 4 3 2 1

springer.com

# Preface

This book is based on research carried out at the Katholieke Universiteit Leuven, Belgium, on transferring data and power over poorly coupled inductive links with near theoretical energy efficiency. This largely mathematical study grew out of the frustration that the link circuits designed with methods available in literature did not perform as calculated. This is attributed to the poor coil coupling typically encountered in our applications. This triggered our eagerness to dive into the theory and find out why things went astray.

Inductive powering has been a reliable and simple method for years to wirelessly transfer power and data over short distances. This engineering discipline originated in biotelemetry with cochlear implants as the commercially most common application, and is now also widely applied in radio-frequency identification (RFID), with high-volume applications like wireless ID tags for asset tracking, key cards for building access control, electronic passports, implanted ID tags for husbandry control etc…

The field of inductive powering splits up along two orthogonal axis: the amount of transferred power and the coil coupling which directly relates to transfer range. Biotelemetry applications typically transfer a few milliwatts over a few centimetres. The RFID applications are characterised by transfer ranges up to half a metre, but run at microwatt power levels where energetic link efficiency is of little importance. The present study aims at transferring several milliwatts over longer ranges with near the theoretical efficiency limit by a careful optimisation of the driven inductive link (chapter 5) and by servo control of the driver output to the actual needs (chapter 6). Although this textbook emphasises applications with weak coil coupling, the described methods and formulae are universal to the complete range of coil coupling factors.

Six chapters make up this book:

- **Chapter 1** is a general introduction on telemetry to situate inductive coupling amongst its competitors.
- **Chapter 2** outlines the basics of inductive powering. It starts off with a summary of magnetic induction theory and then delves into the existing methods to design inductive links. This allows pin pointing the problems that crop up with poor coil coupling. It shows that the approximate formulae that are commonly used to optimise coil sets become invalid at low coupling and also that the design of the coil driver amplifier must be included in the optimisation process.
  This observation dictates the structure of the following chapters.
- A set of exact design formulae that describe the coil link is first derived in **chapter 3**.
- **Chapter 4** then studies the design of primary coil drivers.
- **Chapter 5** finally develops a step-by-step design procedure for driven inductive links that operate as close as possible to their theoretical efficiency limit.
- Basic inductive links may be finicky to handle in many real-world situations. **Chapter 6** therefore expands the optimised driven link with servo loops that automatically adjust the link circuit to changing coupling and load conditions.
- This book also provides three **appendices**. There is a short overview of vector mathematics. A second appendix rounds up some tips and tricks on coil models and measurements. The third appendix is a mathematical exposé on saturating-class-C amplifiers.

Koenraad Van Schuylenbergh
Robert Puers
January 2009.

# Symbols and units

| Symbol | Description | Unit |
|---|---|---|
| $=$ | is equal to | |
| $\equiv$ | is defined as | |
| $\approx$ | is approximately equal to | |
| $\propto$ | is linearly proportional to | |
| **Complex numbers** | | |
| $Re\{\underline{X}\}$ | The real part of the complex number $\underline{X}$ | |
| $Im\{\underline{X}\}$ | The imaginary part of the complex number $\underline{X}$ | |
| $\lvert\underline{X}\rvert$ | The magnitude of the complex number $\underline{X}$ | |
| $\angle\underline{X}$ | The phase angle of the complex number $\underline{X}$ | |
| **Electromagnetic fields** | | |
| $A$ | surface | m |
| $\vec{E}$ | electric field strength | V/m |
| $\vec{D}$ | electric displacement: $\vec{D} = \varepsilon\vec{E} = \varepsilon_0\vec{E} + \vec{P}$ | $C/m^2$ |
| $\vec{P}$ | medium polarisation | $C/m^2$ |
| $\vec{J}$ | electric current density | $A/m^2$ |
| $\rho$ | electric charge density | $C/m^3$ |
| $\Phi_B$ | magnetic flux | Wb $= T.m^2$ |
| $\vec{B}$ | magnetic flux density: $\vec{B} = \mu\vec{H} = \mu_0\left(\vec{H} + \vec{M}\right)$ | T $= \dfrac{N}{A.m}$ |
| $\vec{H}$ | magnetic field strength | A/m |
| $\vec{M}$ | medium magnetisation | A/m |
| $\sigma$ | medium conductivity | S/m $= \dfrac{1}{\Omega.m}$ |
| $\varepsilon$ | medium dielectric constant | F/m |
| $\mu$ | medium permeability | H/m |
| $\mu_r$ | relative medium permeability $\equiv \mu/\mu_0$ | - |
| $R_m$ | reluctance or magnetic resistance | A/Wb |

| Symbol | Description | Unit |
|---|---|---|
| **Waves** | | |
| $t$ | time | s |
| $\lambda$ | wavelength | m |
| $T$ | period | s |
| $f$ | frequency $= 1/T$ | Hz $= s^{-1}$ |
| $\omega$ | angular or radian frequency | rad/s |
| **Voltages** | | |
| $v(t)$ | a voltage signal | V |
| 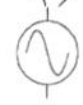 | an ideal sinusoidal voltage source | |
| $\hat{V}$ | peak amplitude of a sinusoidal voltage signal: $$v(t) = \hat{V}.\cos(\omega t + \varphi) = \mathrm{Re}\left\{\hat{V}.e^{j(\omega t+\varphi)}\right\} = \mathrm{Re}\left\{\underline{\hat{V}}.e^{j\omega t}\right\}$$ | V |
| $\underline{\hat{V}}$ | complex peak amplitude of a sinusoidal voltage signal: $$\underline{\hat{V}} = \hat{V}.e^{j\varphi}$$ | V |
| $V$ | RMS amplitude of a sinusoidal voltage signal: $$V = \hat{V}/\sqrt{2}$$ | V |
| $\underline{V}$ | complex RMS amplitude of a sinusoidal voltage signal: $$\underline{V} = \underline{\hat{V}}/\sqrt{2} = V.e^{j\varphi}$$ | V |
| $\varphi$ | phase of a sinusoidal voltage signal | rad |
| **Currents** | | |
| $i(t)$ | a current signal | A |
|  | an ideal sinusoidal current source | |
| $\hat{I}$ | peak amplitude of a sinusoidal current signal: $$i(t) = \hat{I}.\cos(\omega t + \psi) = \mathrm{Re}\left\{\hat{I}.e^{j(\omega t+\psi)}\right\} = \mathrm{Re}\left\{\underline{\hat{I}}.e^{j\omega t}\right\}$$ | A |
| $\underline{\hat{I}}$ | complex peak amplitude of a sinusoidal current signal: $$\underline{\hat{I}} = \hat{I}.e^{j\psi}$$ | A |
| $I$ | RMS amplitude of a sinusoidal current signal: $$I = \hat{I}/\sqrt{2}$$ | A |
| $\underline{I}$ | complex RMS amplitude of a sinusoidal current signal: $$\underline{I} = \underline{\hat{I}}/\sqrt{2} = I.e^{j\psi}$$ | A |
| $\psi$ | phase of a sinusoidal current signal | rad |
| **The primary coil** | | |
| $i_1(t)$ | primary coil current (through the ideal coil of the series R-L-C model) | A |
| $v_1(t)$ | primary coil voltage (across the ideal coil of the series R-L-C model) | V |
| $n_1$ | number of primary coil windings | - |
| $r_1$ | primary coil radius | m |
| $L_1$ | primary coil inductance (R-L coil model) | H $= \frac{V.s}{A}$ |
| $R_{L_1}$ | parasitic primary coil resistance (R-L coil model) | Ω $= V/A$ |
| $Q_{L_1}$ | unloaded primary coil quality factor (R-L coil model) $$Q_{L_1} = \frac{\omega L_1}{R_{L_1}}$$ | - |
| $L_{S_1}$ | primary coil inductance (series R-L-C coil model) | H |
| $R_{S_1}$ | Parasitic primary coil resistance (series R-L-C coil model) | Ω |
| $C_{S_1}$ | Parasitic primary coil capacitor (series R-L-C coil model) | F |
| $Q_{L_{S_1}}$ | unloaded primary coil quality factor (series R-L-C coil model) $$Q_{L_{S_1}} = \frac{\omega L_{S_1}}{R_{S_1}}$$ | - |

| Symbol | Description | Unit |
|---|---|---|
| **The secondary coil** | | |
| $i_2(t)$ | secondary coil current (through the ideal coil of the series R-L-C model) | A |
| $v_2(t)$ | secondary coil voltage (across the ideal coil of the series R-L-C model) | V |
| $n_2$ | number of secondary coil windings | - |
| $r_2$ | secondary coil radius | m |
| $L_2$ | secondary coil inductance (R-L coil model) | H |
| $R_{L_2}$ | parasitic secondary coil resistance (R-L coil model) | Ω |
| $Q_{L_2}$ | secondary coil quality factor (R-L coil model) $$Q_{L_2} = \frac{\omega L_2}{R_{L_2}}$$ | - |
| $L_{s_2}$ | secondary coil inductance (series R-L-C coil model) | H |
| $R_{s_2}$ | parasitic secondary coil resistance (series R-L-C coil model) | Ω |
| $C_{s_2}$ | parasitic secondary coil capacitor (series R-L-C coil model) | F |
| $Q_{L_{S_2}}$ | secondary coil quality factor (series R-L-C coil model) $$Q_{L_{S_2}} = \frac{\omega L_{S_2}}{R_{L_{S_2}}}$$ | - |
| **The coupling** | | |
| $d$ | distance between the coil planes for parallel coils | m |
| $M$ | mutual inductance | H |
| $k$ | coil coupling factor (also called coupling coefficient) $$k = \frac{M}{\sqrt{L_{s_1}.L_{s_2}}}$$ | - |
| $n$ | square root ratio of the coil inductances $$n = \sqrt{\frac{L_{s_2}}{L_{s_1}}}$$ | - |
| **Loads** | | |
| $R_{out}$ | d.c. load at the regulated output of the inductive link | Ω |
| $R_{d.c.}$ | equivalent d.c. resistance of the loaded regulator | Ω |
| $R_{load_2}$ | equivalent a.c. resistor of the rectifier connected to the loaded regulator | Ω |
| **Link voltages** | | |
| $V_{prim}$ | a.c. link input voltage | V |
| $V_{sec}$ | a.c. link output voltage (= voltage across $R_{load_2}$) | V |
| $V_{d.c.}$, $I_{d.c.}$ | the rectified but non-regulated d.c. output of the driven inductive link (= voltage across and current through $R_{d.c.}$) | V, A |
| $V_{out}$, $I_{out}$ | the rectified and regulated d.c. output of the driven inductive link (= voltage across and current through $R_{out}$) | V, A |
| **The primary tank** | | |
| $\omega_{tank}$ | resonance frequency of the primary tank, coupled to the secondary (i.e. amplitude resonance for saturated-class-C drivers and phase resonance for all other drivers). | rad/s |
| $Q_{prim}$ | the quality factor of the primary tank, connected to the driver amplifier, but not coupled to the secondary (including coil losses, on-resistance of the driver transistors, resistive losses in capacitors,...) | - |
| $Z_{eq}$ | the equivalent impedance of the secondary transformed to the primary (series R-L-C model for the primary coil) | Ω |
| $R_{eq}$ | $Z_{eq}$ that has become real at the phase-resonance frequency of the secondary tank (series R-L-C model for the primary coil) | Ω |
| $R^*_{s_1}$ | series resistance of the primary coil coupled to the secondary circuit (R-L-C coil model) $$R^*_{s_1} = R_{eq} + R_{s_1}$$ | Ω |

| Symbol | Description | Unit |
|---|---|---|
| $Q^*_{L_{S_1}}$ | quality factor of the primary coil coupled to the secondary circuit (series R-L-C coil model) $$Q^*_{L_{S_1}} = \frac{\omega L_{S_1}}{R^*_{S_1}} = \frac{\omega L_{S_1}}{R_{eq} + R_{S_1}}$$ | - |
| $R^*_{L_1}$ | series resistance of the primary coil coupled to the secondary circuit (R-L coil model) | Ω |
| $Q^*_{L_1}$ | quality factor of the primary coil coupled to the secondary circuit (R-L coil model) $$Q^*_{L_1} = \frac{\omega L_1}{R^*_{L_1}}$$ | - |
| $C_1$ | primary tank capacitor (class C and D drivers) | F |
| $C_{1ser}$ | primary series capacitor in a class-E driver | F |
| $C_{1par}$ | primary parallel capacitor in a class-E driver | F |
| $C_{1res}$ | resonance capacitor in a (semi-) resonant class-E driver | F |
| **Secondary tank** | | |
| $\omega_{resP}$ | phase-resonance frequency of the uncoupled secondary tank | rad/s |
| $\omega_{resA}$ | amplitude-resonance frequency of the uncoupled secondary tank (i.e. maximal voltage across $R_{load_2}$) <br> $\omega_{resP} = \omega_{resA} \equiv \omega_{res}$ for a series-tuned secondary | rad/s |
| $Q_{sec}$ | quality factor of the loaded, but uncoupled secondary tank | - |
| $C_2$ | secondary tank capacitor | F |
| **The inductive-link driver** | | |
| $V_{cc}$, $I_{cc}$ | the d.c. supply voltage and current of the primary coil driver for the first-order-simplified driver model | V, I |
| $R_{cc}$ | $$R_{cc} \equiv \frac{V_{cc}}{I_{cc}}$$ | Ω |
| $V^*_{cc}$ | real supply voltage that also accounts for the driver losses $$V^*_{cc} \equiv \frac{V_{cc}}{\eta_{driver}}$$ | V |
| $v_{AE}$, $v_S$ | the voltage across the active element or switch of the primary coil driver | V |
| $i_{AE}$, $i_S$ | the current through the active element or switch | A |
| $p$ | the power output capability of the primary coil driver $$p \equiv \frac{P_{link\ in}}{v_{AE_{max}} i_{AE_{max}}} \text{ or } p \equiv \frac{P_{link\ in}}{v_{S_{max}} i_{S_{max}}}$$ | - |
| $D$ | switch duty cycle $$D \equiv \frac{t_{ON}}{t_{ON} + t_{OFF}}$$ | - |
| $t_R$, $t_F$ | switch rise and fall times | s |
| $V_{Ssat}$ | switch saturation voltage (i.e. the extrapolation of the switch voltage-current curve to a zero current) | V |
| $R_{ON}$ | switch-on resistance: $v_S = V_{Ssat} + R_{ON}\, i_S$ | Ω |
| $L_S$ | parasitic series inductance of the switch | H |
| $P_{t_R}$, $P_{t_F}$ | switch dissipation due to non-zero rise and fall times | W |
| $P_{V_{Ssat}}$ | switch dissipation due to non-zero saturation voltage | W |
| $P_{R_{ON}}$ | power dissipation in the switch-on resistance | W |
| $\zeta$ | normalised slope of the switch voltage at turn-on ($\zeta = 0$ for class E and $< 0$ for saturated class C) $$\zeta \equiv \frac{1}{\omega V_{cc}} \left. \frac{dv_S(t)}{dt} \right\|_{turn-on}$$ | - |

| Symbol | Description | unit |
|---|---|---|
| **Link optimisation** | | |
| $\alpha$ | the ratio of $R_{load_2}$ and the impedance of the capacitor $C_2$ <br> $\alpha = \omega.C_2.R_{load_2}$ | - |
| $X$ | abbreviation for $k^2 Q_{L_{S_1}} Q_{L_{S_2}}$ | - |
| $P_{in}$ | d.c. input power of the driven inductive link | W |
| $P_{link\ in}$ | a.c. power into the coil set $= P_{in}.\eta_{driver}$ | W |
| $P_{sec}$ | total power delivered to the secondary $= P_{link\ in}.\eta_{primary}$ | W |
| $P_{link\ out}$ | useful a.c. power output of the coil set $= P_{link\ in}.\eta_{link}$ | W |
| $P_{out}$ | useful d.c. power output of the driven inductive link $= P_{in}.\eta$ | W |
| $\eta_{primary}$ | primary efficiency = ratio of the power transferred to the secondary circuit and the power put into the primary coil | - |
| $\eta_{secondary}$ | secondary efficiency = ratio of the power dissipated in the load $R_{load_2}$ and the total dissipation of the secondary circuit | - |
| $\eta_{link}$ | link efficiency: $\eta_{link} = \eta_{primary}.\eta_{secondary}$ | - |
| $\eta_{rectif}$ | rectifier efficiency | |
| $\eta_{regul}$ | regulator efficiency | - |
| $\eta_{driver}$ | primary coil-driver efficiency $= \frac{P_{link\ out}}{P_{in}}$ | |
| $\eta$ | global efficiency of the driven inductive link <br> $\eta = \eta_{driver}.\eta_{link}.\eta_{rectif}.\eta_{regul}$ | - |

| Symbol | Description | Value |
|---|---|---|
| $\mu_0$ | permeability constant (in vacuum) | $4\pi\ 10^{-7}$ H/m |
| $\varepsilon_0$ | permittivity constant (in vacuum) | $8.85419\ 10^{-12}$ F/m |
| $c$ | speed of light (in vacuum) | $2.99792\ 10^{8}$ m/s |

| Abbr. | Description |
|---|---|
| a.c. | alternating current (which means a non-zero signal frequency) |
| ADC | analogue-to-digital converter |
| AM | amplitude modulation |
| ASK | amplitude shift keying |
| CW | continuous wave |
| DAC | digital-to-analogue converter |
| d.c. | direct current (which refers to a zero frequency signal) |
| EM | electromagnetic |
| EMI | electromagnetic interference |
| FM | frequency modulation |
| FSK | frequency shift keying |
| MO | master oscillator |
| MOPA | master-oscillator-power-amplifier configuration |
| PA | power amplifier |
| PCM | pulse code modulation |
| PDM | pulse duration modulation (a synonym for PWM) |
| PFM | pulse frequency modulation |
| PPM | pulse position modulation |
| PWM | pulse width modulation |
| r.f. | radio frequency |
| RMS | root mean square |
| RX | receiver |
| THD | total harmonic distorsion |
| TX | transmitter |
| VFO | voltage-controlled-frequency oscillator |

# Table of contents

# 1 An introduction on telemetry

The ever-growing technical world surrounding us would be unimaginable without ***connections; pathways for information and/or power from one site to another***. The mutual distances range from micrometers for on-chip connections, to thousands of kilometres for satellite communications. Connections play a massive role in our world. Millions of kilometres of electrical wires and optical fibres are installed worldwide annually, in houses, cars and industrial plants, in electric and electronic appliances, as computer networks or power-grid lines.

The notion of connections as information carriers is not restricted to electrical or optical links. Shafts and pneumatic lines can be seen as their mechanical brethren. The concept of connective systems is not unique to the engineering world either. Our hormone system, for instance, is a beautiful example of a biochemical information network. This system is even matched to a power distribution network that transports adenosine-triphosphate (ATP) via the blood through the body and energises the muscle tissue. In fact, mankind itself is built on communication and connection. Every human interaction is an information exchange through speech, touch, writing, eye contact or body language.

This book studies magnetic induction as a connection medium for short-range telemetry. Telemetry is the engineering branch that occupies itself with wirelessly measuring parameters of objects. The remote telemetry unit is often small compared to the object being measured, in order not to obstruct the object's functionality. The present introduction first reviews common telemetry techniques to situate magnetic induction amongst its competitors and to outline the various data modulation methods available. Examples are often taken from the biomedical field because that's where telemetry originally started. This, nonetheless, doesn't restrict the methods to this field only. RFID (radio frequency identification) is an example of a booming market that shares many operation principles with telemetry.



K. Van Schuylenbergh and R. Puers, *Inductive Powering: Basic Theory and Application to Biomedical Systems*, Analog Circuits and Signal Processing, 1–39. 

## 1.1 WIRELESS CONNECTIONS

The wireless approach is often dictated by boundary conditions between the connected sites that obstruct hard-wire cabling. Radio frequency broadcasting of radio and television are the most common examples, but the list of applications is virtually unlimited: satellite telephone, cellular telephone (GSM, CDMA, 3G, etc.), global positioning satellite (GPS), wireless monitoring of humans and animals (MACKAY, 1970, 1993; AMLANER and MACDONALD, 1980; JEUTTER, 1983a). The obstructions can be various:

- A large mutual distance that makes a cable connection uneconomical (e.g. telephone lines to remote regions) or even not feasible (e.g. connection to a satellite).
- A physical barrier that is too tough to penetrate or that looses its protective function once it perforated (e.g. in situ parameter monitoring in a chemical vessel or in the human body).
- An unknown location of the remote system (e.g. wildlife monitoring).
- The fact that the remote system is moving. Moving cables tend to fail after a while due to metal fatigue. Slip contacts on rotating drums or shafts usually show a poor reliability. The contacts are noisy and their resistance changes with time.

Wireless information transmission requires a local energy source at both sites. It may, however, not always be feasible to connect the remote unit to the mains, especially when it is small and or portable. Sometimes, even batteries can be unwanted because of their size, limited lifetime, toxicity e.g. when used in the human body, or temperature hazard when a circuit fails (e.g. in an explosive environment). This explains the need to have the base unit supply the remote system wirelessly in certain applications.

The wireless communication and/or energising methods are classified in three groups, based on how they physically establish the connection:

- Most wireless links are based on wave propagation:
  - Radio frequency (r.f.) links and optical links both rely on propagating electromagnetic waves (Figure 1.1). The radio frequencies are defined between 30 kHz and 300 GHz, whereas light ranges from far infrared ($10^{12}$ Hz) to ultraviolet ($10^{17}$ Hz).
  - The ultrasound links use acoustical waves with typical frequencies from some 10 kHz to some 10 MHz.
- A second group establishes the connection between both units, via electrical conduction through their intermediate medium.
- The third group couples the units with an alternating but non-radiating, quasi-stationary field:
  - The inductive links use magnetic coupling between two coils; one at each site. Typical frequencies are in the range of 1 kHz–100 MHz.
  - The capacitive links are established via an electrostatic field.[1]

[1] Capacitive links are only mentioned here for the sake of completeness as they only cover millimetre-range distances. However, the principle has been commercialised successfully in isolation amplifiers for the data transfer between the isolated circuit parts (TEXAS INSTRUMENTS, 2005).

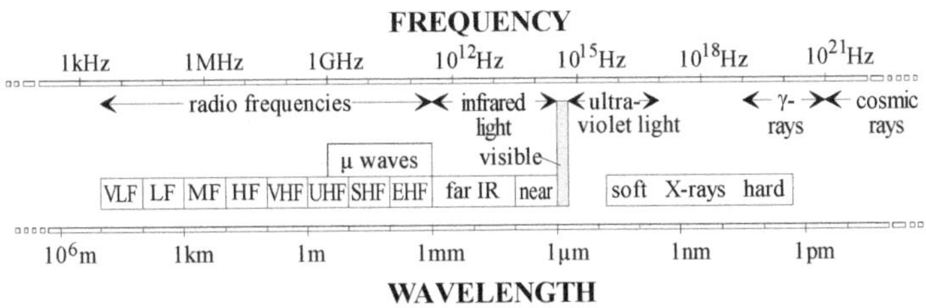


*The electromagnetic spectrum (HALLIDAY and RESNICK, 1977).* *Figure 1.1.*

*THE ELECTROMAGNETIC SPECTRUM (SHANMUGAN, 1979).* *TABLE 1.1.*

| Designation | Frequency | Wavelength | Medium | Applications |
|---|---|---|---|---|
| Very low frequency (VLF) | 3–30 kHz | 100–10 km | Wire pairs, long-wave radio | Audio, telephone, computer terminals, long range and undersea navigation, timing standards |
| Low frequency (LF) | 30–300 kHz | 10–1 km | Wire pairs, long-wave radio | Navigation, AM broadcasting (150–300 kHz), industrial (power line) communication |
| Medium frequency (MF) | 0.3–3 MHz | 1–0.1 km | Coaxial cable, long wave radio | AM broadcasting (540–1,600 kHz) |
| High frequency (HF) | 3–30 MHz | 100–10 m | Coax. cable, short-wave radio | AM broadcasting (6–30 MHz), CB and amateur radio, navigation and aviation |
| Very high frequency (VHF) | 30–300 MHz | 10–1 m | Coaxial cable, short-wave radio | Television band I (40–70 MHz) and III (175–225 MHz), FM broadcasting (88–108 MHz), air traffic control, amateur radio, police, fire brigade,... |
| Ultra high frequency (UHF) | 0.3–3 GHz | 1–0.1 m | Waveguides, μ-wave radio | Television band IV (450–600 MHz) and V (600–960 MHz), GSM, mobile radio, beamed transmission, radar |
| Super high frequency (SHF) | 3–30 GHz | 100–10 mm | Waveguides, μ-wave radio | Beamed transmission, satellite and space communication, radar |
| Extremely high frequency | 30–300 GHz | 10–1 mm | Waveguides, μ-wave radio | Navigation radar, radio astronomy, |
| Light (near IR .. UV) | $10^{14}$–$10^{16}$ Hz | 3 μm–30 nm | Optical fibres | Optical-fibre digital links |

Different link principles are sometimes combined in a single application (HOF *et al.*, 1994): e.g. inductive powering together with an r.f. (FORSTER, 1986) or an infrared data link (MITAMURA *et al.*, 1990).

*TABLE 1.2. A CLASSIFICATION OF MODULATION TECHNIQUES.*

| | |
|---|---|
| Continuous wave carrier modulation (CW): | • Amplitude-type modulation<br>– Amplitude modulation (AM)<br>– Double sideband modulation (DSB)<br>– Single sideband modulation (SSB)<br>• Frequency modulation (FM)<br>• Phase modulation (PM)<br>+ Combinations by subcarrier modulation |
| Pulse modulation | • Analogue pulse modulation:<br>– Pulse-position modulation (PPM)<br>– Pulse-width modulation (PWM)<br>– Pulse-frequency modulation (PFM)<br>• Digital pulse-code modulation (PCM) |

## 1.1.1 MODULATION

A steady radio frequency, acoustic or optical signal doesn't contain a lot of information. Directional antennas can estimate a transmitter's direction from the received signal intensity. Two separate direction-sensitive receivers can point the transmitter's position through triangulation. This makes steady transmission useful for, e.g. animal radio tracking, but that's about it! The transfer of "more intelligent" data requires modulation.

***Modulation is the systematic variation of some attribute of a carrier waveform such as the amplitude, phase or frequency in accordance to the data signal.*** Despite the multitude of modulation techniques, two basic modulation types can be identified: the continuous wave (CW) or analogue modulation where the carrier attribute is modulated in a continuous fashion, and the pulse modulation where the carrier attribute is changed in a discrete abrupt manner (Table 1.2; BLACK, 1953; SCHWARTZ, 1970; SHANMUGAN, 1979; JEUTTER, 1983b).

### 1.1.1.1 Continuous wave carrier modulation

Three modulation types constitute the family of continuous wave-carrier modulations: amplitude modulation (with carrier or with suppressed carrier), frequency and phase modulation.

Amplitude modulation (AM) is a technique where the data signal directly controls the transmitter amplitude: when the data signal goes up, the transmitter output is increased accordingly (Figure 1.2):

$$X_{AM}(t) = [1 + m.X(t)].\cos(\omega_c.t)$$

The data signal $X(t)$ is supposed here to be normalised to a maximal amplitude of one. The modulation index $m$ should be smaller than one to avoid envelope distortion. The carrier frequency $\omega_c$ is at least a decade larger than the highest data frequency so that a simple envelope detector can reconstruct the original data signal (Figure 1.3).

The carrier term in an AM signal contains no information, but represents a substantial component in the signal power. One can thus economise on transmitter power by suppressing this carrier. This kind of amplitude

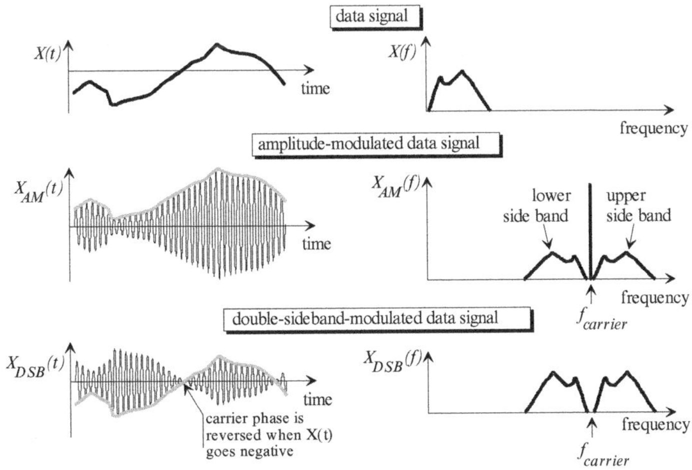


*Amplitude modulation and double-sideband modulation.* *Figure 1.2.*

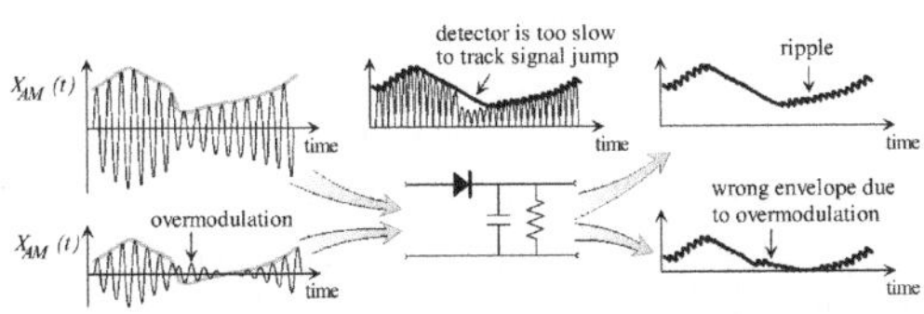


*The principle of envelope detection for amplitude demodulation and the potential distortion causes.* *Figure 1.3.*

modulation with suppressed carrier is called double-sideband modulation (DSB) and is accomplished by multiplying the data signal with a carrier[2]:

$$X_{DSB}(t) = X(t).\cos(\omega_c t)$$

The modulated data is recovered by multiplication with a coherent or synchronous carrier and low-pass filtering:

$$X_{recov.}(t) = X_{DSB}(t).2\cos(\omega_c t) = X(t).2\cos^2(\omega_c t) = X(t) + \underbrace{X(t).\cos(2\omega_c t)}_{\text{rejected by low-pass filtering}}$$

The major problem in the data recovery is to have a coherent local carrier in the receiver. The original carrier is not contained in the DSB signal so it should be generated in the receiver. The higher circuit complexity involved in DSB and SSB rejects them for telemetry applications.

2 DSB modulation still leaves some redundancy, as both side bands are instances of the same information. The next power-economising step is to suppress one of both sidebands to get single sideband modulation (SSB). This, however, further increases the circuit complexity and is therefore never used for small-sized telemetry units.

*TABLE 1.3.* *COMPOSITE AMPLITUDE - AND FREQUENCY-MODULATION TECHNIQUES.*

| | submodulated signal | AM main carrier | FM main carrier |
|---|---|---|---|
| | | AM | FM |
| without subcarrier |  (= input signal) | 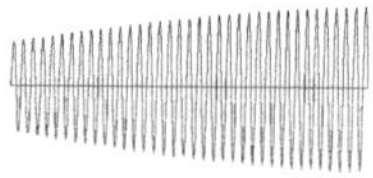 | 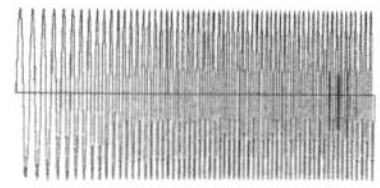 |
| | | AM-AM | AM-FM |
| AM sub-carrier | 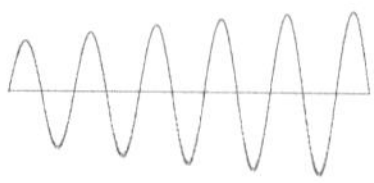 |  | 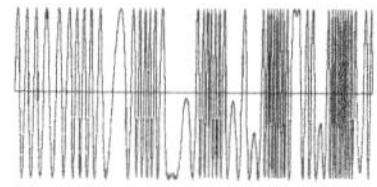 |
| | | FM-AM | FM-FM |
| FM sub-carrier | 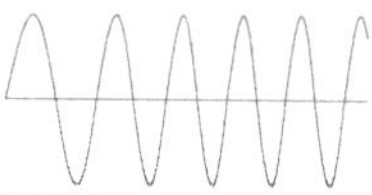 | 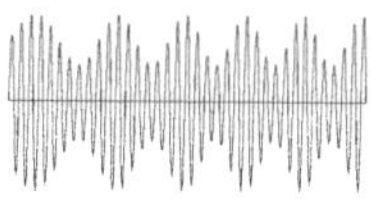 | 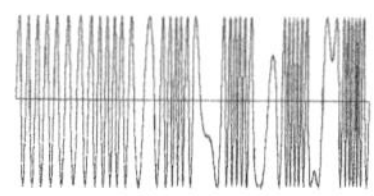 |

Neither AM, nor DSB or SSB can faithfully reproduce slowly varying data or absolute data values, as the reconstructed-data amplitude directly depends on the strength of the received modulated signal. Amplitude fading due to movements of mobile transmitters or changing atmospheric conditions is incorrectly interpreted as a declining data signal.

Frequency modulation (FM) and phase modulation (PM) are both insensitive to such amplitude changes, because they translate the data signal *X(t)* in a frequency shift or a carrier phase shift:

$$X_{FM}(t) = \cos\left[\left[1 + m.X(t)\right].\omega_c.t\right] \quad \text{and} \quad X_{PM}(t) = \cos\left[\omega_c t + \pi.m.X(t)\right]$$

FM and PM are similar in functional form, except for the integration of the data signal in FM:

$$X_{FM}(t) = \cos\left(\omega_c t + m.X(t).\omega_c t\right) = \cos\left[\omega_c t + m.\omega_c \int_{-\infty}^{t} X(\tau).d\tau\right]$$

FM and PM signals are produced by a voltage-controlled-frequency oscillator (VFO) driven by the data signal or its time derivative. The data recovery is performed by the opposite sequence or by a phase-locked-loop circuit (PLL) (SHANMUGAN, 1979, pp. 289–297).

Simultaneous transmission of several data channels through the same medium is done by assigning one carrier frequency to each data channel (frequency multiplexing). As a consequence, increasing the number of output channels multiplies the number of transmitters and the corresponding energy consumption (Figure 1.4a). This becomes particularly problematic with small, battery-powered telemetry units. As the transmitter consumption dominates the overall power drain, this is the place to economise. Furthermore, the small size also limits the circuit complexity.

Subcarrier modulation is a technique to share one carrier frequency and transmitter with several data channels (Table 1.3; Figure 1.4b). Each data signal is first either amplitude or frequency modulated on an individual subcarrier. These contributions are then summed into a composite signal that is then amplitude or frequency modulated on an r.f. carrier (KLEIN

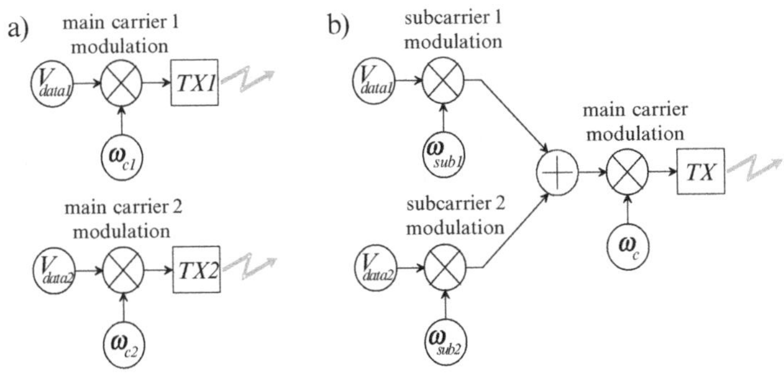


*Two-channel data transmission, without (a) and with subcarrier modulation (b).* *Figure 1.4.*

and DAVIS, 1976; PITSILLIDES *et al.*, 1992). Submodulation is thus also a form of frequency multiplexing, but it takes place on a lower, subcarrier, level.

It should be noted that subcarrier modulation pushes the transmitter frequencies to higher values. The subcarrier frequencies should be at least one decade larger than the highest signal frequency and well separated from each other to allow for easy channel filtering at the receiver. The carrier frequency is in its turn, another one or more decades higher than the maximal subcarrier frequency.

#### 1.1.1.2 Analogue pulse modulation

***Pulse modulation*** involves ***the discrete variation of a carrier attribute (amplitude, frequency or phase) at discrete time intervals corresponding to the modulation pulses.*** The modulation pulses actually represent a form of analogue sampling, which implies that the pulse frequency should be at least twice the upper signal frequency according to Nyquist's Theorem. The pulse frequency is limited at the upper side to about a twentieth of the carrier frequency so that each pulse takes at least about 10 carrier periods.

Three analogue-pulse modulations are distinguished (Table 1.4):

- Pulse-position modulation (PPM)) is a format of consecutive pulse pairs repeated at a fixed rate. Each pair represents a data sample: the first pulse is synchronised to a clock and the second pulse is then positioned such that the time between both is proportional to the momentary data value.
- Pulse-width modulation (PWM), also known as pulse-duration modulation (PDM), is another synchronous sampling method. Each sample value is coded as the width of a single pulse. The rising edges of the consecutive pulses are synchronised to a clock.
- Pulse-frequency modulation (PFM)[3] is an asynchronous format of single pulses. The time between consecutive pulses is proportional or inversely proportional to the momentary data value (both exist) (SANSEN and PUERS, 1984; SANSEN *et al.*, 1984a, b; PUERS and SANSEN, 1985). The sampling frequency thus varies with the data values.

[3] Also often erroneously named pulse-position modulation.

*TABLE 1.4.* *ANALOGUE PULSE MODULATION.*

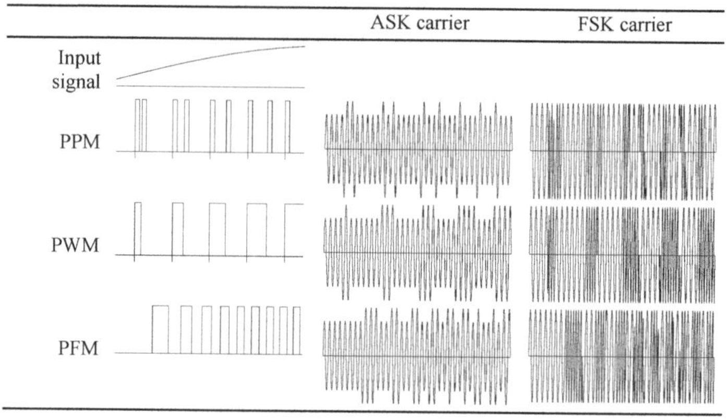


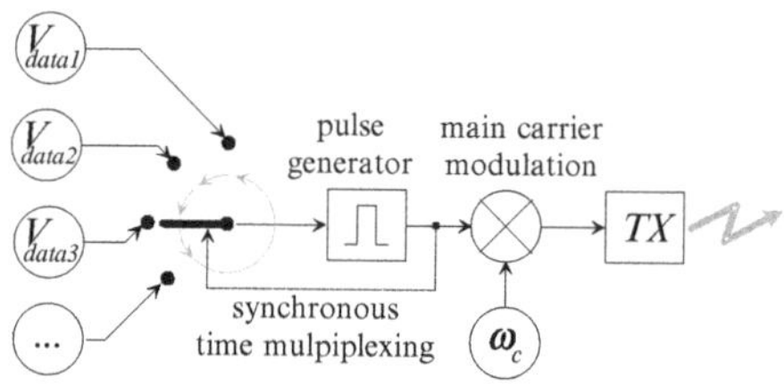


*Figure 1.5.* *Time multiplexing, synchronous with a pulse subcarrier.*

Amplitude-modulating the data pulses on an r.f. carrier is often referred to as amplitude shift keying (ASK) because the carrier amplitude is shifting between two distinct values (a pulse-on value and a pulse-off value). Pulse modulation on an FM or PM carrier is named frequency and phase shift keying (FSK and PSK) for the same reason.

The special kind of ASK where the transmitter is switched fully on and completely off ($m = 1$) by a pulse-modulated signal with a low duty-cycle, is very popular for battery-powered telemetry. The low duty cycle allows the transmitter to run on a higher peak power and hence a larger range for the same average power consumption (LIN and PILLAY, 1974).

Frequency multiplexing was pointed out before as a way to put multiple channels on a single carrier. Time division multiplexing is second way to share a single carrier frequency with multiple data channels. This method combines very well with pulse modulation. The transmitter input is switched from one data channel to the next one after a certain time, e.g. after each data pulse from the pulse-submodulator (Figure 1.5; EVANS, 1989). It may be useful to include a synchronisation pulse at the start of each channel scan to identify the first data channel and to synchronise the receiver with the switching pattern.

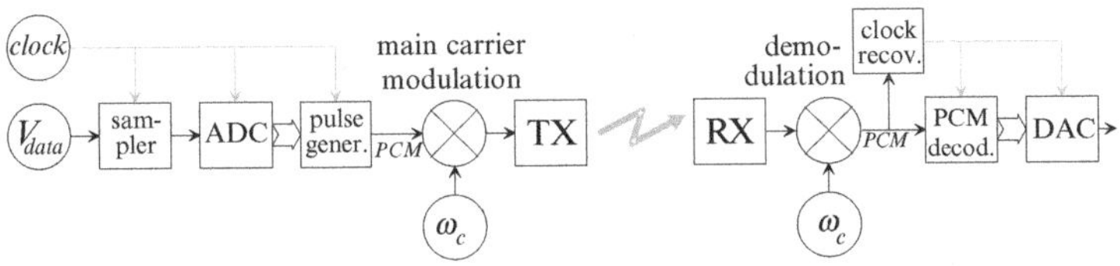


*Digital pulse-code modulation* *Figure 1.6.*

### 1.1.1.3 Digital pulse-code modulation

Digital data transmission can be seen as a natural extension of the above concept of data sampling before transmission. The additional step is that the samples are transmitted in a digitally encoded rather than an analogue format. The transfer procedure then becomes: data sampling, sample conversion to an integer number and transmission of that number as a coded pulse train on an ASK, FSK or PSK carrier (Figure 1.6; CATTERMOLE, 1969). Extra control codes can be inserted in the data stream for error checking, receiver synchronisation or even complete handshake protocols. This conversion procedure from input signal to digital-encoded pulses is often named pulse-code modulation (PCM).

The many different ways to encode a number in a coded pulse train all start from the digital representation of the integer:

$$number = \ldots + digit_n * base^n + \ldots + digit_2 * base^2 + digit_1 * base + digit_0$$

Binary encoding is the simplest method and is therefore preferred for small and lightweight telemetry units. The base value is 2 and there are 2 digits (bits), 0 or 1. Each bit can be translated "as is" to a high level for a "1" or a low level for a "0". A third "nil" level fills the time in between the bits (Figure 1.7 and Figure 1.8).

A binary encoded signal is easy to decode but, as the up and down pulses are not balanced, the method is only suitable for transmission techniques with a good transfer quality for d.c. and low frequency signals. A bipolar encoded pulse signal has no d.c. content (Figure 1.8, 2nd example, SHANMUGAN, 1979, pp. 222–230). There are, however, very few pulse transitions which complicates decoding because it is difficult to recover the original clock from the data stream. Twinned binary or Manchester encoding offers the best of both worlds: a lot of pulse transitions for smooth clock recovery and no d.c. content. Each bit is translated in a double pulse pattern: a high-low pattern for a logical "1" and a low-high pattern for a "0". A major advantage of Manchester encoding is that the "nil" level can be omitted provided that each sequence is preceded by a synchronisation mark. Skipping the "nil" level dramatically simplifies ASK transmission. It is not required anymore to have two output levels in the transmitter and three detection levels in the receiver. The transmitter is just switched on and off, and the receiver only detects if there is a signal or not.

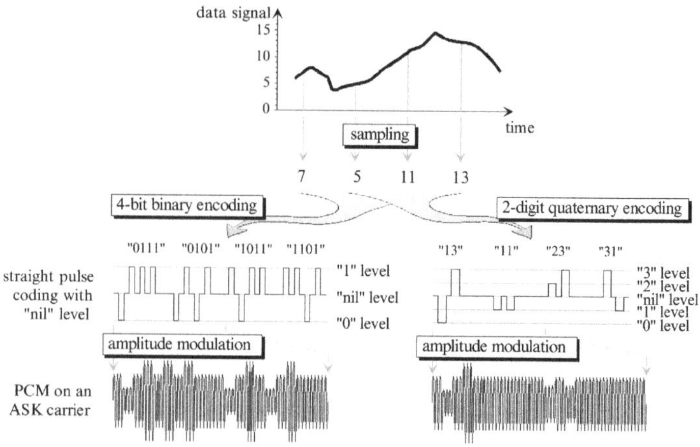


Figure 1.7. *Two examples of PCM. The data signal is sampled at a 4 bit accuracy and then encoded in a 4-bit binary code (base = 2, left) or in a 2-digit quaternary code (base = 4, right). The transmission is on an ASK carrier.*

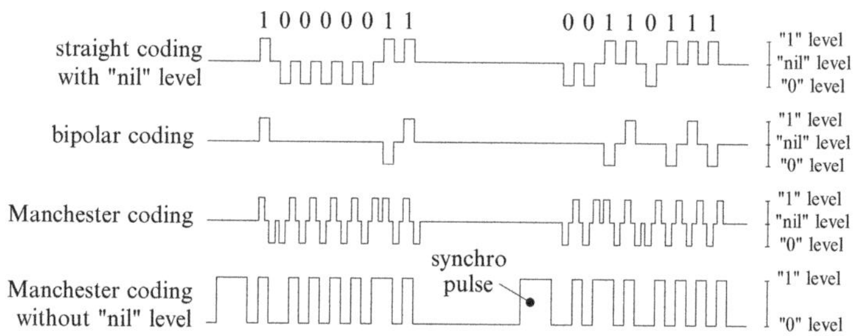


Figure 1.8. *Pulse encoding of bit-streams.*

M-ary coding, with a base value *M* larger than 2, squeezes more than one bit of information in a single data pulse to increase data throughput where the signal bandwidth is limited (the right example in Figure 1.7 has a base value of 4). Telephone-line modems cram 57,600 baud of uncompressed data though a standard 3 kHz telephone line, this way.

Digitising the data transmission matches with the current trend to take signal processing more and more into the digital domain. CMOS circuit technology made high-performance micro-power integrated circuits, including one-chip microcomputers, available at low prices. Silicon micro machining is now an important source of inexpensive, reliable and low-power sensors (MEINDL and FORD, 1984). Both accomplishments made it finally possible to put more intelligence in the remote unit, so it can extract that part of the sensor data that is relevant for transmission (PUERS, 1991; WOUTERS *et al.*, 1991, 1993). The transmission data rates and the corresponding power consumption are reduced accordingly.

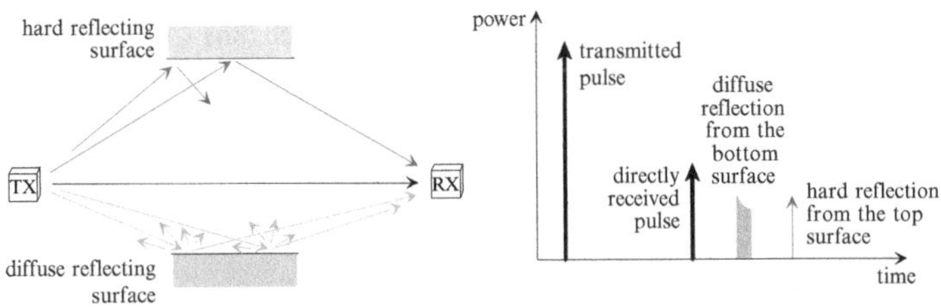


*Figure 1.9.* *The reception of a short wave pulse can be seriously distorted by parasitic reflections and diffusions. Hard reflections produce echoes and diffuse reflections disperse the pulse.*

The following sections summarise the physics of the various connection principles and their potential in terms of data and power transfer.

## 1.1.2 PROPAGATING-WAVE LINKS

Propagating waves are by far the most popular carriers in the telemetry world. Carrier amplitude, phase or frequency modulation is achieved by varying the corresponding attribute of the wave source.

It should be noted, however, that the maximal data frequency is not only limited to about a twentieth of the carrier frequency as explained in the previous paragraph. Parasitic reflections have also their effect on the data bandwidth. Figure 1.9 illustrates how parasitic reflections produce echoes and reverberations that interfere with the directly received signal especially at higher signal frequencies or pulse rates. This phenomenon is further discussed in paragraph 1.1.2.2 on optical links. It is however inherent to all wave-based communication systems.

Waves transport energy along with their propagation: e.g. a sheet of paper will vibrate when held in the radiation field of a sound source. It is thus possible for the remote unit to capture some of the radiated wave energy and convert it into electrical energy to supply the remote circuits.

### 1.1.2.1 Radio-frequency links

Radio-frequency links transfer power and information over propagating electromagnetic waves. Electromagnetic waves arise from the interaction between magnetic fields that induce electric fields and *vice versa*. In general, the two vector quantities, the electric and the magnetic field, are related to each other by Maxwell's equations:

$$\vec{\nabla} \times \vec{E} = -\frac{\partial \vec{B}}{\partial t}$$ *Equation 1.1.*

$$\vec{\nabla} \times \vec{H} = \frac{\partial \vec{D}}{\partial t} + \vec{J}$$ *Equation 1.2.*

$$\vec{\nabla}.\vec{D} = \rho$$ *Equation 1.3.*

$$\vec{\nabla}.\vec{B} = 0$$ *Equation 1.4.*

where $\vec{D}$ represents the electric displacement, $\vec{E}$ the electric field strength, $\vec{J}$ the free electric current density (excluding the atomic currents in magnetic materials), $\vec{B}$ the magnetic flux density, $\vec{H}$ the magnetic field strength and $\rho$ the free electric-charge density (excluding the charges induced in dielectric substances). All are function of position and time. The electric displacement vector and the electric-field strength are linked by the medium dielectric constant $\varepsilon$, whereas the medium permeability $\mu$ associates the magnetic flux density to the field strength:

*Equation 1.5.*
$$\vec{D} = \varepsilon\vec{E} = \varepsilon_0\vec{E} + \vec{P}$$

*Equation 1.6.*
$$\vec{B} = \mu\vec{H} = \mu_0\left(\vec{H} + \vec{M}\right)$$

$\vec{P}$ is the built-in medium polarisation, caused by the electric field. $\vec{M}$ is the built-in medium magnetisation, caused by the magnetic field.

Equation 1.1.Equation 1.1. states that time-varying magnetic fields induce electric fields. Its integrated version is known as Faraday's Law and will pop up again in Chapter 2. It is easily derived via Stokes' Theorem (cf. Appendix A1). Equation 1.2.Equation 1.2. points out that magnetic fields are automatically produced by any electric current or time-varying electric field. Equation 1.3.Equation 1.3. means that electric charges generate an electric field while Equation 1.4.Equation 1.4. states that magnetic charges that may bring about magnetic fields, do not exist. The Maxwell equations are to be solved for each specific electromagnetic problem, be it static or dynamic, pure electric, magnetic or a complex combination of these. This can become quite complicated especially when $\varepsilon$, $\mu$ or $\sigma$ are field dependant for non-linear materials, direction dependant for anisotropic materials or frequency dependant.

Nevertheless, a simple combination of the Maxwell equations already proofs the occurrence of electromagnetic waves in free space.[4] Taking the curl of both sides in equation 1.2 yields

$$\vec{\nabla} \times \vec{\nabla} \times \vec{E} = -\frac{\partial\left(\vec{\nabla} \times \vec{B}\right)}{\partial t} \Leftrightarrow \nabla^2\vec{E} = \mu_0\varepsilon_0\frac{\partial^2\vec{E}}{\partial t^2}$$

This is the general differential equation for a vector wave, in this case the electric field, in a three dimensional space. The wave propagates at the speed $c$, the speed of light, independent of the frequency:

*Equation 1.7.*
$$c = \frac{1}{\sqrt{\mu_0\varepsilon_0}}$$

A similar wave equation is produced for the magnetic field, which demonstrates identical wave behaviour at the same wave velocity $c$:

$$\nabla^2\vec{B} = \mu_0\varepsilon_0\frac{\partial^2\vec{B}}{\partial t^2}$$

This should not be surprising. Maxwell's equations state that it is impossible to have an electric disturbance without having a magnetic one also, and *vice versa.* Every electromagnetic wave has an electric portion and a magnetic portion travelling along together. To have one without the other would be analogous to having a water wave in which there is motion without displacement, or displacement without motion. Although there are separate equations for $\vec{E}$ and $\vec{B}$, they represent physically inseparable quantities.

[4] The connotation "free" points to a space where no free or built-in charge or current distributions exist (hence $\varepsilon = \varepsilon_0$, $\mu = \mu_0$ and $\sigma = 0$).

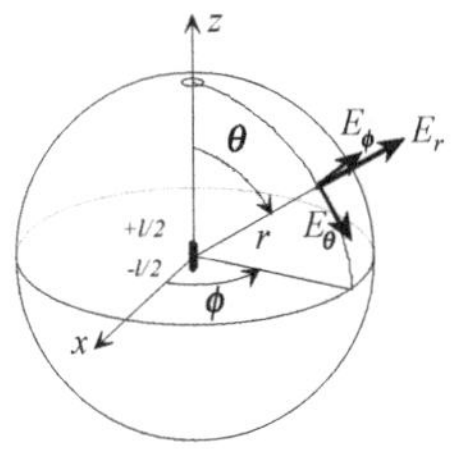


*A spherical co-ordination system around a short current carrying wire.* *Figure 1.10.*

Consider now a short wire in free space, carrying an alternating current *i:*

$$i = I.\ sin\ (\omega.t)$$

The wire length *l* is assumed to be much smaller than the wavelength $\lambda$. Solving the Maxwell equations in spherical co-ordinates produces (SKILLING, 1948, pp. 165–167 and Figure 1.10):

$$E_r = -\sqrt{\frac{\mu}{\varepsilon}}\frac{I.l.\cos\theta}{r\lambda}\left[\frac{1}{4\pi^2}\frac{\lambda^2}{r^2}\cos\left(2\pi\frac{r}{\lambda}-\omega t\right)+\frac{1}{2\pi}\frac{\lambda}{r}\sin\left(2\pi\frac{r}{\lambda}-\omega t\right)\right]$$

$$E_\theta = \sqrt{\frac{\mu}{\varepsilon}}\frac{I.l.\sin\theta}{2r\lambda}\left[-\frac{1}{4\pi^2}\frac{\lambda^2}{r^2}\cos\left(2\pi\frac{r}{\lambda}-\omega t\right)-\frac{1}{2\pi}\frac{\lambda}{r}\sin\left(2\pi\frac{r}{\lambda}-\omega t\right)\right.$$
$$\left.+\cos\left(2\pi\frac{r}{\lambda}-\omega t\right)\right]$$

$$E_\phi = H_r = H_\theta = 0$$

$$H_\phi = \frac{I.l.\sin\theta}{2r\lambda}\left[-\frac{1}{2\pi}\frac{\lambda}{r}\sin\left(2\pi\frac{r}{\lambda}-\omega t\right)+\cos\left(2\pi\frac{r}{\lambda}-\omega t\right)\right]$$

These expressions are not important as such, but they help understand the distinction between two different regions of interest. ***At distances r much smaller than a wavelength $\lambda$ from the current-carrying wire***, the highest-order terms in $\lambda/r$ dominate and the expressions are reduced to quasi-stationary equations of two oscillating $\vec{E}$ and $\vec{H}$-fields. This is the so-called ***near field*** solution. The fields near the wire are mainly non-radiating and are called the ***induction fields***.

***At distances well away from the wire, r is much larger than $\lambda$*** and the terms in $\lambda/r$ can be neglected:

$$E_r = E_\phi = H_r = H_\theta = 0$$

$$E_\theta = \sqrt{\frac{\mu}{\varepsilon}}\frac{I.l.\sin\theta}{2r\lambda}\cos\left(2\pi\frac{r}{\lambda}-\omega t\right)$$

$$H_\phi = \frac{I.l.\sin\theta}{2r\lambda}\cos\left(2\pi\frac{r}{\lambda}-\omega t\right)$$

This approximation describes two $\vec{E}$ and $\vec{H}$-fields that propagate away from the wire and is therefore often referred as ***the radiation field***. Note that the magnetic and electric components are mutually perpendicular.

The energy radiating from a transmitting antenna directly relates to its dimensions. A popular unit to express an antenna's capability to radiate electromagnetic power is the ***radiation resistance***. It is defined as ***the average radiated power divided by the square of the effective value of the current in the antenna lead.***

The radiation resistance $R_{rad}$ of the short wire carrying a uniform current, described on the previous page, equals:

$$R_{rad} = \frac{2\pi}{3} \frac{l^2}{\lambda^2} \sqrt{\frac{\mu}{\varepsilon}}$$

The uniform current in the wire implies that it is included in a closed electric circuit. However, the small dipole antenna with open ends is much more common. Such an antenna carries a linear decreasing current towards its ends. This makes it look like a wire with a uniform current of half its actual length. Its radiation resistance becomes:

$$R_{rad} = \frac{\pi}{6} \frac{l^2}{\lambda^2} \sqrt{\frac{\mu}{\varepsilon}}$$

The half-wave dipole antenna in free space has a sinusoidal current pattern. Its radiation resistance is always 73.1 Ω regardless the frequency.

The expressions for the radiation resistance indicate that efficient wave generation involves antenna dimensions in the order of the electromagnetic wavelength. This explains why the very low frequency bands are not used for telecommunications (Figure 1.1). It also explains the drive to higher frequencies for the small remote systems envisaged here. But there is a trade-off between both trends, as conductive or magnetic media more easily absorb higher frequencies.

At the receiving end, an electromagnetic wave passing a conductive wire induces an electrical field in that wire. The passing wave distorts by the presence of this antenna; its electric and magnetic fields are weakened as the electric field in a perfect conductor is zero, and current flows in the antenna depending on its termination.

There is a second, more electric way of looking at receiving antennae. This paradigm is also more practical for computations. It assumes that electromotive force is induced in the antenna by the magnetic field of the undistorted passing wave. As a result, current flows in the antenna, producing a second electromagnetic field that superimposes on the undistorted wave and radiates from the antenna. The current flow is limited by the distributed capacitance and inductance of the conductor, by its Ohmic loss resistance and by its radiation resistance.

The antenna design used in biotelemetry applications is far more complicated than indicated here. But, the fundamental relations between wavelengths, antenna dimensions and antenna performance remain. The design complexity arises from the electrical-conductive tissue surrounding the antennae (KING 1977, 1978) and the wave behaviour at the interface between the body and the free space. This latter topic, and the way intermediate layers can improve the wave transition are discussed in DIAZ *et al.* (1987).

Radio waves usually only carry information. It is, however, perfectly possible to power the remote circuit with energy captured from the inci-

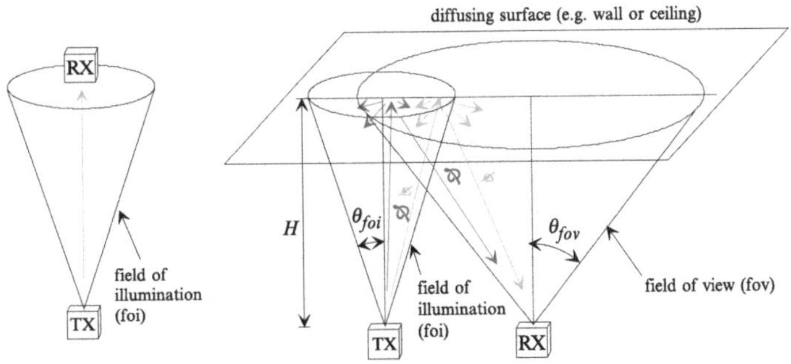


Figure 1.11. *Direct (left) and indirect (right) radiation (copied from ANNOVAZZI-LODI and DONATI, 1988). The signal bandwidth of indirect, diffuse radiation links is limited because of the slightly different propagation delays for light coming from different spots on the illuminated surface.*

dent electromagnetic waves. The waves induce an electrical voltage on the antenna terminals that can be rectified to supply an electronic circuit. In fact, the headset of a vintage crystal radio is powered by the radio energy captured by the receiver antenna.

Efficient power transfer asks for an optimal antenna design with dimensions comparable to the wavelengths used, and a minimal energy spill by carefully focusing the transmitted waves in the direction of the remote unit. In practice, the dimensional constraints of telemetry devices limit this technique to short wavelengths and short distances (BAUMANN *et al.*, 1987). The Electronic Product Code (EPC) RFID tags used by the WalMart discount department stores to manage their supply chain operate in the 860–960 MHz UHF band according to this principle (EPCGLOBAL INC., 2008).

### 1.1.2.2 Optical links

Light is electromagnetic wave energy with frequencies between 300 GHz and $10^{17}$ Hz. The main distinction between r.f. and optical links is how the radiation is generated and captured. Antennas and resonant cavities are typical r.f. techniques, whereas optical links use lamps, LED's and laser-light sources. The optical receivers are commonly built of light-sensitive transistors or diodes (KIMMICH, 1982).

In fact, most wireless optical data links can be seen as fibreless versions of conventional optical-fibre links. The data is commonly transmitted as analogue or digital pulse codes that on-off key the light source (100% ASK). The pulses are transmitted as light flashes that propagate through the transparent medium. The major advantage of optical data transfer is its high level of immunity against electromagnetic-interference (EMI).

Two wireless optical principles can be identified:

- direct-view radiation, where the transmitter and the receiver are in direct optical contact (Figure 1.11, left and KAWAHITO *et al.*, 1994).
- diffuse radiation where the transmitted light first scatters against a wall or ceiling before reaching the receiver (Figure 1.11, right and ANNOVAZZI-LODI and DONATI, 1988).

The maximal data rate for direct view radiation is theoretically limited by the carrier frequency. This would bring extremely high data rates within reach, as the frequency of light exceeds thousands of gigahertz. However, parasitic reflections and diffusions induce unequal propagation delays, echoes and disperse the impulse response (Figure 1.9).

ANNOVAZZI-LODI and DONATI (1988), calculated the -3 dB signal bandwidth $BW_{-3\,dB}$ for optical transfer by first-order diffuse radiation[5]:

$$\mathrm{BW}_{-3\mathrm{dB}} = \frac{0.22c}{H}\frac{\cos\theta}{1-\cos\theta}$$

where c is the speed of light, H the distance to the reflecting surface and $\theta = Max\left(\theta_{foi}, \theta_{fov}\right)$. For a $\theta$ of $\pi/6$, the product $BW_{-3\,dB} . H = 427\ MHz.m$, which indicates that the signal bandwidth is reduced to some 10 MHz, for indoor applications.

Optical powering of remote units from the base side is feasible with solar cells (EDELL, 1996). A typical capacity is in the order of 1 mW/cm$^2$, assuming exposure to bright sunlight and a 10% cell efficiency.

#### 1.1.2.3 Ultrasound links

Ultrasounds are plain acoustic waves with a frequency above the audible range of 20 Hz–17 kHz. Sound is a small pressure disturbance superimposed on the constant ambient pressure that propagates by the back and forth exchange of potential and kinetic energy of adjacent medium sections. This explains the need for a physical medium, in contrast to the electromagnetic waves that also propagate in vacuum.

Ultrasound transmission is very attractive for marine telemetry. Acoustic waves propagate much better in water than r.f. waves, that are absorbed by the relative high electrical conductivity of surrounding water (SLATER and BELLET, 1969). LIEW and CHAN (1993) obtained ranges of 1–2 km in seawater, while tracking green turtles with commercial ultrasonic temperature-depth and swim speed transmitters. ROCHELLE (1974) achieved a 400-m range in fresh-water lakes and rivers at frequencies between 70 and 90 kHz with an acoustic output of only 14 mW. This sharply contrasts with the 3-m range obtained by KUDO and UEDA (1980) in an aquatic tank with a 4.5 mW r.f. telemeter at 78 MHz.

A carefully designed transducer can avoid reflection of ultrasound energy at the transducer-medium interface. Horn loading of the transmitter and receiver transducers, a technique adopted from loudspeaker engineering, can improve their matching with the medium characteristics. It is also possible to enhance the transducer-to-medium transition by the interposition of intermediate layers. Coupling gel is applied between the skin and the transducer medical ultrasonography for this very same reason.

---

[5] "First order" indicates that the light pulses reflect just once before reaching the receiver. The power contributions for multiple diffusions rapidly decrease with increasing order of diffusion and are neglected here.

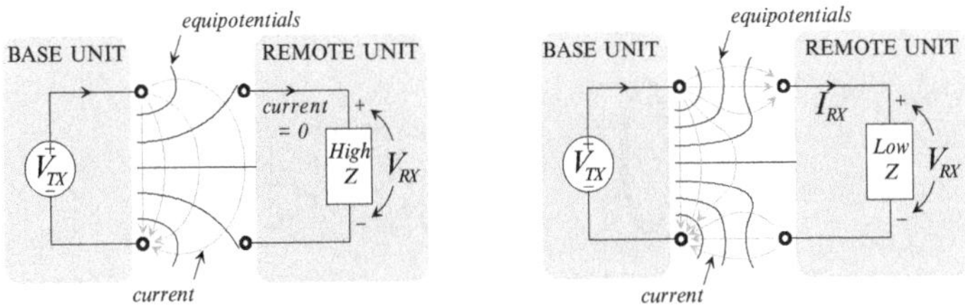


*The conduction-link principle. The left drawing shows a high-impedance receiver that draws no current and leaves the generated current field unaffected. The remote unit on the right, extracts energy from the current field to receive the data or to supply its circuits.* *Figure 1.12.*

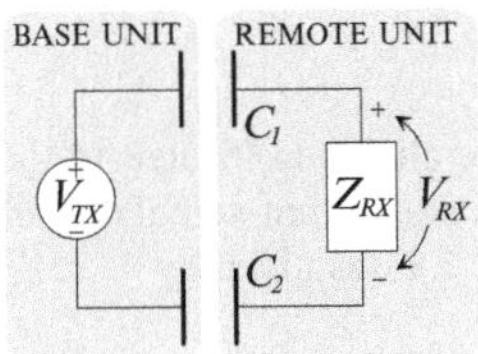


*The capacitive-link principle.* *Figure 1.13.*

Acoustic powering of the remote unit is possible, though rarely used. TSUJI *et al.* (1991, 1993) excited a miniaturised tuning fork implanted in a human body. It had a temperature-sensitive resonance frequency and acted as a pure mechanical temperature telemeter. It was excited at 40 kHz from the outside of the body for 0.4 s. Its ringing response was then measured with the same probe during the next 0.6 s. HOLMER (1973, 1974) reported on the use of a piezoelectric crystal to convert ultrasound mechanical energy into electricity, to supply a remote amplifier. In this case, the base and the remote unit were linked with a glass rod, to optimise their acoustic coupling.

### 1.1.3 CONDUCTION LINKS

The electrical conductivity of media like water or body tissue, seriously interferes with r.f. transmission. On the other hand, this property can be used to establish a direct galvanic connection through the medium (Figure 1.12 and UTSUYAMA *et al.*, 1988). Special precautions are to be taken if the data and/or energy is transferred through living tissue. The signal should be kept below the sensation and stimulation levels and modulated onto an a.c. carrier because sustained d.c. signals are harmful. Electrical current flows as electrons in metal and as ions in fluids with redox reactions occurring at the metal-fluid interface where one conduction mode converts to the other one. This conversion process releases metal ions in the fluid, i.e. body tissue. A sustained d.c. current leaves a net amount of such toxic metal ions in the body. One should thus place blocking capacitors between the current source and both drive electrodes. The carrier frequency should be chosen in the range of lowest sensation.

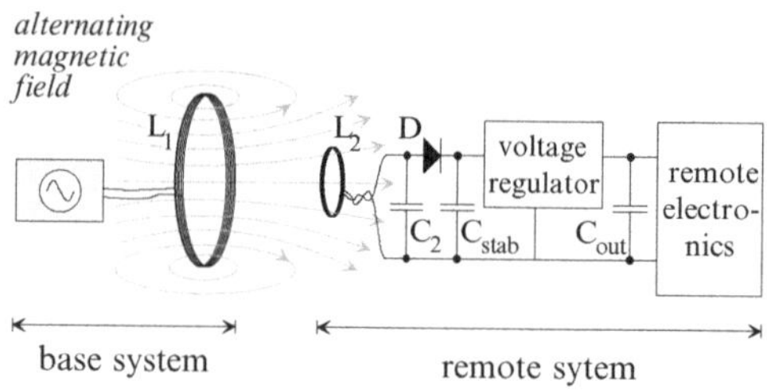


*Figure 1.14.* *The principle of inductive powering.*

### 1.1.4 QUASI-STATIONARY OR INDUCTION FIELD LINKS

#### 1.1.4.1 Capacitive links

A capacitive link is built of capacitors with one electrode located in the base unit and the other one in the remote unit. The output signal $V_{RX}$ of the most simple link example depicted in Figure 1.13, is given by

$$V_{RX} = V_{TX} \frac{Z_{RX}}{Z_{RX} + \frac{1}{j\omega}\left(\frac{1}{C_1} + \frac{1}{C_2}\right)}$$

This expression indicates the need for high carrier frequencies $\omega$ and high capacitor values $C_1$ and $C_2$ to yield substantial output levels.

Big capacitors require large electrode areas and small mutual distances, which explains why capacitive links are only found inside isolation amplifiers and were never used for telemetry (TEXAS INSTRUMENTS, 2005).

#### 1.1.4.2 Magnetic links

Magnetic or inductive power links, also named transcutaneous energy transfer systems (TET) or skin transformers, are basically d.c.–d.c. converters built around a loosely coupled transformer. The technique is based on the magnetic coupling between a remote coil $L_2$ and a primary, base coil $L_1$ that is driven by an alternating current (Figure 1.14). Both coils form a coreless transformer. Resonance of the remote coil with a series or parallel capacitor $C_2$ is most often used for higher link efficiency (cf. Chapter 2; SCHUDER *et al.*, 1961; KO *et al.*, 1977; DONALDSON and PERKINS, 1983). The remote-coil voltage is rectified and stabilised to a d.c. voltage for powering the rest of the remote electronics. A voltage regulator ensures a constant output voltage regardless of coil movements or load variations.

Inductive power links were originally developed in the sixties and the seventies, to power artificial heart-assist devices and auditory prostheses. The heart-assist systems demanded continuous power transfers up to 50 W. And even higher power rates, up to 1 kW, were used to recharge implanted batteries (SCHUDER *et al.*, 1971).

***The ratio between the flux captured by the remote coil and the generated flux is the coil coupling factor*** and was typically 10–20% with these earlier applications. This value was well predictable and constant over time. There is, however, a trend to apply inductive links also in less favourable conditions. Coupling factors often drop below 1% nowadays, and vary over about a decade in a very unpredictable matter (HEETDERKS, 1988). Some representative examples are:

- Multichannel functional electrical stimulation (FES) of paralysed muscles where each stimulation channel consists of a complete micro-implant that locally stimulates the muscle (LOEB *et al.*, 1991; NARDIN *et al.*, 1996)
- 125 and 134.2 kHz transponders for item management and husbandry control (ISO 11785:1996; ISO 14223-1:2003; ISO/IEC 18000-2:2004 and WOUTERS *et al.*, 1993)
- 13.56 MHz proximity ID cards, credit cards, and biometric passports (ISO/IEC 14443-2:2001; ISO/IEC 18000-3:2008)
- Small telemonitoring implants (MUELLER and NAGLE, 1996)

The power-transfer efficiency is dominated by the coil coupling factor (cf. Chapters 2 and 3). The resistive coil losses play a second important role in the link's performance. Power-transfer efficiencies of the order of 70% and more have been reported for coupling factors of at least 15%. This efficiency drops rapidly down to 10% for a coupling factor of only 2%. The latter corresponds to a parallel and coaxial set-up with circular loop coils and a mutual distance of about twice the remote-coil diameter. This already indicates that induction telemetry can only cover fairly small distances, compared to r.f., ultrasound or optical telemetry.

The link's power capacity is directly related to the link efficiency in combination with the maximal allowed base-side dissipation. A capacity of a few mW is readily achieved, even for very low coil coupling.

The transfer frequency is determined by the available space at the remote side and by the characteristics of the surrounding materials. Higher frequencies result in smaller coils, but from a certain frequency on, the resistive coil losses and the losses in the surrounding materials and the base-coil driver become important. The transfer frequency rarely exceeds 30 MHz. The 100 kHz range is more typical.

Nearby permeable (like iron, ferrite or μ-metals) or conductive materials (like metals) can seriously alter the magnetic field distribution and the corresponding link performance. For instance, when the remote coil $L_2$ is embedded in a metal package, eddy currents are induced in the surrounding package metal and shield $L_2$ from the applied magnetic field (DONALDSON, 1992; GESELOWITZ *et al.*, 1992). As a result, no voltage can be induced in $L_2$. The higher the frequency, the more dramatic this effect becomes. Apart from the shielding effect, the eddy currents also cause additional resistive losses that increase with the frequency. This indicates why the powering frequency is to be lowered to some kilohertz when working with conductive material in the neighbourhood. But, nearby materials not always degrade link performance. They can also have a positive effect. Applying a highly permeable core for one or both coils can enhance the coil coupling (MITAMURA *et al.*, 1990; MATSUKI *et al.*, 1990). The coil cores "attract" the magnetic flux and forces it to pass through the coil centres.

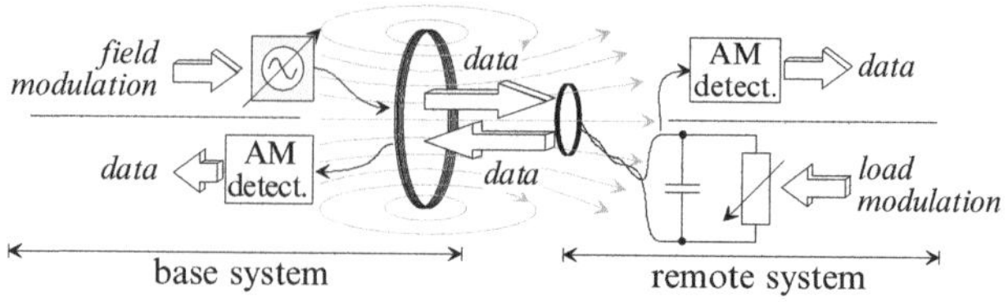


*Figure 1.15.* *Data transfer over an inductive link. The data are transferred to the remote unit by modulating the applied magnetic field and in the opposite direction, by varying the load.*

Note that the magnetic coupling field is supposed to be quasi stationary. This implies that the coil dimensions are small compared to the electromagnetic wavelength at the given transfer frequency. Radiation actually represents an additional link loss as it is not contributing to the energy transfer to the remote device.

Although inductive links were originally conceived for power transfer, they feature the possibility for transmitter-free bi-directional data communication (Figure 1.15). Information can be transferred to the remote device by modulating the powering field (SAWAN *et al.,* 1992). Straightforward amplitude demodulation of the power signal across the remote coil recovers the data. A high data rate will require high power transfer frequency (at least about 20 times the data rate). The remote module can transmit data in the opposite direction, back to the base side, by modulating its power consumption (DONALDSON, 1986; NEUKOMM *et al.*, 1988; TANG *et al.,* 1995) or changing the tuning characteristics of the secondary coil and its resonance capacitor (TOWE, 1986).

This so called backscattering modulates the field around the remote coil and thus also around the primary coil because of the coil coupling, thereby varying the primary coil voltage. The achievable signal bandwidth is again limited to about a twentieth of the powering frequency. The quality factors of the primary and secondary circuits also impact the data throughput. A resonating tank with higher quality factor, responds more slowly to signal variations and reduces the maximal data rate.

There is a second method to relay remote data to the base site provided that the rectifier capacitor or a small battery in the remote unit stores enough energy to run the remote circuit briefly without powering field. In a tic-tac fashion, the base powers the remote intermittently and the remote relays its information back over its powering coil in the dead periods.

## 1.2 HOW TO CHOOSE BETWEEN LINK TYPES?

This paragraph walks through the various aspects that determine what link is best suited for your application:

- Device dimensions
- The impact of the link operation on its surroundings, and
- The influence of the environment on the link performance

A summary table is provided at the end of this chapter (Table 1.8).

### 1.2.1 TRANSDUCER DIMENSIONS

All wireless links have ***a device*** at the transmitter side (e.g. antenna, ultrasound speaker, LED, coil, ...) ***that converts electric signals into electromagnetic or acoustic waves, light, magnetic or electric fields***. The ***reverse*** process is carried out in the receiver. Both conversion devices are called ***transducers***.

A general requirement for <u>all wave-based links</u> is that the transducer dimensions have to be comparable to the wavelength for efficient wave generation or reception (cf. radiation resistance) This is why small devices call for high carrier frequencies.

<u>Magnetic links</u> know an optimal radius $r_1$ for a base coil, given the remote coil radius $r_2$ and the mutual distance $D$ (KO *et al.*, 1977):

$$r_1 = \sqrt{D^2 + r_2^2}$$

*Equation 1.8.*

This means that the external coil radius equals about the distance to be covered (for small remote coils). On the other hand, the remote-coil dimensions determine the power capacity: the smaller the remote coil, the worse the coil coupling and the lower the link efficiency. It remains, though, possible to employ millimetre-sized remote coils for microwatt power transfers or for data transmission only (WOUTERS *et al.,* 1991).

The <u>capacitive-link</u> electrodes should have a large area compared to their mutual distance, to yield a substantial capacitor value.

The drive electrodes of a <u>conduction link</u> should be positioned apart from each other, and left and right from the sensing electrodes to enclose them in the generated field. The sensing electrodes are to be well separated from each other to receive a significant fraction of the "transmitter" voltage.

### 1.2.2 INFLUENCE OF THE LINK ON ITS ENVIRONMENT

#### 1.2.2.1 R.f. electromagnetic fields

Electromagnetic fields are often absorbed by media, which produces heat. This is especially important when the generated electromagnetic fields run through fuels, explosive media (FICCHI, 1971, pp. 231–234) or living tissue (JOHNSON and GUY, 1972; LERNER, 1980; POLK and POSTOW, 1995; GANDHI, 1987a; FOSTER and MOULDER, 1996). With the growing use of r.f fields in communication, power distribution, industrial and medical applications, people are more and more exposed to r.f. signals (STUCHLY and MILD, 1987).

Electromagnetic radiation consists of quanta with an energy that is proportional to their frequency. Quanta of ionizing radiation, like ultraviolet light or X-rays, possess a high enough energy to ionize atoms or molecules. Such radiation, when absorbed by tissue, creates free radicals that are a major source of cell damage. Electromagnetic energy at radio frequencies is non-ionising: their frequency and hence quantum energy is orders of magnitude too small to disrupt even the weakest chemical bond. The hazards from electromagnetic fields or electrical currents at radio frequencies exhibit clear thresholds and do not depend on cumulative exposure, quite different from the case of ionising radiation.

When considering possible biological effects of electric fields, it is important to distinguish between ***field strengths outside the subject, and those within the body. The relation between both is called the coupling*** and varies greatly with the frequency. A 1 kV/m electric field at 50 Hz will induce an electric field of only 1 mV/m in the body, which is well below the fields from natural sources, such as the heart. But, the coupling increases with the frequency, up to about 70 MHz where the body has an electrical resonance and absorbs most strongly. The magnitudes of the incident electric field and the field within the body remain comparable above this frequency.

Body tissues are basically non-magnetic. The magnetic fields inside the body are identical to those outside it and there is very little direct interaction between magnetic fields and tissues. SEIDEL *et al.* (1968) reported on visual sensations of magnetic fields (magnetophosphenes) when a person's head is exposed to alternating 60 Hz magnetic flux densities of about 20 mT. There are also indications that some animals can sense the earth's magnetic field and use that for orientation. Apart from these, no further direct interaction is known. However, the indirect interaction with the body is self-evident: time-varying magnetic fields induce electric fields and thus eddy currents that may evoke biological effects (GANDHI *et al.*, 1986). UENO *et al.* (1986) reported on pain sensation and changes in skin blood flow, probably due to eddy currents, when a hand was exposed to the 3.8 kHz magnetic fields of industrial induction heaters.

***Thermal effects of r.f. electromagnetic fields on biological tissue***

Thermophysiological effects of r.f. absorption arise from tissue heating (ADAIR, 1987; FOSTER, 1996). A useful benchmark in assessing thermal effects is the basal metabolic rate of man, which is about 1 W/kg of body mass. If the body is subjected to heating from an external source at a much greater rate, thermal tissue damage can occur. Some idea of the thermal tolerance of the human body is given by the amount of metabolic heat that it generates. An ordinary person playing tennis has an average metabolic rate of 2 W/kg; an athlete can produce 16 W/kg. An electric field $E$ in a medium of conductivity $\sigma$ generates power at a rate of $\sigma E^2$. The electric field strength should be roughly kept below 10–100 V/m for a maximal dissipation of 1 W/kg.

Assessing r.f. exposure hazards is a complex problem. The tolerable levels depend of the r.f. environmental conditions, the rate of heat removal, and the physical condition of the subject. The r.f. environmental conditions are the power density, signal frequency, modulation type and duty cycle. The location within the body where the heat is produced, ambient temperature, humidity and wind velocity greatly influence the rate of heat removal. The r.f. signal frequency determines the penetration depth and the location where the absorption occurs within the body (Table 1.5). Above 3 GHz, only the outer tissue layers are heated and a large fraction of the energy is reflected. The skin layers respond quickly to temperature changes so that the warming effects on the skin are recognised in time to avoid any hazard. Also, the skin's outward heat dissipation is excellent.

*ELECTROMAGNETIC PROPERTIES OF HUMAN TISSUES (JOHNSON and GUY, 1972; GUY et al., 1974).* *TABLE 1.5.*

| | | **Muscle, skin and tissues with high water content** | | | | **Fat, bone and tissues with low water content** | | | |
|---|---|---|---|---|---|---|---|---|---|
| Freq. (MHz) | $\lambda$ in air (m) | Relative $\varepsilon_{high}$[a] | $\sigma_{high}$ (mS/m) | Penetr. depth[b] (mm) | $\lambda$ in tissue (m) | Relative $\varepsilon_{low}$[a] | $\sigma_{low}$ (mS/m) | Penetr. depth[b] (mm) | $\lambda$ in tissue (m) |
| 1 | 300 | 2,000 | 400 | 913 | 4.36 | | | | |
| 10 | 30 | 160 | 625 | 216 | 1.18 | | | | |
| 27.12 | 11.06 | 113 | 612 | 143 | 0.681 | 20 | 10.9–43.2 | 1,590 | 2.41 |
| 40.68 | 7.38 | 97.3 | 693 | 112 | 0.513 | 14.6 | 12.6–52.8 | 1,180 | 1.87 |
| 100 | 3 | 71.7 | 889 | 66.6 | 0.270 | 7.45 | 19.1–75.9 | 604 | 1.06 |
| 200 | 1.5 | 56.5 | 1,280 | 47.9 | 0.166 | 5.95 | 25.8–94.2 | 392 | 0.597 |
| 300 | 1 | 54 | 1,370 | 38.9 | 0.119 | 5.7 | 31.6–107 | 321 | 0.410 |
| 433 | 0.693 | 53 | 1,430 | 35.7 | 0.0876 | 5.6 | 37.9–118 | 262 | 0.288 |
| 750 | 0.400 | 52 | 1,540 | 31.8 | 0.0534 | 5.6 | 49.8–138 | 230 | 0.168 |
| 915 | 0.328 | 51 | 1,600 | 30.4 | 0.0446 | 5.6 | 55.6–147 | 177 | 0.137 |
| 1,500 | 0.200 | 49 | 1,770 | 24.2 | 0.0281 | 5.6 | 70.8–171 | 139 | 0.0841 |
| 2,450 | 0.122 | 47 | 2,210 | 17.0 | 0.0176 | 5.5 | 96.4–213 | 112 | 0.0521 |
| 3,000 | 0.100 | 46 | 2,260 | 16.1 | 0.0145 | 5.5 | 110–234 | 97.4 | 0.0425 |
| 5,000 | 0.060 | 44 | 3,920 | 7.88 | 0.0089 | 5.5 | 162–309 | 66.7 | 0.0263 |

[a]*The relative dielectric constant is normalised to the dielectric constant of vacuum.*
[b]*The penetration depth is defined here as the distance that the propagating wave will travel before the power density decreased by a factor of* $e^{-2}$.

Deep tissue heating occurs roughly between 200 MHZ and 3 GHz, a band that includes the various frequencies used for portable and data communication. The cellular frequency bands centre around 450, 800, 850, 900, 1,800, 1,900, 2,110–2,200 and 2,500–2,690 MHz depending of the country and communication standard (GSM, CDMA, 3G, etc.). Cordless DECT telephones operate around 1,900 MHz. Wireless local area networks (WLAN according to IEEE 802.11 standards) run at 2.4 or 5-5.8 GHz. Note that 2,450 MHz is used in domestic microwave ovens for dielectric heating of food and is usually a rather noisy frequency.

Deep tissue heating is hazardous because the brain cannot be warned of the dangerous rise in tissue temperature in time. The physical condition of the individual controls his ability to dissipate excess heat. This includes general health, especially the circulatory condition, the physical labour in which the person is engaged, and the garments worn at the time of exposure.

Microwave hyperthermia used for the treatment of localised malignancies, clearly demonstrates the influence of the blood circulation. Healthy tissue is better perfused with blood than cancerous tissue, so it is possible by controlled and focused deep heating of tissue, to heat tumours to temperatures above 44.5°C and kill their cells. At the same time, the surrounding tissue is protected because a significant increase in local blood flow rapidly carries the heat away.

### *Nonthermal effects of r.f. electromagnetic fields on biological tissue*

Electromagnetic fields induce electric currents in biological tissue, due to its fairly high conductivity. Low-frequency currents above a threshold level of about 10 A/m$^2$ or an electric field in the tissue of 100 V/m, excite cell membranes causing muscle contractions and heart rhythm disorders.

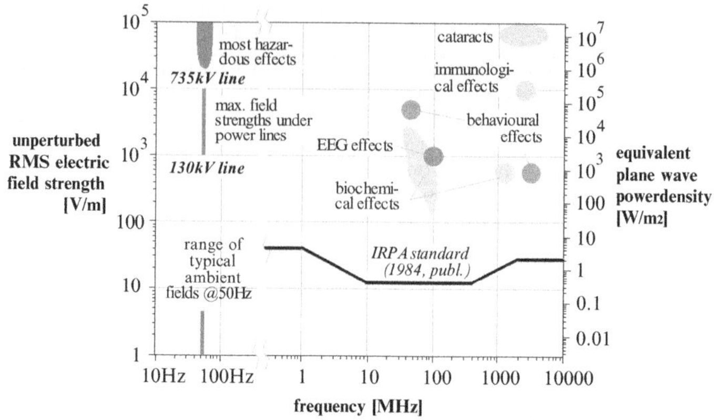


*Figure 1.16. Some biological effects of electromagnetic radiation that resulted from animal experiments (LERNER, 1980). The field strengths are measured outside the body.*

This electroshock threshold increases rapidly with the frequency above about 100 Hz. This makes that at low frequencies, shock is the limiting hazard, while at higher frequencies, burns and heat stress occur at lower field strength than shock. Electrosurgical devices for cutting, coagulation or desiccation with output powers up to 150 W, therefore utilise electric currents of 300 kHz to 1.5 MHz, to avoid electroshock hazards (BRONZINO, 1995).

Because the body is very weakly coupled to external fields below 100 Hz, electroshock normally requires direct contact with conductors.

Alternating magnetic fields induce eddy currents in the tissue that can produce shock; but this requires very high field strengths.

While the thermal and the electroshock effects of electromagnetic radiation are well established, other effects are less understood, more speculative and controversial. Either are their effects so small that they haven't been demonstrated yet in a reliable way, or they are not really relevant to the problem, as they take place at levels way above those that already induce dangerous thermal effects (Figure 1.16; FOSTER, 1996).

### *Exposure standards*

Major exposure standards are generally designed to limit whole-body exposure to 0.08–0.4 W/kg, and partial body exposure to 4 W/kg for any point of the body (ICNIRP 1998; IEEE STD. C95.1-1999, 1992; GANDHI, 1987b; REPACHOLI, 1987).

The standard limits are also function of the frequency and the exposure time. The lowest values are around the body's resonance at 70 MHz where most power is absorbed (Figure 1.17). The standard values at low frequencies, e.g. below 10 MHz, are not based on absorption levels but on the levels of electric field strength that minimise possible r.f. shocks when making contact with conducting objects.

Higher exposure levels are allowed for short or intermittent exposure.

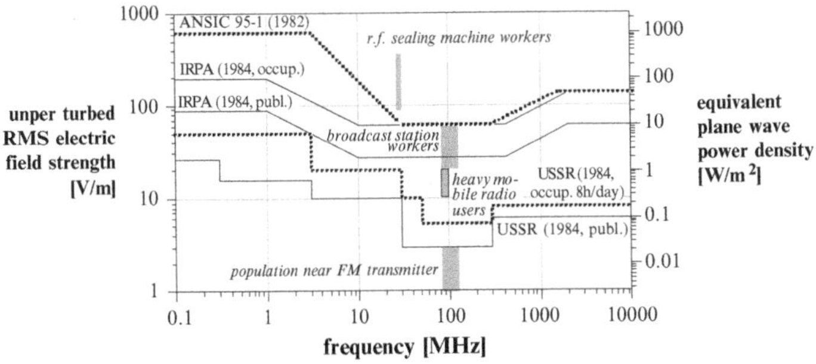


Figure 1.17. *The international IRPA and two national standards for electromagnetic radiation exposure together with some actual exposure levels (REPACHOLI, 1987). The field strengths were measured outside the body. Note how operators of r.f. sealers actually face a real exposure hazard.*

The standard limits are also different for occupational and for general public exposure. Workers are presumed to be healthy adults, aware of the risks, exposed only during working periods, and trained to take any necessary safety precautions. The general public on the other hand, contains the full spectrum of sensitivities (babies, young, aged, sick, etc.). They may be unaware of any exposure, and are potentially exposed 24 h per day, every day.

***Electromagnetic interference and immunity***

It would not be accepted that electromagnetic signals from telemetry devices would induce noise in other equipment or interfere with normal radio communications. Also *vice versa*, telemetry links should be immune to electromagnetic noise generated by other equipment.

The explosive growth of communication and other electronic devices the past 20 years required new regulations that control their electromagnetic output levels and spectra. The latest Electromagnetic Compatibility (EMC) Directive 2004/108/EC repeals the previous EMC Directive 89/336/EEC and led to a multitude of EN standards (2008/C 280/05). Radio equipment and telecommunications terminal equipment are covered by a different Directive 1999/5/EC and corresponding EN standards (2008/C 280/06). Medical devices also have separate EMC Directives, the 90/385/EEC and 93/42/EEC and about a dozen amendments. The corresponding EN standards are listed in 2008/C 304/06.

Acoustic and optical telemetry lack an electromagnetic data carrier and have therefore an excellent electromagnetic immunity and compatibility. This intrinsic property is one of their strongest features.

### 1.2.2.2 Electric currents through the body

Low-frequency electrical currents can excite muscles to contract, as explained above in the section on non-thermal r.f. effects. A current that passes through a body from one hand to the other, and which amplitude exceeds the so called let-go threshold, stimulates the muscles at such a level that the victim is left unable to release the conductors (Figure 1.18).

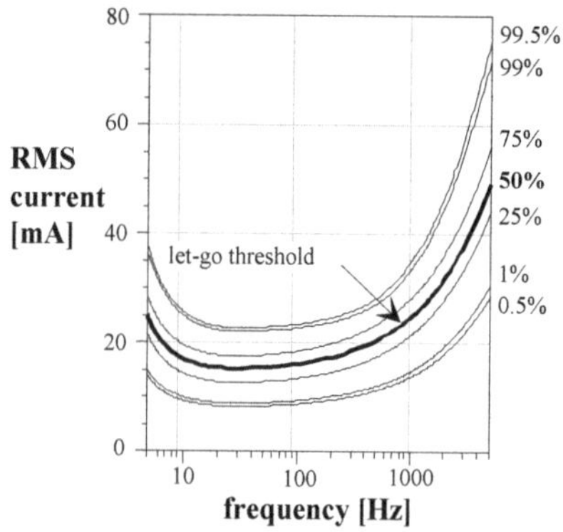


*Figure 1.18. Percentile curves of the "cannot-let-go current" and the definition of the "let-go threshold" as the curve at 50% (DALZIEL, 1968).*

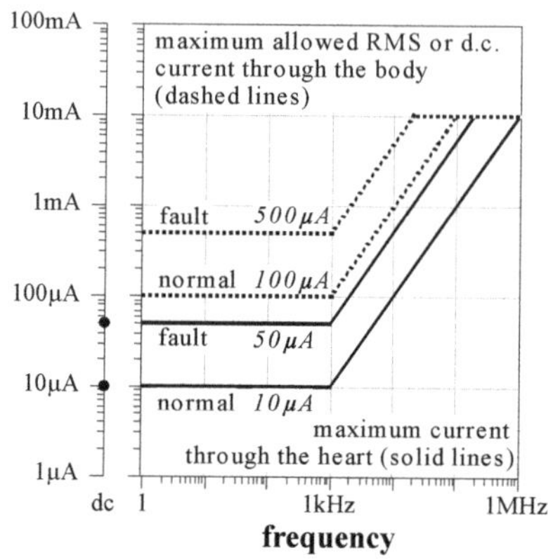


*Figure 1.19. Maximal allowed RMS or d.c. levels for electrical current through the body or the heart, for normal and single faulty conditions (EN 60601-1-1:2001). The upper limits for d.c. currents in the body are taken equal to the maximal currents in the heart (10 μA for the normal and 50 μA for the single fault condition). This is because sustained d.c. currents dissolve metal ions from the electrodes and as such contaminate the tissue.*

The most lethal electroshock effect is ventricular fibrillation. Relatively small currents drawn through the heart muscle, can already disrupt the electric management that controls its pacing, and cause it to contract in a very unorganized way thereby stopping its pumping function. Only a sudden, well-controlled and high current pulse can reset the heart muscle after which it restarts its normal beating.

Two modes of electroshocks are commonly identified. The macroshock comes from contacts on the body exterior. As the current spreads over the body, only a small part actually reaches the heart. The current density created at the heart is 50 $\mu A/mm^2/A$ entering the chest (EN 60601-1-1:2001). Fibrillation is possible above 100-300 mA (Table 1.6). Contacts applied directly onto the heart, like catheters, are more dangerous to induce a so-called microshock. There is a 1% chance that a current of 50 μA through the heart causes ventricular fibrillation. This probability increases to 5% for 100 μA and to 100% for 500 μA.

D.c. body currents are already hazardous below macroshock levels, because the ions from the metal electrodes dissolve in the tissue and move along with the current. This gradually poisons the surrounding tissue. The EN 60601 standard is therefore more stringent for d.c. body currents.

TABLE 1.6. EFFECTS OF 60 HZ ELECTRIC CURRENT ON AN AVERAGE HUMAN THROUGH THE BODY TRUNK (MACROSHOCK) (FREIDLANDER, 1971).

| Current intensity (1 s. contact) (mA) | Physiological effects |
|---|---|
| 1 | Threshold of perception |
| 5 | Accepted as a harmless maximum |
| 10–20 | Let-go current before sustained muscular contraction |
| 50 | Pain, possible fainting, exhaustion, mechanical injury. Heart and respiratory functions continue. |
| 100–300 | Ventricular fibrillation. Respiratory centre remains |
| 6,000 | Sustained myocardial contraction, followed by normal heart rhythm. Temporary respiratory paralysis. Burns if current density is high. |

### 1.2.2.3 Optical effects

The photon energy of UV light is sufficient to effect chemical reactions. It can disrupt weak bonds or form new ones, thereby chemical degrading a wide range of substances.

Optical radiation plays a very significant role in mankind. On the whole, man is very well adapted to the sun's optical and infrared radiation and depends on it for warmth, the processing of food through photosynthesis, and as the principle source for sensory information through the eyes. Optical hazards to man are few, and are mostly confined to high intensities form artificial sources such as lasers.

Infrared light acts like electromagnetic r.f. energy of ultra high frequency, in the sense that it only induces heating of superficial skin and tissue layers. That is why most r.f. exposure standards extend to the far infra-red frequencies. Infrared light is invisible and only observed by the warming effect on the skin for levels above the human perception threshold of 0.67 mW/cm$^2$ (whole body exposure).

A sensation of "very warm or hot" was experienced at an average power density of 21.7 ± 4.0 mW/cm$^2$ for an exposure area of 40.6 cm$^2$ on the dorsum of the right hand, while a similar sensation occurred for a lower exposure area of 9.6 cm$^2$ for a larger power density of 55.9 ± 4.9 mW/cm$^2$. The sensation of "very warm or hot" for larger areas, such as the whole body exposure, may well occur at 8.7 mW/cm$^2$ which is quite close to the ANSI-recommended safety level of 5 mW/cm$^2$ (= 50 W/m$^2$) which may indicate a potential problem with the ANSI r.f. safety standard (GANDHI, 1987b).

More severe precautions are necessary when exposing tissues to ultraviolet rays. The high photon energy creates tissue reactions that are more complex than simple heating: superficial cell damage (sunburn) and augmented risk for cancers. The skin plays an important role in protecting the underlying tissues from UV damage. It therefore absorbs UV radiation for wavelengths above 0.3 μm. The inefficiency in this protection gives rise to sunburn.

*TABLE 1.7. THE INFLUENCE OF ENVIRONMENTAL PARAMETERS ON THE LINK PERFORMANCE.*

| **Link Principle** | **Field is determined by:** Main parameter | Add. param. | **Obstructions** | **Absorptions** | **Environmental Noise** |
|---|---|---|---|---|---|
| r.f. | Dielectric constant $\varepsilon$, permeability $\mu$ | Conductivity $\sigma$ | Reflector, conductor | EM absorption | High: other r.f. equipmt., fast digit. electronics, sparks, fluorescent lamps, high power electric appliances, electric motors |
| Ultrasound | Compressibility, specific mass | Viscosity | Reflector | Acoustic damping | Typically low |
| Optical | Refraction index | – | Reflector | Imperfect transparency | Ambient light, IR noise is normally low |
| Conduction | Conductivity $\sigma$ | – | (Grounded) conductor | Resistivity | EMI, power line, low with proper modulation |
| Capacitive | Dielectric constant $\varepsilon$ | – | Grounded conductor | Dielectric losses | EMI, power line, low with proper modulation |
| Induction | Permeability $\mu$ | – | Conductor | Magnetic losses | EMI, power line, low for tuned coils |

#### 1.2.2.4 Ultrasound effects

Ultrasound levels should be limited, because ultrasound can distress animals that are sensitive to higher pitches than humans. The impact of high sound levels can also physically damage nearby objects or living tissue. Hazardous levels for tissues depend on the exposure time. Maximal allowable levels are between 100 W/cm$^2$ during 100 µs and 100 mW/cm$^2$ for an infinite exposure (ULRICH, 1974).

### 1.2.3 INFLUENCE OF THE ENVIRONMENT ON THE LINK PERFORMANCE

The space around and between the telemetry units has a fourfold influence on the link performance (Table 1.7):

- The intrinsic medium properties control the characteristics of applied fields. Electromagnetic waves have different wavelengths, velocities and amplitudes depending on the medium conductivity, permeability or dielectric constant. These three parameters also determine the magnitudes of the quasistatic fields used in conduction, induction or capacitive links. The medium density and mechanical compressibility control its acoustic wave behaviour.
- Obstacles placed between the transmitter and the receiver can have an obstructing effect by either reflecting the transmitter signal away from the receiver, or by shielding the transmitter or receiver. Conductors larger than a good fraction of a wavelength act as reflectors of r.f. signals. Abrupt media transitions can be reflective for electromagnetic (incl. light) or sound waves, depending on the difference in media characteristics and the angle of incidence. Their interface surface should also measure at least a good fraction of a wavelength for this effect to occur.

  Shielding effects find their origin in the fact that an electric or alternating magnetic field cannot exist inside a perfect conductor. Static or dynamic electric fields acting on a conductor, induce charges on its outer

surface that cancel out the original field inside the conductor. Alternating magnetic fields induce eddy currents that also compensate the applied fields. The better the conductor, the more effective the applied fields are neutralized. Shielding makes r.f. or induction telemetry problematic if the devices are put inside metal parts (car), surrounded by conductive media or packaged in full metal cans.
Note that the obstructing effects have nothing to do with absorption, because there is no dissipation required. They do occur in lossless situations with ideal reflectors and conductors.

- Absorption is the third environmental effect. One can distinguish the direct absorption, which directly relates to the applied fields, and the indirect absorption, which originates from the parasitic effects induced by the applied fields. Direct absorption of non-radiating fields is found with high-permeable substances that exhibit hysteresis losses for alternating magnetic fields, and with insulators that show dielectric losses for alternating electric fields. An example of indirect absorption is the Ohmic dissipation by eddy currents that are induced by an alternating magnetic field.
  Direct and indirect absorption causes waves to gradually lose their energy as they pass through a medium. A homogeneous absorbing medium with no scattering obeys the simple one-dimensional exponential law of absorption called the Lambert-Beer Law:
  $$I(x) = I(x=0).e^{-\alpha x}$$
  $I(x)$ represents the intensity of a plane wave travelling in the $x$-direction and $\alpha$ the absorption per meter. When the wavelength approaches the size of an object or medium inhomogeneity, scattering occurs and substantial deviations from the Lambert-Beer Law are observed.
- Finally, the telemetry signals can be masked by environmental noise.

### 1.2.3.1 Influence on electromagnetic links

Electromagnetic properties of technical materials are readily found in textbooks on electromagnetic and electrical engineering. Reliable biological data are somewhat more difficult to obtain. Biological tissues have electromagnetic properties that highly depend on the frequency. Cell membranes act as insulating layers with a high capacity below roughly 100 MHz. The dielectric constant $\varepsilon$ is high and the conductivity $\sigma$ is low. At higher frequencies, the capacitor reactance of the cells decreases so that the currents also flow in the intracellular medium (Table 1.5; NEUKOMM, 1978). An extensive review of electromagnetic tissue properties is given by SCHWAN and FOSTER (1980). Other general data are found in (PETHIG, 1979; Table I in SOWINSKI and VAN DEN BERG, 1992; Table III in ORCUTT and GANDHI, 1988; Table I in SULLIVAN *et al.*, 1987, 1988; Table I in HOQUE and GANDHI, 1988; Table I in STUCHLY *et al.*, 1987; and in Table I in LIN, 1976). Specialised data on muscle tissue were published by GIELEN *et al.* (1984, 1986) and some on cortical bone by SAHA and WILLIAMS (1992).

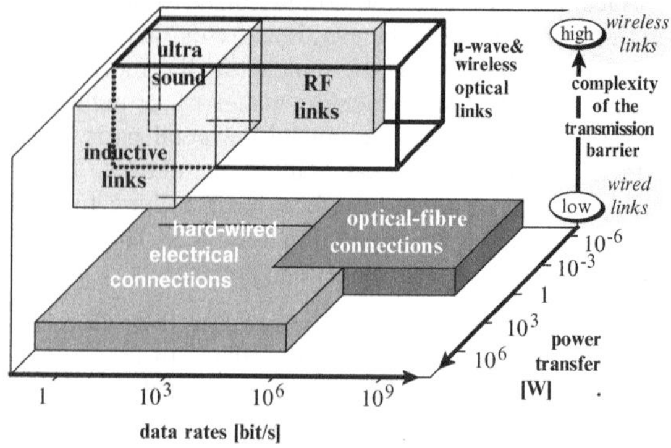


*Figure 1.20. A comparison of interconnection techniques, as a function of their data and power transfer capabilities and how easy they pass complex transmission barriers.*

### 1.2.3.2 Influence on optical links

Infrared optical telemetry is popular for various reasons related to its interaction with the surroundings. Though it is mainly useful indoors, due to its somewhat limited transmission range and the high ambient light levels outdoors, it shows some very interesting characteristics there. Metallic structures and electric cabling in modern buildings seriously hamper medium range (10 m) indoor r.f. telemetry, because of the distortions, reflections and shielding effects they cause. Most wall and ceiling covers are however, well reflective for IR rays. Hidden metal constructions or electric installations have no influence. The IR signals are not absorbed but reflected back into the room, which keeps signal levels high and spatially uniform. Inserting an obstacle between the transmitter and the receiver does not obstruct the data transfer. Unlike with r.f. telemetry, the diffraction phenomena associated to the reflections do not restrain the signal reception, as IR light has very short wavelengths. The receiver cells have orders-of-magnitude larger sizes and diffractions are thus not perceived. Transmission boundaries are also well defined, as there is no spillover to adjacent rooms. This is very practical if several identical telemetry systems operate simultaneously, but makes telemetry from one room to another impossible. Note that glass is opaque for IR.

IR communication with devices implanted in the body is possible as soft tissue is fairly transparent for IR light above 700 nm. Visible light and UV are seldom used as they don't penetrate the body that well and because of their potential hazards (JOHNSON and GUY, 1972). Light absorption of soft tissues is strongly related to the optical behaviour of skin and blood. There is light scattering in tissues because of the micrometer sizes of cells and blood cells. Blood shows a strong absorption band centred at 415 nm with minor absorption peaks in the 550 nm region. The absorption is low above 700 nm. The absorption of blood is function of its oxygenation. Oxyhemoglobin has lower absorption in the red portion of the spectrum compared to hemoglobin. Thus blood looks red when oxyhemoglobin molecules are predominant.

*QUALITATIVE COMPARISON OF THE WIRELESS, RADIATION LINKS FOR TELEMETRY APPLICATIONS (THE DATA TRANSMITTER OUTPUT IS ASSUMED TO BE SOME mW TO ACCOMMODATE SMALL UNITS).* *TABLE 1.8.*

| | r.f. Links | Ultra sound | Optical links |
|---|---|---|---|
| **Transfer range (from base to remote site)** | Determined by:<br>→TX output power<br>→Media losses<br>→Antennae gains<br>→Antenna-to-medium matching<br>→Shielding effects<br>→Noise in the used frequency band<br>e.g. high in air (km), small in water (m) | Determined by:<br>→TX output power<br>→Media losses<br>→Transducer-to-medium matching<br>→Reflections at media interfaces<br>→Ambient noise in the used frequency band<br>e.g. high in water (km), medium in air (10 m) | Determined by:<br>→TX output power<br>→Media transparency<br>→Reflections at media interfaces<br>→Ambient light or other light sources<br>e.g. medium in air (10 m) |
| **Transfer range (remote→base)** | Identical as for base to remote site | Identical as for base to remote site | Identical as for base to remote site |
| **Appropriate modulation techniques** | →AM, FM, PM<br>→AM/AM, FM/AM, PM/AM, AM/FM, FM/FM, PM/FM, AM/PM, FM/PM, PM/PM<br>→any ASK, FSK or PSK | →AM, FM,<br>→AM/AM, FM/AM, PM/AM, AM/FM, FM/FM, PM/FM,<br>→any kind of ASK or FSK | →AM<br>→Any kind of ASK |
| **Max. data signal freq. or pulse rate** | $\leq f_{carrier}/20$<br>without reflections | $\leq f_{carrier}/20$<br>without reflections | (a) Direct view transfer<br>$\leq f_{carrier}/20$<br>(b) Diffuse transfer: some 10 Mpulse/s indoor |
| **Number of independent data channels** | 1 per carrier freq.<br>or<br>1 per subcarrier | 1 per carrier freq.<br>or<br>1 per subcarrier | 1 per carrier (= light colour). Subcarrier modulation is not adopted for optical transfer |
| **Data transfer mode** | →Half duplex on 1 carrier. (by time multiplexing)<br>→Full duplex on 2 carriers | →Half duplex on 1 carrier (by time multiplexing)<br>→Full duplex on 2 carriers | →Half duplex on 1 carrier (by time multiplexing)<br>→Full duplex on 2 carriers |
| **Max. power transfer capacity to remote unit** | →Some Watts over some cm (with μ-wave techniques)<br>→Some μW over larger distances | →Small amounts of mechanical power<br>→Can be transformed into electrical energy with piezo-crystals | Limited by:<br>→Illumination level<br>→Solar cell size ($\leq$1 mW/cm$^2$) |
| **Transducer dimensions** | Base and remote antennae: one or several wavelengths | Ultrasound microphone and speaker: some cm$^3$ | →Light emitter: some mm$^3$ for an LED<br>→Light detector: some mm$^3$ for an photo- diode or -transistor<br>→Solar cell $\geq$ 0.1 m$^2$/W (0.5 mm thickness) |
| **Media requirements** | →r.f. Transparent<br>→Low electrical conductivity and low magnetic permittivity to avoid shielding | →Low acoustic damping<br>→No sharp interfaces to avoid echoes | →Optical transparent<br>→No sharp interfaces to avoid reflections |
| **Environmental impact** | →Eddy currents in nearby conductors (including body tissue)<br>→EM absorption (in body tissue)<br>→EMI to other radio communications | →Animal disturbance<br>→Physical damage on nearby objects or living tissue<br>→Excellent EMC | →Heating by absorption<br>→UV degradation of chemical substances<br>→UV hazards on living tissue<br>→excellent EMC |

*TABLE 1.8. (CONTINUED)* *QUALITATIVE COMPARISON OF THE WIRELESS, RADIATION LINKS FOR TELEMETRY APPLICATIONS (THE DATA TRANSMITTER OUTPUT IS ASSUMED TO BE SOME* **mW** *TO ACCOMMODATE SMALL UNITS).*

| | **conduction links** | **capacitive links** | **induction links** |
|---|---|---|---|
| **Transfer range (base to remote)** | Small (= the size of the generated current field) | Very small (something like a tenth of the electrode size) | Small (≈ the base coil radius) |
| **Transfer range (remote to base)** | Identical as for base to remote site | Identical as for base to remote site | Small (= the size of the remote coil magnetic field = a few times the remote coil diameter) |
| **Appropriate modulation techniques** | → AM, FM, PM<br>→ AM/AM, FM/AM, PM/AM, AM/FM, FM/FM, PM/FM, AM/PM, FM/PM, PM/PM<br>→ any kind of ASK, FSK or PSK | → AM, FM, PM<br>→ AM/AM, FM/AM, PM/AM, AM/FM, FM/FM, PM/FM, AM/PM, FM/PM, PM/PM<br>→ any kind of ASK, FSK or PSK | From base to remote:<br>→ AM, FM,<br>→ AM/AM, FM/AM, PM/AM, AM/FM, FM/FM, PM/FM,<br>→ any kind of ASK, FSK<br>From remote to base:<br>→ AM,<br>→ AM/AM, FM/AM, PM/AM,<br>→ any kind of ASK |
| **Max. data signal freq. or pulse rate** | $\leq f_{carrier}/20$ | $\leq f_{carrier}/20$ | $\leq f_{\text{magn. field}}/20$ |
| **Number of independent data channels** | 1 per carrier freq.<br>or<br>1 per subcarrier | 1 per carrier freq.<br>or<br>1 per subcarrier | One |
| **Data transfer mode** | → Half duplex on 1 carrier (by time multiplexing)<br>→ Full duplex on 2 carriers | → Half duplex on 1 carrier (by time multiplexing)<br>→ Full duplex on 2 carriers | → Half duplex (by time multiplexing) |
| **Max. power transfer capacity to remote unit** | → Limited by the maximal allowed current level in the medium | → Limited by the maximal allowed voltage across the link capacitors | → Some 10 W for good coil coupling<br>→ Several mW for low coil coupling |
| **Transducer dimensions** | → Electrode size is not important (few $mm^3$)<br>→ The electrode placement is the dominant factor (well separated) | Capacitor size ≈ at least ten times the desired range | → Base coil radius ≈ desired range<br>→ Remote coil: some cm (for transfers in the Watt range) or even mm (mW) |
| **Media requirements** | → High electrical conductivity | → High permittivity, low loss angle, low electrical conductivity | → Low electrical conductivity to avoid eddy currents |
| **Environmental impact** | → Ohmic dissipation in the medium<br>→ EM absorption (in body tissue...)<br>→ Electro-shock risk at too high current levels | → Eddy currents in conductors between the capacitor electrodes (including body tissue)<br>→ EM absorption (in body tissue) | → Eddy currents in nearby conductors (including body tissue)<br>→ EM absorption by body tissue<br>→ EMI to other radio communications |

The optical skin properties are controlled by scattering and absorption. The transmittance shows peaks for wavelengths around 1, 1.6 and 2.15 μm. The skin heavily absorbs UV radiation below 0.3 μm, to protect the underlying tissues against hazardous UV exposure. Beyond 1.4 μm, the skin transmission is governed by the Lambert-Beer Law and is very similar to that of water.

#### 1.2.3.3 Influence on ultrasound links

Ultrasound is most appropriate for telemetry through electrically conductive media where r.f. transmission is impaired, like metals, but also sea water or body tissues that are conductive due to their heavy ion load. Ultrasounds of MHz frequencies are very popular and well accepted for medical imaging and blood flow measurements. Their wavelength in most soft tissues is 1.5 mm at 1 MHz, and decreases proportionally with the frequency. The wave velocity of sounds and ultrasounds in soft body tissues, goes from 1,450 to 1,520 m/s. The attenuation lies around 0.5 dB/cm/MHz for most soft tissues (BRONZINO, 1995).

## 1.3 CONCLUSIONS

Telemetry opened a wide array of measurement applications the past few decades on items that were previously inaccessible. A multitude of telemetry principles evolved throughout the years, based on electromagnetic radiation, optical or acoustic coupling or magnetic induction (Figure 1.20). Analogue and digital carrier modulation allowed condensing the information flow. Signal digitising with handshaking and error correction pushed transfer reliability to levels that rival hardwired connections.

As data transmission requires energy, the powering of telemetry units is an important issue, especially when dimensions are critical. Battery powering is often considered when mains access is not available. But, the remote units may be too small to encompass sufficient battery capacity, the battery toxicity or the risk of hazardous heat production in case of circuit failure, may render them incompatible with critical biological or explosive environments. Fortunately, power can be transferred over short distances along with the data, to energise the remote unit.

Wireless data transmission and remote unit powering, induces parasitic effects in nearby objects. This ranges from simple absorption and heat dissipation, to complex reactions in biological tissue. The applied energy levels are strictly controlled by several national and international standards, to rule out dangerous exposures that may cause immediate damage, as well as lower-level exposures that may have long-term effects. In the opposite way, surrounding objects and materials also affect the link performance:

- The physical medium properties can alter signal characteristics, like wavelength or amplitude
- Obstructions between transmitter and receiver can just do that, obstruct the link
- Signal energy may be absorbed while travelling through the medium; and
- The telemetry signal may be masked by environmental noise

## 1.4 REFERENCES[6]

**1999/5/EC,** "Radio equipment and telecommunications terminal equipment and the mutual recognition of their conformity," *Official Journal of the European Communities,* no. L 91/10, Mar. 9, 1999 (available on-line at http://ec.europa.eu/enterprise/rtte/ dir99-5.htm)

[6] The essential references on this chapter are marked with an asterisk.

**90/385/EEC**, "Active implantable medical devices," *Official Journal of the European Union*, no. L 189/17, June 20, 1990 (available on-line at http://ec.europa.eu/enterprise/newapproach/standardization/harmstds/reflist/implmedd.html)

**93/42/EEC**, "Medical devices," *Official Journal of the European Union*, no. L 169/1, June 12, 1993 (available on-line at http://ec.europa.eu/enterprise/newapproach/standardization/harmstds/reflist/meddevic.html)

**2004/108/EC**, "Electrical equipment – electromagnetic compatibility (EMC)," *Official Journal of the European Union*, no. L 390/24, Dec. 31, 2004 (available on-line at http://ec.europa.eu/enterprise/electr_equipment/emc/directiv/ dir2004_108.htm)

*Guide for the EMC Directive 2004/108/EC*, May 21, 2007 (available on line at http://ec.europa.eu/enterprise/electr_equipment/emc/guides/emcguide_may2007.pdf)

**2008/C 280/05**, "Commission communication in the framework of the implementation of Directive 2004/108/EC of the European Parliament and of the Council on the approximation of the laws of the Member States relating to electromagnetic compatibility and repealing Directive 89/336/EEC – Publication of titles and references of harmonized standards under the directive," *Official Journal of the European Union*, no. C 280/14, Nov. 4, 2008 (available on-line at http://ec.europa.eu/enterprise/ newapproach/standardization/harmstds/reflist/emc.html)

**2008/C 280/06**, "Commission communication in the framework of the implementation of the Directive 1999/5/EC of the European Parliament and of the Council on radio equipment and telecommunications terminal equipment and the mutual recognition of their conformity – Publication of titles and references of harmonized standards under the directive," C 280/33 Official Journal of the European Union, Nov. 4, 2008 (available on-line at http://ec.europa.eu/enterprise/newapproach/standardization/harmstds/reflist/emc.html)

**2008/C 304/06**, "Commission communication in relation to Council Directive 90/385/EEC on medical devices – Publication of titles and references of harmonized standards under the directive," *Official Journal of the European Union*, no. C 304/8, Nov. 27, 2008 (available on-line at http://ec.europa.eu/enterprise/newapproach/ standardization/harmstds/reflist/meddevic.html)

**Adair E.R.**, "Thermophysiological effects of electromagnetic radiation," *IEEE Eng. Med. Biol.*, vol. EMB-6, no. 1, pp. 37–41, 1987.

**Amlaner C.J. Jr.** and D.W. Macdonald (Eds.), *A handbook on biotelemetry and radio tracking – Proceedings of an International Conference on Telemetry and Radio Tracking in Biology and Medicine*, University of Oxford, March 20–22, 1979, Pergamon, Oxford, England, 1980.

***Annovazzi-Lodi V.** and S. Donati, "An optoelectronic interconnection for bidirectional transmission of biological signals," *IEEE Trans. Biomed. Eng.*, vol. BME-35, pp. 595–606, 1988.

**Baumann S.B.**, W.T. Joines and E. Berman, "Feasibility study of batteryless temperature transponders using miniature microwave cavity resonators," *IEEE Trans. Biomed. Eng.*, vol. BME-34, pp. 754–757, 1987.

**Black H.S.**, *Modulation theory*, D. Van Nostrand, Princeton, NJ, 1953.

***Bronzino J.D.** (Ed), *The biomedical engineering handbook*, CRC Press, Boca Raton, FL, ISBN 0-8493-8346-3, 1995.

**Texas Instruments, "**ISO124 – Precision lowest-cost isolation amplifier**"**, *datasheet SBOS074C*, Texas Instruments Inc., www.ti.com, Sept. 1997, revised Sept. 2005.

**Cattermole K.W.**, *Principles of pulse code modulation*, Iliffe Books, London, 1969.

**Dalziel C.F.**, "Reevaluation of lethal electric currents," *IEEE Trans. IGA.*, vol. IGA-5, no. 5, 1968.

**Diaz S.B.**, B.A. Ovide and V.H.P. Pintos, "Increasing electromagnetic energy transmission from one medium to another by the interposition of intermediate layers," *Med. Biol. Eng. Comput.*, vol. 25, pp. 347–349, 1987.

***Donaldson N. de N.**, "Passive signalling via inductive coupling," *Med. Biol. Eng. Comput.*, vol. 24, pp. 223–224, 1986.

**Donaldson N. de N.**, "Effect of the metallic seal of a hermetic enclosure on the induction of power to an implant," *Med. Biol. Eng. Comput.*, vol. 30, pp. 63–68, 1992.

***Donaldson N. de N.** and T.A. Perkins, "Analysis of resonant coupled coils in the radio frequency transcutaneous links," *Med. Biol. Eng. Comput.*, vol. 21, pp. 612–627, 1983.

**Edell D.J.**, "Chronically implantable neural information transducers," *Proc. 18th. Annual Int. Conf. IEEE-EMBS*, Amsterdam, The Netherlands, paper no. M3-3 on CD-ROM Proceedings, 1996.

**EN 60601-1-1:2001,** "Medical electrical equipment. General requirements for safety. Collateral standard. Safety requirements for medical electrical systems. Section 1.1 Collateral standard: Safety requirements for medical electrical systems," July 2001.

**EN 60601-1-2:2007**, "Medical electrical equipment. General requirements for basic safety and essential performance. Collateral standard. Electromagnetic compatibility. Requirements and tests," Sept. 2007.

**EPCglobal Inc.** *EPCglobal Tag Data Standards Version 1.4* June 2008 (available online at http://www.epcglobalinc.org/standards/tds/)

**Evans N.E.,** "Baseband transmit and receive processors for two-channel PWM telemetry," *Med. Biol. Eng.*, vol. 27, pp. 215–220, 1989.

***Ficchi R.F.** (Ed.), *Practical design for electromagnetic compatibility*, Hayden Book Company, New York, 1971.

**Forster I.C.**, "Preliminary development of a radiotelemetry system for biological applications," *Med. Biol. Eng. Comput.*, vol. 24, pp. 281–291, 1986.

***Foster R.K.**, "Electromagnetic field effects and mechanisms," *IEEE Eng. Med. Biol.*, vol. EMB-15, no. 4, pp. 50–56, 1996.

***Foster R.K.** and J.E. Moulder (Eds.), "Questioning biological effects of EMF," *IEEE Eng. Med. Biol.*, vol. EMB-15, no. 4, pp. 23–102, 1996.

**Friedlander G.D**, "Electricity in hospitals: elimination of lethal hazards," *IEEE Spectrum.*, pp. 40–51, Sept. 1971.

***Gandhi O.P.** (Ed.), "Biological effects of electromagnetic radiation," *IEEE Eng. Med. Biol.*, vol. EMB-6, no. 1, pp. 14–58, 1987a.

**Gandhi O.P.**, "The ANSI radio frequency safety standard: its rationale and some problems," *IEEE Eng. Med. Biol.*, vol. EMB-6, no. 1, pp. 22–25, 1987b.

**Gandhi O.P.**, J.Y. Chen and A. Riazi, "Currents induced in a human being for plane-wave exposure conditions 0-50 MHz and for RF sealers," *IEEE Trans. Biomed. Eng.*, vol. BME-33, pp. 757–767, 1986.

**Geselowitz D.B.,** Q.T.N. Hoang and R.P. Gaumond, "The effects of metals on an transcutaneous energy transmission system," *IEEE Trans. Biomed. Eng.*, vol. BME-39, pp. 928–934, 1992.

**Gielen F.L.H.**, W. Wallinga-de Jonge and K.L. Boon, "Electrical conductivity of skeletal muscle tissue: experimental results form different muscles in vivo," *Med. Biol. Eng. Comput.*, vol. 22, pp. 569–577, 1984.

**Gielen F.L.H.**, H.E.P. Cruts, B.A., Albers, K.L. Boon, W. Wallinga-de Jonge and H.B.K. Boom, "Model of electrical conductivity of skeletal muscle based on tissue structure," *Med. Biol. Eng. Comput.*, vol. 24, pp. 34–40, 1986.

**Guy A.W.**, J.F. Lehmann and J.B. Stonebridge, "Therapeutic applications of electromagnetic power," *Proc. IEEE*, vol. 62, no. 1, pp. 55–75, 1974.

**Halliday D.** and R. Resnick, *Physics - Parts 1 and 2*, Wiley, New York, ISBN 0-471-02456-2, 1977.

**Heetderks W.J.**, "RF powering of millimeter- and submillimeter-sized neural prosthetic implants," *IEEE Trans. Biomed. Eng.*, vol. BME-35, pp. 323–327, 1988.

**Hof A.L.,** G.J.J. Bonga, F.G.J. Swarte and L. de Pater, "Modular PPM telemetry system with radio, infra-red and inductive loop transmission," *Med. Biol. Eng.*, vol. 32, pp. 107–112, 1994.

**Holmer N.-G.**, "Isolation amplifier energized by ultrasound," *IEEE Trans. Biomed. Eng.*, vol. BME-21, pp. 329–333, 1974.

**Holmer N.-G.** and K. Lindström, "Highly isolated power supply energised by ultrasound," *Med. Biol. Eng.*, pp. 233–235, 1973.

**Hoque M.** and O.P. Gandhi, "Temperature distributions in the human leg for VLF-VHF exposures at the ANSI-recommended safety levels," *IEEE Trans. Biomed. Eng.*, vol. BME-35, pp. 442–449, 1988.

**ICNIRP** International Commission on Non-Ionizing Radiation Protection, "Guidelines for limiting exposure to time-varying electric, magnetic, and electromagnetic fields (up to 300 GHz)," *Health Physics*, vol. 74, no. 4, pp. 494–522, 1998 (available on-line at http://www.icnirp.de/documents/emfgdl.pdf).

**IEEE std. C95.1-1991**, *IEEE standards for safety levels with respect to human exposure to radio frequency electromagnetic fields, 3 kHz to 300 GHz*, IEEE, New York, ISBN 1-55937-179-X, 1992.

**IEEE std. C95.1-1999**, *IEEE standard for safety levels with respect to human exposure to radio frequency electromagnetic fields, 3 kHz to 300 GHz,* The institute of electrical and electronics engineers, Inc., 345 East 47th Street, New York, NY 10017–2394, USA, 1999.

**ISO 11785:1996**, *Radio frequency identification of animals – Technical concept*, 1996.

**ISO 14223-1:2003**, *Radio Frequency Identification of Animals, Advanced transponders – Part 1: air interface*, 2003.

**ISO/IEC 14443-2:2001**, *Identification cards – Contactless integrated circuit cards – Proximity cards— Part 2: Radio frequency power and signal interface,* 2001.

**ISO/IEC 18000-2:2004** *Information technology – Radio frequency identification for item management – Part 2: Parameters for air interface communications below 135 kHz*, 2004.

**ISO/IEC 18000-3:2008**, *Information technology – Radio frequency identification for item management – Part 3: Parameters for air interface communications at 13,56 MHz*, 2008.

***Jeutter D.C.** (Ed.), "Biomedical Telemetry," *IEEE Eng. Med. Biol.*, vol. EMB-2, no. 1, pp. 10–50, 1983a.

***Jeutter D.C.**, "Overview of biomedical telemetry techniques," *IEEE Eng. Med. Biol.*, vol. EMB-2, no. 1, pp. 17–23, 1983b.

***Johnson C.C.** and A.W. Guy, "Nonionizing electromagnetic wave effects in biological materials and systems," *Proc. IEEE*, vol. 60, pp. 692–720, 1972.

**Kawahito S.**, S. Ueda, M. Ishida, T. Nakamura, S. Usui and S. Nagaoka, "A CMOS integrated circuit for multichannel multiple-subject biotelemetry using bidirectional optical transmissions," *IEEE Trans. Biomed. Eng.*, vol. BME-41, pp. 400–406, 1994.

**Kimmich H.P.,** "Biotelemetry, based on optical transmission," *Biotelem. Patient Monit.*, vol. 9, pp. 129–143, 1982.

**King R.W.P.,** "Embedded bare and insulated antennas," *IEEE Trans. Biomed. Eng.*, vol. BME-24, pp. 253–260, 1977.

**King R.W.P.,** "Transmission and reception with embedded antennas," *IEEE Trans. Biomed. Eng.*, vol. BME-25, pp. 397–400, 1978.

**Klein F.F.** and D.A. Davis, "A low-powered 4-channel physiological radio telemetry system for use in surgical patient monitoring," *IEEE Trans. Biomed. Eng.*, vol. BME-23, pp. 478–481, 1976.

***Ko W.H.**, S.P. Liang and C.D.F. Fung, "Design of radio-frequency powered coils for implant instruments," *Med. Biol. Eng. Comput.*, vol. 15, pp. 634–640, 1977.

**Kudo Y.** and K. Ueda, "The measurement of the electric activity of the Olfactory Bulb in free swimming carp (*Cyprinus Carpio* L.) by underwater telemetry systems," *IEEE Trans. Biomed. Eng.*, vol. BME-27, pp. 694–698, 1980.

**Lerner E.J.**, "RF radiation: biological effects," *IEEE Spectrum*, pp. 51–59, Dec. 1980.

**Liew H.-C.** and E.-H. Chan, "Biotelemetry of green turtles (Chelonia mydas) in Pulau Redang, Malaysia, during the internesting period," in: *Biotelemetry XII (Proc. 12th Int. Symposium on Biotelemetry)*, Ancona, Eds. P. Mancini and S. Fioretti, Editrice Universitaria Litografia Felici, Pisa, Italy, pp. 157–163, 1993.

**Lin W.C.** and S.K. Pillay, "A micropower pulsewidth-modulation-pulse-position-modulation two-channel telemetry system for biomedical applications," *IEEE Trans. Biomed. Eng.*, vol. BME-21, pp. 273–280, 1974.

**Lin J.C.**, "Interaction of two cross-polarized electromagnetic waves with mammalian cranial structures," *IEEE Trans. Biomed. Eng.*, vol. BME-23, pp. 371–375, 1976.

**Loeb G.E.**, C.J. Zamin, J.H. Schulman and P.R. Troyk, "Injectable microstimulator for functional electrical stimulation," *Med. Biol. Eng. Comput.*, vol. 29, pp. NS13-NS19, 1991.

***Mackay R.S.**, *Bio-medical telemetry – sensing and transmitting biological information from animals and man*, 2nd edn., Wiley, New York, ISBN 0-471-56030-8, 1970.

***Mackay R.S.**, *Bio-medical telemetry – sensing and transmitting biological information from animals and man*, 2nd edn., IEEE Press, 445 Hoes Lane, PO Box 1331, Piscataway, NJ, ISBN 0-7803-1028-4, 1993.

**Matsuki H.,** M. Shiiki, K. Nadehara, T. Yamamoto and K. Murakami, "Transcutaneous energy transmission using amorphous magnetic fiber for an artificial heart system," *Int. J. Appl. Electromagnet. Mater.* 1, pp. 201–205, 1990.

**Meindl J.D.** and A.J. Ford, "Implantable telemetry in biomedical research," *IEEE Trans. Biomed. Eng.*, vol. BME-31, pp. 817–823, 1984.

**Mitamura Y.**, E. Okamoto, A. Hirano and T. Mikami, "Development of an implantable motor-driven assist pump system," *IEEE Trans. Biomed. Eng.*, vol. BME-37, pp. 146–156, 1990.

**Mueller J.S.** and H.T. Nagle, "Feasibility of inductive powering of miniature low-power biotelemetry for use with microfabricated biomedical sensors," in: *Biotelemetry XIII (Proc. 13th Int. Symposium on Biotelemetry)*, Williamsburg, VI, Eds. C. Cristalli, C.J. Amlaner Jr. and M.R. Neuman, pp. 372–377, 1996.

**Nardin M.,** B. Ziaie, J. Von Arx, A. Coghlan, M. Dokmeci and K. Najafi, "An inductively powered microstimulator for functional neuromuscular stimulation," in: *Biotelemetry XIII (Proc. 13th Int. Symposium on Biotelemetry)*, Williamsburg, VI, Eds. C. Cristalli, C.J. Amlaner Jr. and M.R. Neuman, pp. 99–104, 1996.

**Neukomm P.A.**, "Artefacts from RF and MW telemetry: estimation of safety aspects and review on biological effects," in: *Biotelemetry IV (Proc. 4th Int. Symposium on Biotelemetry)*, Garmisch-Partenkirchen, Germany, Eds. H.-J. Klewe and H.P. Kimmich, Klewe/Kimmich, Döring-Druck, Braunschweig, Germany, pp. 51–54, 1978.

**Neukomm P.A.**, H. Kündig, H. Baggenstos and K. Zerobim, "Passive Telemetry by absorption modulation – A new principle for long term transabdominal monitoring of pressure and EMG of the uterus of cows," in: *Biotelemetry X (Proc. 10th Int. Symposium on Biotelemetry)*, Ed. C.J. Amlaner Jr., University of Arkansas Press, Fayetteville, AR, pp. 487–496, 1988.

**Orcutt N.** and O.P. Gandhi, "A 3-D impedance method to calculate power deposition in biological bodies subjected to time varying magnetic fields," *IEEE Trans. Biomed. Eng.*, vol. BME-35, pp. 577–583, 1988.

**Pethig R.**, *Dielectric and electronic properties of biological materials*, Wiley, Chichester, ISBN 0-471-99728-5, 1979.

**Pitsillides K.F.**, J.D. Symons and J.C. Longhurst, "Biotelemetry of cardiovascular hemodynamic measurements in miniswine," *IEEE Trans. Biomed. Eng.*, vol. BME-39, pp. 982–986, 1992.

***Polk C.** and E. Postow (Eds.), *Handbook of biological effects of electromagnetic fields*, 2nd edn., CRC Press, Boca Raton, FL, ISBN 0-8493-0641-8, 1995.

**Puers B.** and W. Sansen, "A dedicated processing and control IC for pressure telemetry," *Proc. IEEE CICC 1985*, Portland, OR, pp. 144–146, 1985.

**Puers B.**, P. Wouters, K. Van Schuylenbergh, M. De Cooman, W. Sansen, V. Goedseels, R. Geers, B. Truyen, K. Goosens, H. Ville, "An intelligent data-acquisition and control unit for animal husbandry welfare using telemetry links," in: *Biotelemetry XI (Proc. 11th Int. Symposium on Biotelemetry)*, Yokohama, Eds. A. Uchiyama and C.J. Amlaner Jr., Waseda University Press, Tokyo, Japan, pp. 141–145, 1991.

**Repacholi M.H.**, "Radiofrequency electromagnetic field exposure standards," *IEEE Eng. Med. Biol.*, vol. EMB-6, no. 1, pp. 18–21, 1987.

**Rochelle J.M.**, "Design of gateable transmitter for acoustic telemetering tags," *IEEE Trans. Biomed. Eng.*, vol. BME-21, pp. 63–66, 1974.

**Saha S.** and P.A. Williams, "Electric and dielectric properties of wet human cortical bone as function of frequency," *IEEE Trans. Biomed. Eng.*, vol. BME-39, pp. 1298–1304, 1992.

**Sansen W.** and B. Puers, "Pressure senor signal processing," *Proc. IEEE Symp. Biosensors,* Los Angeles, CA, pp. 116–118, 1984.

**Sansen W.**, B. Puers and R. Govaerts, "Design and realisation of an ultra-miniature hybrid for a pressure telemetry capsule," *Proc. Int. Microelectronics Conf. 1984,* Tokyo, Japan, pp. 406–409, 1984a.

**Sansen W.,** R. Puers and R. Vereecken, "Realization of a pressure telemetry capsule for cystometry," *Proc. IEEE Front. Eng. Comput. Health Care,* Los Angeles, CA, pp. 711–714, 1984b.

**Sawan M.**, F. Duval, M.M. Hassouna, J.-S. Li, M.M. Elhilali, J. Lachance, M. Leclair, S. Pourmehdi and J. Mouïne, "Computerized transcutaneous control of a multichannel implantable urinary prosthesis," *IEEE Trans. Biomed. Eng.*, vol. BME-39, pp. 600–609, 1992.

***Schuder J.C.,** H.E. Stephenson and J.F. Townsend, "High-level electromagnetic energy transfer through a closed chest wall," *IRE Int. Conv. Record.*, vol. 9, pp. 119–126, 1961.

**Schuder J.C.**, J.H. Gold and H.E. Stephenson, Jr., "An inductively coupled RF system for the transmission of 1 kW of power through the skin," *IEEE Trans. Biomed. Eng.*, vol. BME-18, pp. 265–273, 1971.

***Schwan H.P.** and K.R. Foster, "RF-field interactions with biological systems: electrical properties and biophysical mechanisms," *Proc. IEEE*, vol. 68, no. 1, pp. 104–113, 1980.

**Schwartz M.**, *Information transmission, modulation, and noise – A unified approach to communication systems*, 2nd ed., McGraw-Hill, New York, 1970.

**Seidel D.,** M. Knoll and J. Eichmeier, "Stimulation of subjective appearance of light (phosphenes) in man by alternating magnetic fields," *Pflugers Archiv*, vol. 229, p. 11, 1968.

**Shanmugan K.S.**, *Digital and analog communication systems*, Wiley, New York, ISBN 0-471-03090-2, 1979.

**Skilling H.H.**, *Fundamentals of electric waves*, Wiley, New York/London, 1948.

**Slater A.** and S. Bellet, "An underwater temperature telemetry system," *Med. Biol. Eng.*, vol. 7, pp. 633–639, 1969.

**Sowinski M.J.** and P.M. van den Berg, "A three-dimensional iterative scheme for an electromagnetic inductive applicator," *IEEE Trans. Biomed. Eng.*, vol. BME-39, pp. 1255–1264, 1992.

**Stuchly M.A.** and K.H. Mild, "Environmental and occupational exposure to electromagnetic fields," *IEEE Eng. Med. Biol.*, vol. EMB-6, no. 1, pp. 15–17, 1987.

**Stuchly M.A.,** A. Kraszewski, S.S. Stuchly, G.W. Hartsgrove and R.J. Spiegel, "RF energy deposition in a heterogeneous model of man: near-field exposures," *IEEE Trans. Biomed. Eng.*, vol. BME-34, pp. 945–950, 1987.

**Sullivan D.M.**, Borup D.J. and O.P. Gandhi, "Use of finite-difference time-domain method in calculating EM absorption in human tissues," *IEEE Trans. Biomed. Eng.*, vol. 34, pp. 148–157, 1987.

**Sullivan D.M.**, Gandhi O.P. and A. Taflove, "Use of finite-difference time-domain method in calculating EM absorption in man models," *IEEE Trans. Biomed. Eng.*, vol. 35, pp. 179–185, 1988.

**Tang Z.**, B. Smith, J.H. Schild and P.H. Peckham, "Data transmission from an implantable biotelemeter by load-shift keying using circuit configuration modulator," *IEEE Trans. Biomed. Eng.*, vol. BME-42, pp. 524–528, 1995.

**Towe B.C.**, "Passive biotelemetry by frequency keying," *IEEE Trans. Biomed. Eng.*, vol. BME-33, pp. 905–909, 1986.

**Tsuji T.**, T. Ohshima, D. Hashimoto and T. Togawa, "Telemetry of human organ temperature with quartz crystal resonator using ultrasonic detection," in: *Biotelemetry XI (Proc. 11th Int. Symposium on Biotelemetry)*, Yokohama, Japan, Eds. A. Uchiyama and C.J. Amlaner Jr., Waseda University Press, Tokyo, Japan, pp. 378–380, 1991.

**Tsuji T.**, T. Fujimoto, T. Izuto and T. Togawa, "Telemetry of organ temperature with quartz crystal resonator using ultrasonic detection for postoperative patient monitoring," *Proc. 12th Int. Symposium on Biotelemetry*, Ancona, Italy, Eds. P. Mancini and S. Fioretti, Editrice Universitaria Litografia Felici, Pisa, Italy, pp. 378–380, 1993.

**Ueno S.,** P. Lövsund and P.A. Öberg, "Effects of alternating magnetic fields and low-frequency electric currents on human skin blood flow," *Med. Biol. Eng. Comput.*, vol. 24, pp. 57–61, 1986.

**Ulrich W.D.**, "Ultrasound dosage for nontherapeutic use on human beings – extrapolations from a literature survey," *IEEE Trans. Biomed. Eng.*, vol. BME-21, pp. 48–51, 1974.

**Utsuyama N.,** H. Yamaguschi, S. Obara, H. Tanaka, S. Fukuta, J. Nakahira, S. Tanabe, E. Bando and H. Miyamoto, "Telemetry of human electrocardiograms in aerial and aquatic environments," *IEEE Trans. Biomed. Eng.*, vol. BME-35, pp. 881–884, 1988.

**Wouters P.**, M. De Cooman, K. Van Schuylenbergh, B. Puers and W. Sansen, "Injectable biotelemetry CMOS chip for identification and temperature measurement system," *Proc. ESSCIRC '91*, Milano, Italy, Sept. 11–13, pp. 45–48, 1991.

**Wouters P.**, M. De Cooman, S. Vergote and B. Puers, "Simultaneous telemetric monitoring of body temperature and activity of herding mammals," in: *Biotelemetry XII (Proc. 12th Int. Symposium on Biotelemetry)*, Ancona, Italy, Eds. P. Mancini and S. Fioretti, Editrice Universitaria Litografia Felici, Pisa, Italy, pp. 128–136, 1993.

# 2 The concepts of inductive powering

This chapter delves into the basics of inductive powering and reviews the different techniques that are currently available in literature. The fundamentals of inductive powering are explained after a brief recapitulation of magnetic-induction theory. This section also includes an overview of the different blocks that build up a driven-inductive link; being the coil set, the rectifier, the regulator and the primary coil driver.

Inductive link designers initially strove to optimise the link efficiency. Later on, the link's tolerance to coupling variations and its bandwidth for data transmission were also considered. This spurted a myriad of design methods that may actually confuse rather than help the aspirant designer. This chapter therefore reviews and classifies the various optimisation techniques. It turns out that no more than four main groups encompass all available designs techniques. An evaluation of each group's significance for weakly coupled applications complete this review.

This book puts an emphasis on link design in the case of weak coupling, as that is where the common design methods fail: instead of the holistic system approach that's called for with weak coupling, the primary coil driver, the primary and secondary tanks and the rectifiers and regulator are all considered separately. Chapters 3–5 therefore completely revise inductive link theory and propose a single procedure that is both workable, not approximating and as general as possible to keep it valid over a wide range of applications, including but not limited to weak and very weak coupling.



K. Van Schuylenbergh and R. Puers, *Inductive Powering: Basic Theory and Application to Biomedical Systems*, Analog Circuits and Signal Processing, 41–76. 

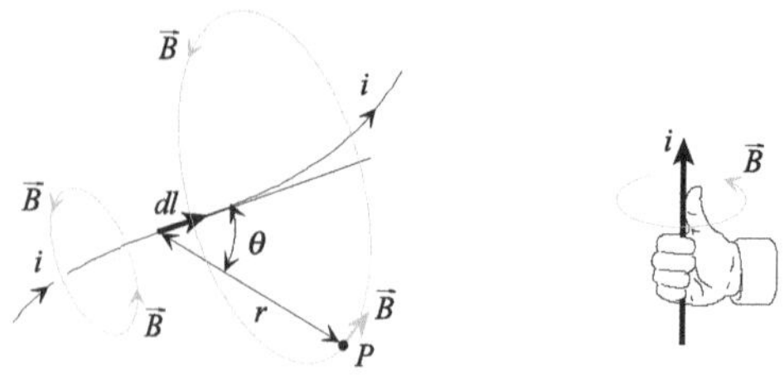


*Figure 2.1.* *The magnetic-flux density around a current conductor. The field direction is found with the "right-hand rule": Grasp the wire with the right hand, the thumb pointing in the direction of the current. The fingers will curl around the wire in the field's direction.*

# 2.1 INDUCTION THEORY

## 2.1.1 MAGNETIC FIELDS

A magnetic flux density $\vec{B}$ [T] can be generated by an electric current. The field direction is described by the "right-hand rule" as depicted in Figure 2.1. The field amplitude is calculated from ***Ampère's Law*** that states that the loop integral of the $\vec{B}$-field equals the net current $i$ enclosed by the loop:

$$\oint \vec{B}.d\vec{l} = \mu_0.i$$

The parameter $\mu_0$ is the ***permeability*** in vacuum ($\mu_0 = 4.\pi.10^{-7}$ *H/m*). Some materials (mainly iron and nickel and a variety of alloys of these and other elements) have built-in microscopic permanent magnets that line up to externally applied fields like compass needles, thereby setting up a magnetisation field $\vec{M}$ that amplifies the applied field. The higher material permeability $\mu$ should be used instead of $\mu_0$ in the expression above to include this so called ***ferromagnetic effect. The relative permeability $\mu_r$ is defined as the ratio between the permeability and the permeability in vacuum*** (Table 2.1). The ferromagnetic effect is non-linear. It shows hysteresis for varying fields and it saturates above a certain field strength because all the built-in magnets are aligned then.

The magnetic-flux density $\vec{B}$ in a point $P$ is also calculated by integration of the ***Biot-Savart Law***. Each current element $i.d\vec{l}$ establishes a field contribution $d\vec{B}$ at the point P (Figure 2.1):

$$dB = \frac{\mu_0}{4\pi}\frac{i.dl.\sin\theta}{r^2},$$

where $r$ is the distance from the length element $d\vec{l}$ to the point $P$. The vector $d\vec{B}$ runs perpendicular to the plane defined by $P$ and $d\vec{l}$.

The ***magnetic flux $\Phi_B$*** that flows through any arbitrary surface $A$ encircled by a given closed loop, is defined as

$$\Phi_B \equiv \iint_A \vec{B}.d\vec{A} \qquad [Wb = T.m^2]$$

*THE RELATIVE PERMEABILITY OF SOME POPULAR MATERIALS.* *TABLE 2.1.*

| Material | Relative permeability $\mu_r$ |
|---|---:|
| Vacuum | 1 |
| Air | 1 |
| Aluminium | 1 |
| Copper | 1 |
| Cobalt | 250 |
| Nickel | 600 |
| Soft steel | 2,000 |
| Iron (0.2% impurities) | 5,000 |
| Silicon iron (4% Si) | 7,000 |
| 78 Permalloy (78% Ni) | 100,000 |
| μ-Metal | 100,000 |
| Pure iron (0.05% ) | 200,000 |
| Supermalloy (5% Mo, 79% Ni) | 1,000,000 |

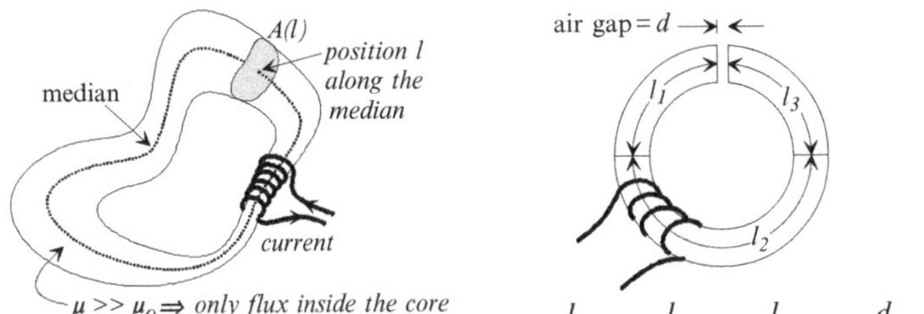


$$R_m = \frac{l_1}{\mu_1.A_1} + \frac{l_2}{\mu_2.A_2} + \frac{l_3}{\mu_3.A_3} + \frac{d}{\mu_0.A_d}$$

*The Ampère Law in terms of reluctances. The example on the right is an air-gapped toroid core made of three different materials.* *Figure 2.2.*

This value $\Phi_B$ depends only on the given loop and not on the chosen surface because magnetic-flux lines are always closed.

It is quite cumbersome to calculate the flux of a magnetic field that flows through several different media with the Ampère or the Biot-Savart Laws. Reformulating the Ampère's Law in terms of ***reluctances*** makes it more adapted to such applications:

$$\oint \frac{\vec{B}}{\mu} d\vec{l} = i \Rightarrow \Phi_B \oint \frac{dl}{\mu A(l)} = i$$

$$\Rightarrow \Phi_B = \frac{i}{R_m} \text{ where } R_m = \oint \frac{dl}{\mu.A(l)}$$

$R_m$ is called the ***reluctance*** or the ***magnetic resistance*** of the magnetic loop [A/Wb]. *A(l)* is the cross-section perpendicular on the field vectors at the position *l* in the loop. The total reluctance of a composite magnetic circuit can now be easily split into a sum of reluctances (Figure 2.2, right). Reluctances have properties similar to electrical resistances: the classic formulae for parallel and series resistors are valid, magnetic fields follow tracks of minimal reluctance, the biggest reluctance, like an air gap, is dominant for the total reluctance of a magnetic series circuit, and so on.

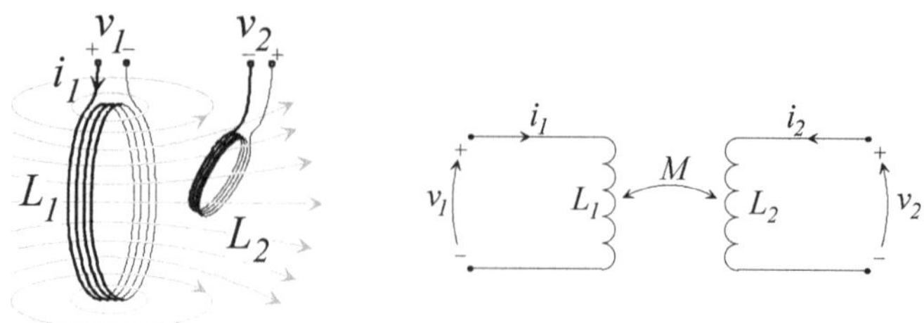


*Figure 2.3. Two mutually coupled coils and their electronic network symbol.*

***Faraday's Law*** states that a changing magnetic flux through a loop-shaped conductor, induces an electromotive force (or emf) in that loop.

$$\oint \vec{E}.d\vec{l} = \frac{d\Phi_B}{dt} \qquad [V]$$

The flux changes can be generated by moving the loop within a non-uniform field or by changing the current that produces the field. The emf is polarised such that if one would connect the loop-conductor ends, the loop current would cause an extra magnetic field that counteracts the original flux changes (this effect is called ***Lenz's Law***).

Note that the magnetic-flux density $\vec{B}$ in the Laws of Ampère, Biot-Savart and Faraday, equals the net magnetic field formed by the original field plus the induced fields.

### 2.1.2 MUTUALLY COUPLED COILS

If two coils are in each other's vicinity, a current $i_1$ in one coil $L_1$ sets up a flux $\Phi_B$ through the second coil $L_2$ (Figure 2.3). If this flux is changed by changing the $L_1$ current, an induced emf appears in the second coil $L_2$ according to Faraday's Law. The emf's in the consecutive coil windings are added so that the voltage $v_2$ across the coil terminals becomes:

$$v_2(t) = n_2.\frac{d\Phi_B(t)}{dt}$$

The parameter $n_2$ is the number of $L_2$ coil windings. The flux $\Phi_B$ is linear proportional to the coil current $i_1$ (cf. Biot-Savart).

$$v_2(t) \propto n_2.\frac{di_1(t)}{dt} \Leftrightarrow v_2(t) = M_{12}.\frac{di_1(t)}{dt}$$

A current $i_2$ in the second coil $L_2$ will induce in the very same way, a voltage $v_1$ on the coil $L_1$:

$$v_1(t) = M_{21}.\frac{di_2(t)}{dt}$$

It can be shown that $M_{12} = M_{21} = M$. This parameter ***M*** [H] is called the ***mutual inductance*** between the coils. It is linearly proportional to the product $n_1.n_2$ and is further function of the geometry of the coils, their mutual position and the surrounding materials that influence the magnetic-field distribution. There are two effects involved in this. Neighbouring ferromagnetic substances reshape the magnetic field so that flux lines follow pathways of minimal reluctance. Nearby electric conductors become virtual coil loops in which currents are induced. These ***eddy currents*** counteract the field that induced them and cause Ohmic dissipation in the conductive material.

However, it's not really necessary to have two coils to produce induction effects. An inductive emf also appears in a single coil if the current in that coil is changed. This emf is polarised such that it counteracts the external source that made the coil current change.

$$v_1(t) = L_1.\frac{di_1(t)}{dt}$$

This effect is called ***self-induction***. The coil inductance $L_1$ [H] is directly proportional to the square of the number of coil windings $n_1^2$ and is function of the coil geometry and of surrounding materials that influence the field distribution.

Both induction effects always act simultaneously. This makes that a coil emf has two contributions: one from the current in the coil itself and one from each current in any other, mutually coupled coil. In this way, the network equations for the two-coil set-up shown in Figure 2.3, become:

$$v_1(t) = L_1\frac{di_1(t)}{dt} + M\frac{di_2(t)}{dt}$$ *Equation 2.1.*

$$v_2(t) = M\frac{di_1(t)}{dt} + L_2\frac{di_2(t)}{dt}$$ *Equation 2.2.*

The ***coupling coefficient or coupling factor k*** and the ***square root of the inductance ratio n*** are two popular dimensionless variables to describe mutually coupled coils:

$$k \equiv \frac{M}{\sqrt{L_1.L_2}}; \quad n \equiv \sqrt{\frac{L_2}{L_1}}$$

***The coupling coefficient is equal to the fraction of the flux generated by the first coil that flows through the second coil (and vice versa).***

It is common in electronics to study a circuit's behaviour on a single-frequency sinusoidal signal. The circuit behaviour for other signals is then computed afterwards, considering their Fourier-series expansion. The sine expressions are mostly written in their complex-exponential form (cf. Euler's Law) to further ease the calculations:

$$v(t) = \hat{V}.\cos(\omega t + \varphi) = \mathrm{Re}\left\{\hat{V}.e^{j(\omega t+\varphi)}\right\} = \mathrm{Re}\left\{\underline{\hat{V}}.e^{j\omega t}\right\}$$

Reformulating the network equations of the double-coil set-up in terms of the complex signal amplitudes, results in:

$$\underline{V_1} = j\omega L_1.\underline{I_1} + j\omega M.\underline{I_2}$$ *Equation 2.3.*

$$\underline{V_2} = j\omega M.\underline{I_1} + j\omega L_2.\underline{I_2}$$ *Equation 2.4.*

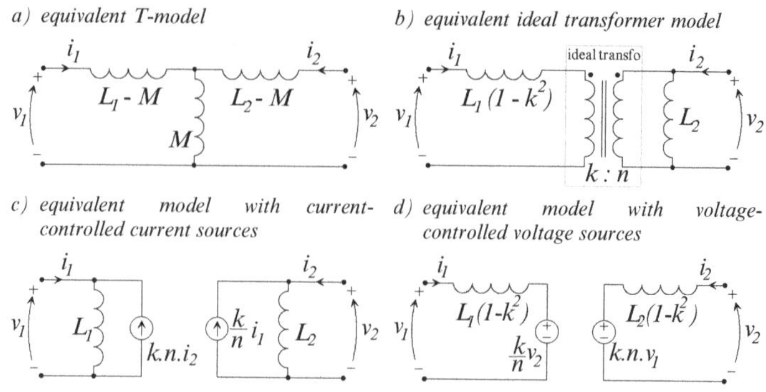


*Figure 2.4. Some equivalent circuit models of two mutually coupled coils.*

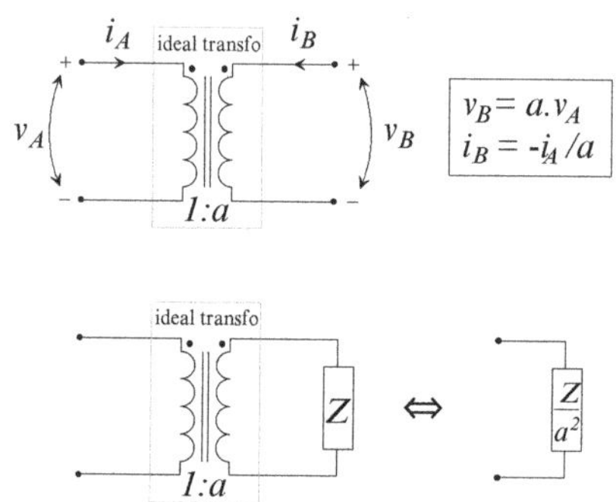


*Figure 2.5. The transformation equations of an ideal transformer.*

### 2.1.3 EQUIVALENT NETWORK MODELS

The mutual coupling complicates circuit computations. It is therefore more practical to replace the coupled coils in a circuit description by an equivalent network model. ***An equivalent network model is defined as another network that is described by the same network equations.*** This means that equivalent models are not unique. The following four models showed to be most convenient for link calculations. Their equivalence is easily proven mathematically.

- The equivalent T-model with three inductances separates the non-coupled part of the inductances from the mutual inductance and models the mutual inductance $M$ as a self-inductance with the same value (Figure 2.4a). ***The non-coupled inductance components $L_1$ - M and $L_2$ - M are named the leak inductances.***
- The equivalent model with an ideal transformer also splits the non-coupled parts of the inductances from the mutual inductance but models the mutual coupling as an ideal transformer (Figure 2.4b) with the coupling factor $k$ as the number of primary windings and the square root of the inductance ratio $n$ as the number of secondary windings.

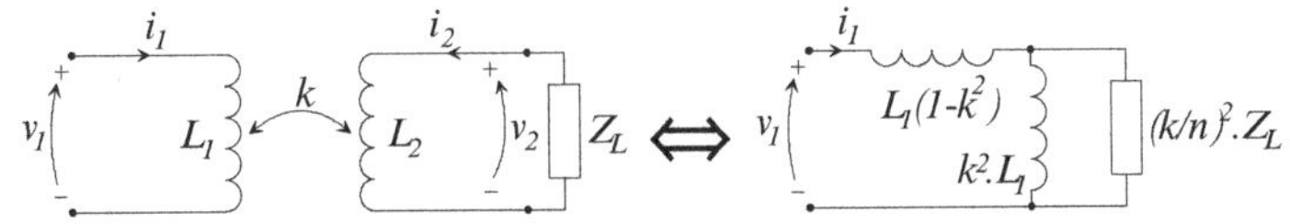


*The equivalent impedance of an inductively coupled circuit, reduced to the primary side. The equivalent model from Figure 2.4 and the ideal transformer equations from Figure 2.5 were used for the reduction.* *Figure 2.6.*

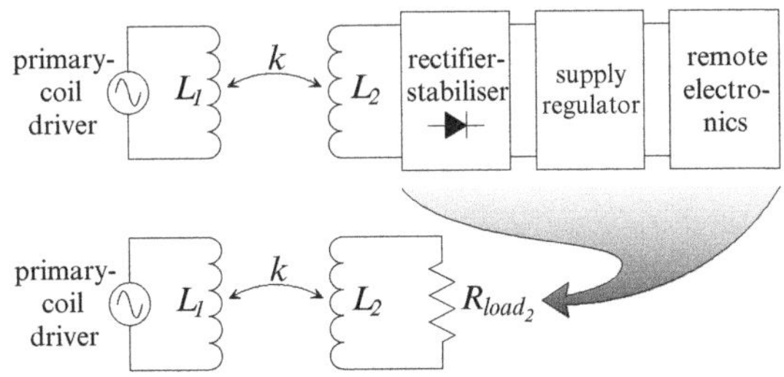


*Basic block diagram of an inductive power link. The rectifier/stabiliser, the supply regulator and the rest of the remote electronics can be represented as an equivalent a.c. resistor $R_{load_2}$.* *Figure 2.7.*

This model is very practical to transform secondary components into equivalent primary impedances (Figure 2.5 and Figure 2.6).

- The two other equivalent models use current-dependant current sources (Figure 2.4c) or voltage-dependant voltage sources (Figure 2.4d) to represent the mutual inductance. Combinations of both are also possible: e.g. a current-controlled current source in the primary and a voltage-controlled voltage source in the secondary.

## 2.2 INDUCTIVE POWERING

### 2.2.1 POWER TRANSFER

Chapter 1 already explained that coupled coils are used to transfer energy to remote electronics. The a.c. voltage induced in a secondary coil is then rectified to supply some remote circuitry (Figure 2.7). A voltage regulator is also added to even the variation on the induced voltage caused by changing coupling and loading conditions.

The power consumption of the remote electronics, the rectifier and the regulator are often represented by an ***equivalent a.c. resistor*** $R_{load_2}$ in inductive link design. Note that the value of this resistor is function of the amplitude of the received voltage: the larger this voltage, the more the regulator cuts and the more it dissipates.

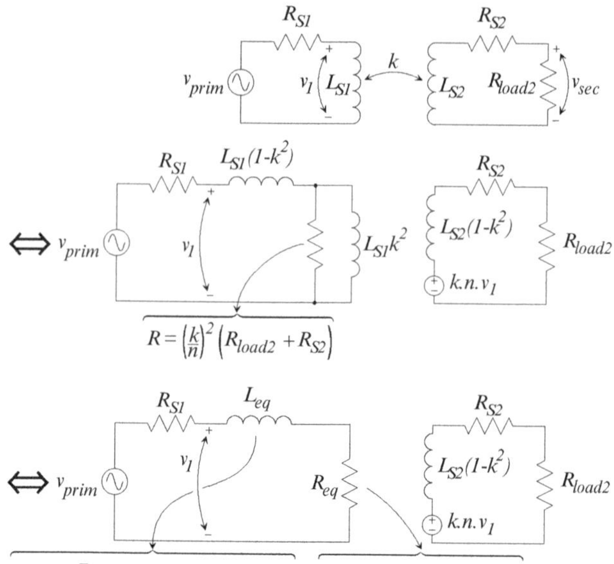


$$R = \left(\frac{k}{n}\right)^2 \left(R_{load2} + R_{S2}\right)$$

$$L_{eq} = L_{S_1} \cdot \left[ \frac{R^2 + \omega^2 L_{S_1}^2 k^4 \left(1 - k^2\right)}{R^2 + \omega^2 L_{S_1}^2 k^4} \right] \qquad R_{eq} = R \left[ \frac{\omega^2 L_{S_1}^2 k^4}{R^2 + \omega^2 L_{S_1}^2 k^4} \right]$$

*Figure 2.8.* *The equivalent link circuits for the calculation of the link efficiency. It is a mix of the models in Figure 2.4.d and Figure 2.6.*

Figure 2.8 demonstrates how the transfer efficiency from input source $V_{prim}$ to the equivalent a.c. load is calculated. The coil-loss resistors $R_{S_1}$ and $R_{S_2}$ have been added to model the link losses.[1] The link efficiency is computed from the resistive dividers $R_{S_1}$–$R_{eq}$ in the primary[2] and $R_{load_2}$–$R_{S_2}$ in the secondary:

$$\eta_{link} = \left( \frac{R_{eq}}{R_{eq} + R_{S_1}} \right) \left( \frac{R_{load_2}}{R_{load_2} + R_{S_2}} \right) = \left( \frac{k}{n} \right)^2 \frac{R_{load_2} \omega^2 L_{S_1}^2 k^4}{\left(R + R_{S_1}\right) \omega^2 L_{S_1}^2 k^4 + R^2 R_{S_1}}$$

---

1 The parasitic coil capacitors are omitted here for simplicity. The transfer frequency is thus assumed to be much smaller than the coil-resonance frequencies. Coil parasitics and equivalent coil models are discussed in detail in Appendix A2.

2 Both resistors conduct the same current. Hence, the power dissipated in $R_{eq}$ is $I^2 R_{eq}$. The total dissipation for both is $I^2 \left(R_{S_1} + R_{eq}\right)$ and the ratio is simply the ratio of resistances.

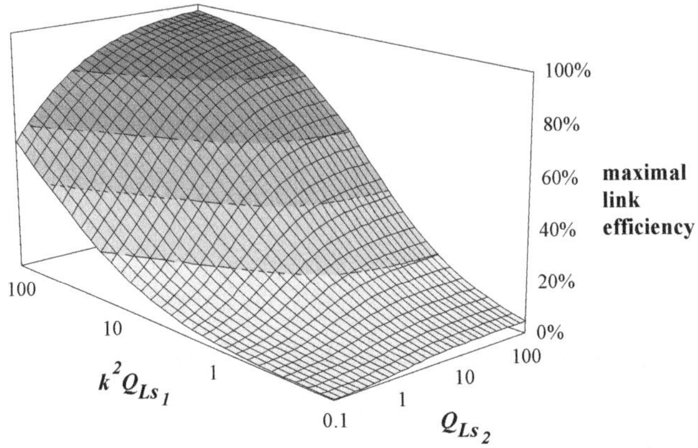


*Maximum link efficiency in function of $k^2 Q_{L_{S_1}}$ and $Q_{L_{S_2}}$ on a log-log scale.* *Figure 2.9.*

This expression becomes easier to interpret by substituting the ***load factor a*** and the coil quality factors $Q_{L_{S_1}}$ and $Q_{L_{S_2}}$[3]:

$$\eta_{link} = \frac{1}{1 + \dfrac{a}{Q_{L_{S_2}}} + \dfrac{a}{k^2 Q_{L_{S_1}}} + \dfrac{1}{k^2 a Q_{L_{S_2}}} + \dfrac{2}{k^2 Q_{L_{S_1}} Q_{L_{S_2}}} + \dfrac{a}{k^2 Q_{L_{S_1}}^2 Q_{L_{S_2}}}},$$

$$\text{where } Q_{L_{S_1}} \equiv \frac{\omega L_{S_1}}{R_{S_1}},\ Q_{L_{S_2}} \equiv \frac{\omega L_{S_2}}{R_{S_2}} \text{ and } a \equiv \frac{\omega L_{S_2}}{R_{load_2}}.$$

This link efficiency becomes maximal for a load factor $a_{\eta_{link_{max}}}$ which means that there is a secondary coil inductance $L_{S_2}$ that optimally fits to the given a.c. load $R_{load_2}$.

$$a_{\eta_{link_{max}}} = \frac{Q_{L_{S_2}}}{\sqrt{1 + Q_{L_{S_2}}^2 + k^2 Q_{L_{S_1}} Q_{L_{S_2}}}}.$$

The maximal link efficiency is only function of two parameters: $k^2 Q_{L_{S_1}}$ and $Q_{L_{S_2}}$ (Figure 2.9):

$$\eta_{link_{max}} = \frac{k^2 Q_{L_{S_1}} Q_{L_{S_2}}}{2 + k^2 Q_{L_{S_1}} Q_{L_{S_2}} + 2\sqrt{1 + Q_{L_{S_2}}^2 + k^2 Q_{L_{S_1}} Q_{L_{S_2}}}}.$$

This maximal efficiency increases with the coil coupling and quality factors. It remains, though, impractically low for realistic quality factors (in the range of 100) and weak to moderate coupling (k below 5%).

[3] This load factor $a$ was originally defined by FORSTER (1981).

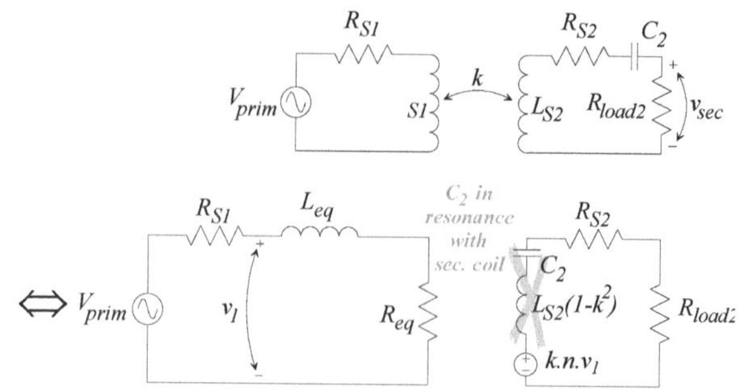


*Figure 2.10.* *Cancelling of the secondary leak inductance* $L_{S_2}\left(1-k^2\right)$ *with a series capacitor* $C_2$.

### 2.2.2 SECONDARY TANK RESONANCE

One of the reasons of poor link efficiency at low coupling comes from the secondary leak inductance $L_{S_2}\left(1-k^2\right)$. This inductance is much larger than the useful load $R_{S_2}$ at weak coupling and demands for a high induced voltage $k.n.v_1$. A higher $v_1$ involves higher primary coil current and hence higher losses. It is therefore common in inductive link design to cancel the secondary leak inductance with a capacitor $C_2$ (Figure 2.10).

The maximal link efficiency for such a series-tuned link is calculated in Chapter 3 and equals[4]:

$$\eta_{link_{max}} = \frac{k^2 Q_{L_{S_1}} Q_{L_{S_2}}}{\left(1+\sqrt{1+k^2 Q_{L_{S_1}} Q_{L_{S_2}}}\right)^2}$$

This maximal link efficiency is only function of one single parameter $k^2 Q_{L_{S_1}} Q_{L_{S_2}}$. Figure 2.11 clearly shows that secondary resonance indeed improves the link efficiency. This diagram, though, may be a bit misleading, because it erroneously suggests that a lower secondary coil quality factor corresponds to a higher link efficiency in case of a non-resonant secondary. The reason is that $X = k^2 Q_{L_{S_1}} Q_{L_{S_2}}$ is taken as horizontal co-ordinate. A lower secondary coil quality factor for the same $X$, automatically implies a higher primary coil quality factor. The corresponding reduction in the primary dissipation overcompensates the increased secondary loss.

---

[4] Expanding the denominator reveals a remarkable similarity with the expression for the link efficiency of non-tuned links, only the $Q^2_{L_{S_2}}$ -term in the denominator is missing:

$$\eta_{link_{max}} = \frac{k^2 Q_{L_{S_1}} Q_{L_{S_2}}}{2 + k^2 Q_{L_{S_1}} Q_{L_{S_2}} + 2\sqrt{1+k^2 Q_{L_{S_1}} Q_{L_{S_2}}}}$$

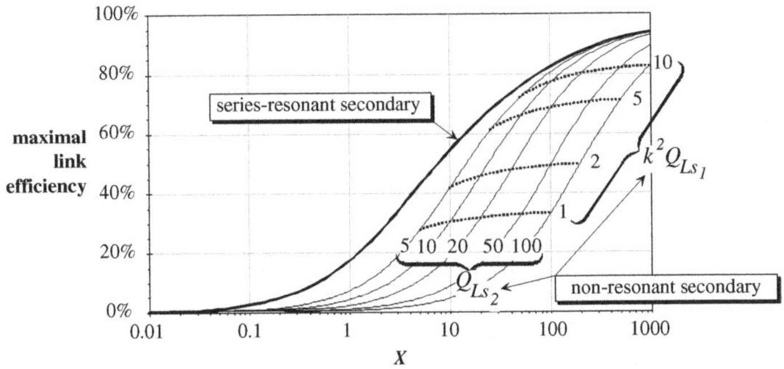


*The maximal link efficiency for links with series-resonant secondary compared to the link efficiency of non-resonant links as a function of X ($\equiv k^2 Q_{L_{S_1}} Q_{L_{S_2}}$ ) and the coil quality factors.* *Figure 2.11.*

Some lines of constant $k^2 Q_{L_{S_1}}$ have therefore been added to indicate that increasing $Q_{L_{S_2}}$ for the same $k^2 Q_{L_{S_1}}$ really increases the efficiency.

Note that high-efficiency, series-resonant links can only be realised for small $R_{load_2}$ as a high $R_{load_2}$ dampens the secondary tank resonance too much.[5] A series-resonant link has thus a current-source output characteristic.

Inductive links with a series-resonant secondary have a current output. They only match to small a.c. load resistors $R_{load2}$.

A voltage-source type output is achieved by parallel resonance of the secondary. The idea of cancelling the leak inductance remains, but the tank capacitor is now placed in parallel to the secondary coil.

Inductive links with a parallel-resonant secondary have a voltage output. They only suitable for large a.c. load resistors $R_{load2}$.

A parallel-tuned set-up has the additional advantage that the parasitic coil capacitor is absorbed in the resonance capacitor $C_2$ allowing higher transfer frequencies and thus smaller coils (Figure 2.12). It is even feasible to omit the separate $C_2$ and use the parasitic coil capacitor as resonance capacitor (LOEB *et al.*, 1991).

---

[5] This also follows from the link expressions in Chapter 3. The maximal link efficiency is found for $L_{S_2} = \dfrac{R_{load_2}}{\alpha\omega}$ and $\alpha = \dfrac{\sqrt{1 + k^2 Q_{L_{S_1}} Q_{L_{S_2}}}}{Q_{L_{S_2}}}$ according to Equation 3.43. and Equation 3.44. The product $k^2 Q_{L_{S_1}} Q_{L_{S_2}}$ is in the order of 1 for moderate to weakly coupled links which makes *a* in the order of $1/Q_{L_{S_2}}$ . $R_{load_2}$ is thus in the order of $\omega L_{S_2} / Q_{L_{S_2}}$ , which is small because high link efficiencies need a large $Q_{L_{S_2}}$ .

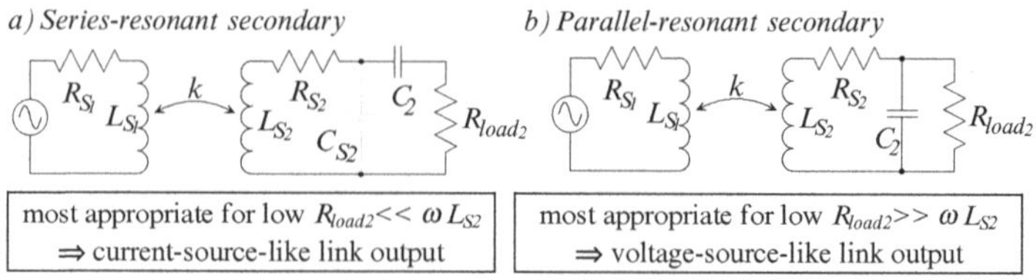


*Figure 2.12.* *(a) Series secondary tuning assumes that the parasitic coil capacitor $C_{S2}$ is much smaller than the resonance capacitor $C_2$.*
*(b) Parallel secondary tuning allows including the parasitic coil capacitor in $C_2$.*

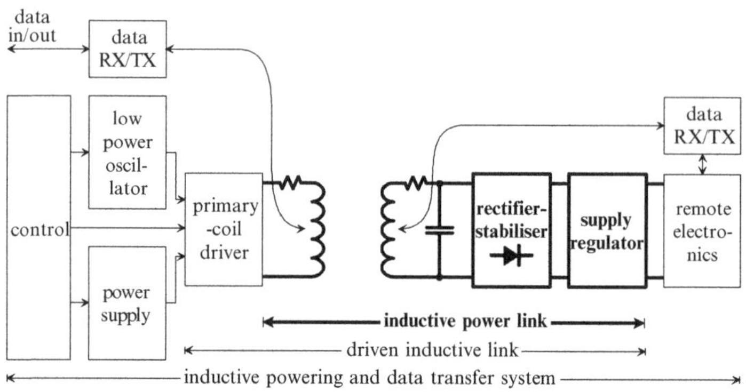


*Figure 2.13.* *From a set of inductively coupled coils to a complete inductive power and data transfer system. The definitions indicated here are used throughout this text.*

## 2.3 THE DRIVEN INDUCTIVE LINK

Two inductively coupled coils need additional circuitry before they can power a remote circuit (Figure 2.13). A considerable a.c. current is to be driven into the primary coil to generate the magnetic fields. The signal received at the secondary side needs rectification and regulation to get a stable d.c. output regardless coupling variations. The determination of the transfer frequency and the driver output power may also need some kind of control. Finally, the data blocks at primary and secondary side take care of the data handling over the link.

### 2.3.1 RECTIFIERS

There are quite a few schemes available to convert an a.c. voltage into a d.c. voltage. This wide selection offers a convenient way to find an a.c. load that better fits the link requirements because, given a certain d.c. load, they all produce other equivalent a.c. loads. An optimal choice cannot be set on forehand, but depends on the actual application. A comprehensive overview therefore follows now.

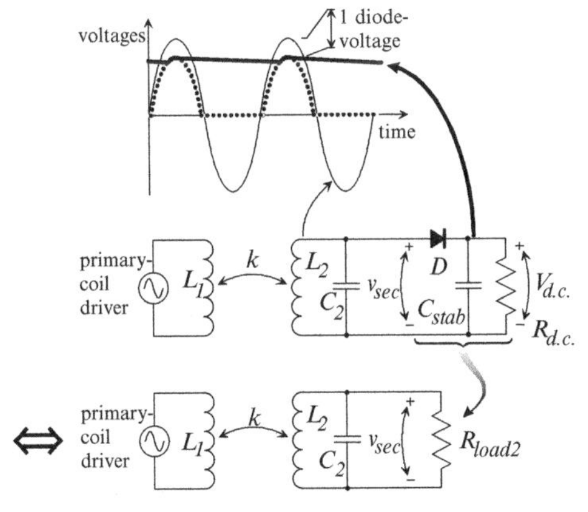


*Half-wave rectification and the definition of the equivalent a.c. load resistor $R_{load_2}$.* *Figure 2.14.*

### 2.3.1.1 Half-wave rectification

The half-wave rectification illustrated in Figure 2.14 is the simplest method. The diode $D$ cuts the negative part off the a.c. signal on $L_2$ and the stabiliser capacitor $C_{stab}$ smoothes the result to a constant d.c. voltage across the load $R_{d.c.}$.

The equivalent a.c. resistor and the rectifier efficiency are calculated as follows. The power dissipation in the secondary is the sum of the power drain in the resistor $R_{d.c.}$ and the dissipation in the diode:

$$P = P_{R_{d.c.}} + P_{diode} = \frac{V_{d.c.}^2}{R_{d.c.}} + \frac{1}{T}\int_0^T v_{diode}.i_{diode}\,dt$$

$$= \frac{V_{d.c.}^2}{R_{d.c.}} + V_{diode}I_{d.c.} = \frac{V_{d.c.}^2}{R_{d.c.}} + \frac{V_{d.c.}V_{diode}}{R_{d.c.}} = \frac{V_{d.c.}}{R_{d.c.}}\left(V_{d.c.} + V_{diode}\right)$$

The d.c. voltage $V_{d.c.}$ equals the a.c. amplitude of the voltage $v_2$ at the rectifier input, minus one diode drop, assuming that the stabilisation capacitor $C_{stab}$ is big enough to neglect any $V_{d.c.}$ ripple. Hence,

$$P = P_{R_{load_2}} = \frac{V_{\sec}^2}{R_{load_2}} = \frac{\hat{V}_{\sec}^2}{2R_{load_2}} = \frac{\left(V_{d.c.} + V_{diode}\right)^2}{2R_{load_2}}$$

$$\Leftrightarrow \begin{cases} R_{load_2} = \dfrac{R_{d.c.}}{2}\left(1 + \dfrac{V_{diode}}{V_{d.c.}}\right) & \text{Equation 2.5.} \\[2ex] \eta_{rectif} \equiv \dfrac{P_{R_{d.c.}}}{P_{R_{d.c.}} + P_{diode}} = \dfrac{V_{d.c.}}{V_{d.c.} + V_{diode}} & \text{Equation 2.6.} \end{cases}$$

It is important to note that the half-wave rectifier of Figure 2.14 accepts a voltage type input and produces a voltage type output. It is thus not suitable for the series-resonant links with their current output, because it cuts the negative phase of the secondary tank current.

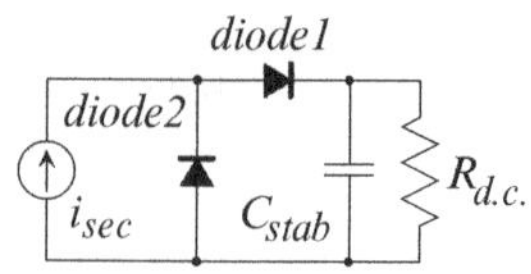


*Figure 2.15. A class-D half-wave rectifier with a current type input and a voltage type output (from KAZIMIERCZUK and JÓZWIK, 1990a).*

A fly-back diode should therefore be added in parallel to $C_2$ in this case, to form a class-D rectifier (Figure 2.15). The positive part of the current $i_{sec}$ flows through the top diode to $C_{stab}$ and $R_{d.c.}$, the negative part flows through the shunt diode.

The power dissipation in the secondary is the power drain in the resistor $R_{d.c.}$ plus the dissipation in the two rectifier diodes:

$$P = P_{R_{d.c.}} + P_{diode1} + P_{diode2} = \frac{V_{d.c.}^2}{R_{d.c.}} + \frac{V_{diode}}{T}\left(\int_0^{T/2} |i_{sec}|\,dt + \int_{T/2}^{T} |i_{sec}|\,dt\right)$$

The d.c. output current $I_{d.c.}$ is the average of the current through diode1:

$$I_{d.c.} = \frac{1}{T}\int_0^{T/2} |i_{sec}|\,dt = \frac{\hat{I}_{sec}}{T}\int_0^{T/2} \sin(\omega t)\,dt = \frac{\hat{I}_{sec}}{\pi}$$

$$\Leftrightarrow P = \frac{V_{d.c.}^2}{R_{d.c.}} + 2V_{diode}I_{d.c.} = \frac{V_{d.c.}^2}{R_{d.c.}} + 2\frac{V_{d.c.}V_{diode}}{R_{d.c.}} = \frac{V_{d.c.}^2}{R_{d.c.}}\left(1 + \frac{2V_{diode}}{V_{d.c.}}\right)$$

The dissipation in the equivalent a.c. model is:

$$P = P_{R_{load_2}} = I_{sec}^2 R_{load_2} = \frac{\hat{I}_{sec}^2 R_{load_2}}{2} = \frac{\pi^2 I_{d.c.}^2 R_{load_2}}{2} = \frac{\pi^2 V_{d.c.}^2 R_{load_2}}{2R_{d.c.}^2}$$

$$\Leftrightarrow \begin{cases} R_{load_2} = \dfrac{2R_{d.c.}}{\pi^2}\left(1 + \dfrac{2V_{diode}}{V_{d.c.}}\right) & \text{Equation 2.7.} \\[2ex] \eta_{rectif} \equiv \dfrac{P_{R_{d.c.}}}{P_{R_{d.c.}} + P_{diode1} + P_{diode2}} = \dfrac{V_{d.c.}}{V_{d.c.} + 2V_{diode}} & \text{Equation 2.8.} \end{cases}$$

There exist six types of half-wave rectifiers, each with a different combination of current or voltage input and current or voltage output (KAZIMIERCZUK and JÓZWIK, 1990a). However, the two rectifiers described here are the only ones suited for inductive powering.

### 2.3.1.2 Full-wave rectification with voltage doubling

***Full-wave rectification*** means that ***the negative phase of the input signal is flipped upwards and also transferred to the d.c. load at the output.*** The time between the current peaks that flow into the stabiliser capacitor is 50% shorter compared to half-wave rectification, and the output ripple is reduced accordingly.

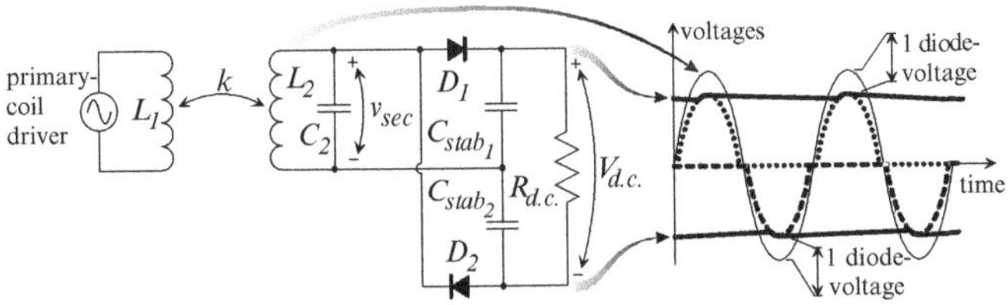


*Full-wave rectification with voltage doubling (from ZIERHOFER and HOCHMAIR, 1992).* *Figure 2.16.*

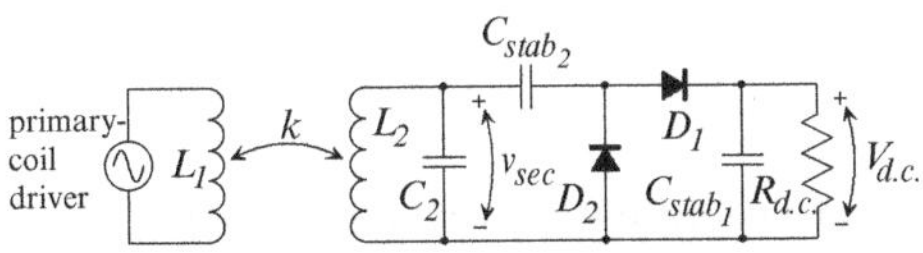


*Full-wave rectification with voltage doubling (from KO, 1977).* *Figure 2.17.*

The a.c. equivalent and efficiency are again calculated as before. The secondary power dissipation is the power drain in the resistor $R_{d.c.}$ plus the dissipation in both rectifier diodes (Figure 2.16 and Figure 2.17):

$$P = P_{R_{d.c.}} + P_{diode1} + P_{diode2} = \frac{V_{d.c.}^2}{R_{d.c.}} + \frac{V_{diode}}{T}\left(\int_0^{T/2} i_{diode1}\, dt + \int_{T/2}^{T} i_{diode2}\, dt\right)$$

The stabiliser capacitors $C_{stab1}$ and $C_{stab2}$ in Figure 2.16. are connected in series and the d.c. current $I_{d.c.}$ flows simultaneously through both of them. $I_{d.c.}$ is thus equal to the average of the current that charges $C_{stab1}$ and also equal to the average of the current that charges $C_{stab2}$. Each integral divided by the period T in the previous expression, is the average of such a capacitor current. Their sum is thus equal to $2I_{d.c.}$ which also counts for the circuit in Figure 2.17:

$$P = \frac{V_{d.c.}^2}{R_{d.c.}} + 2.V_{diode} I_{d.c.} = \frac{V_{d.c.}}{R_{d.c.}}\left(V_{d.c.} + 2V_{diode}\right)$$

Both rectifiers work as a voltage-doubler. The d.c. voltage $V_{d.c.}$ is twice the amplitude $\hat{V}_2$ of the rectifier input, minus two diode drops $V_{diode}$.

$$P = P_{R_{load_2}} = \frac{\hat{V}_{sec}^2}{2R_{load_2}} = \frac{\left(V_{d.c.} + 2V_{diode}\right)^2}{8R_{load_2}}$$

$$\Leftrightarrow \begin{cases} R_{load_2} = \dfrac{R_{d.c.}}{8}\left(1 + \dfrac{2V_{diode}}{V_{d.c.}}\right) \\ \eta_{rectif} \equiv \dfrac{P_{R_{d.c.}}}{P_{R_{d.c.}} + P_{diode1} + P_{diode2}} = \dfrac{V_{d.c.}}{V_{d.c.} + 2V_{diode}} \end{cases}$$

*Equation 2.9.*

*Equation 2.10.*

The full-wave rectifiers have also a complete family with current and/or voltage in- and outputs (KAZIMIERCZUK and JÓZWIK, 1990a).

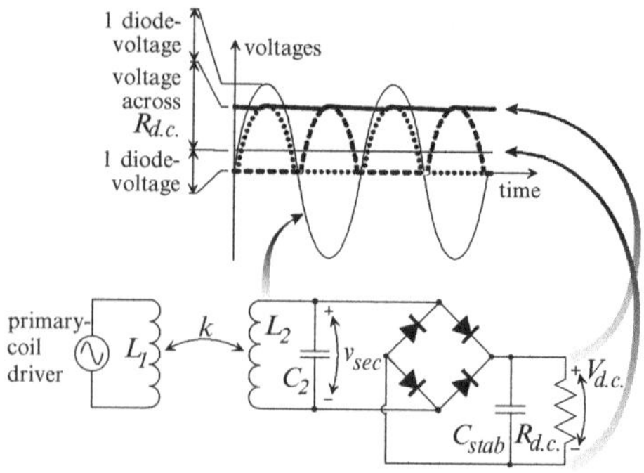


*Figure 2.18. A full-bridge rectifier.*

### 2.3.1.3 Full-wave rectification without voltage doubling

The classic bridge rectifier depicted in Figure 2.18 is also a full-wave rectifier but it lacks the voltage-doubling effect. During the negative phase, the bridge diodes reverse the coil direction with respect to the load.

The secondary dissipation is the power drain in $R_{d.c.}$ and the diodes:

$$P = P_{R_{d.c.}} + P_{diode1} + P_{diode2} + P_{diode3} + P_{diode4}$$

$$= \frac{V_{d.c.}^2}{R_{d.c.}} + \frac{2V_{diode}}{T}\left(\int_0^{T/2} i_{diode1}\,dt + \int_{T/2}^{T} i_{diode3}\,dt\right) = \frac{V_{d.c.}^2}{R_{d.c.}} + 2V_{diode}I_{d.c.}$$

$$= \frac{V_{d.c.}^2}{R_{d.c.}} + 2V_{diode}\frac{V_{d.c.}}{R_{d.c.}} = \frac{V_{d.c.}}{R_{d.c.}}\left(V_{d.c.} + 2V_{diode}\right)$$

The sum of integrals divided by the period T in the above expression, is the average of the diode currents. It equals the d.c. output current $I_{d.c.}$.

The d.c. voltage $V_{d.c.}$ equals the amplitude $\hat{V}_2$ of the rectifier input, minus two diode drops $V_{diode}$:

$$P = P_{R_{load_2}} = \frac{V_{sec}^2}{R_{load_2}} = \frac{\hat{V}_{sec}^2}{2R_{load_2}} = \frac{\left(V_{d.c.} + 2V_{diode}\right)^2}{2R_{load_2}}$$

*Equation 2.11.*

$$\Leftrightarrow \begin{cases} R_{load_2} = \dfrac{R_{d.c.}}{2}\left(1 + \dfrac{2V_{diode}}{V_{d.c.}}\right) \\ \eta_{rectif} \equiv \dfrac{P_{R_{d.c.}}}{P_{R_{d.c.}} + \sum P_{diode_i}} = \dfrac{V_{d.c.}}{V_{d.c.} + 2V_{diode}} \end{cases}$$

*Equation 2.12.*

### 2.3.1.4 Class-E rectifiers

A new class of rectifiers has come about in the area of high-efficient d.c./d.c. converters. Though they haven't been applied for inductive links yet, they may offer new opportunities and deserve more attention in the future. The resonant class-E rectifiers are summarised here mainly to point to the pioneering work of Ronald J. Gutmann and Marian K. Kazimierczuk (GUTMANN, 1980; KAZIMIERCZUK, 1990; KAZIMIERCZUK and JÓZWIK, 1989a, b, 1990a, b; BÎRCA-GALATEANU and COCQUERELLE, 1995).

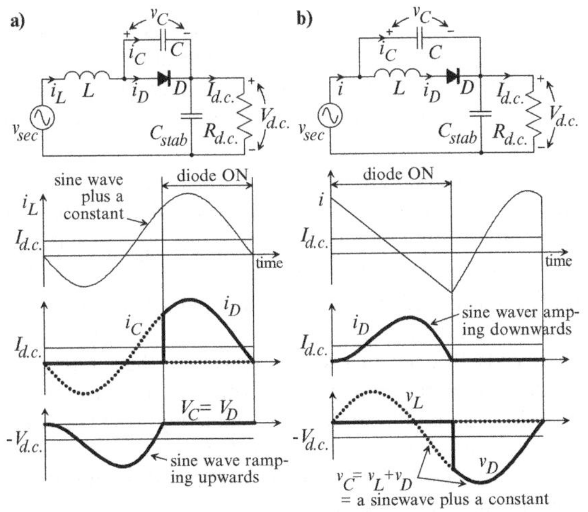


*The class-E zero-voltage-switching (a) and zero-current-switching rectifier (b).* *Figure 2.19.*

The class-E rectifier employs the same principles as the class-E power inverter in order to obtain high power-conversion efficiency at high frequencies (e.g. above 1 MHz). The above expressions for conventional rectifiers assumed ideal diode switching (apart from the voltage drop $V_{diode}$ during conduction). This works fine for moderate frequencies up to a few 10 kHz, but at frequencies above, e.g. 1 MHZ, diode switching is not instantaneous anymore and switching losses come in to play. Parasitic oscillations are also more likely to occur at higher a.c. frequencies.

Switching losses are caused by dissipating the charge stored in diode-junction capacitances or the inductive energy stored in parasitic inductances. The reverse diode-recovery current is flowing while the diode voltage is already building up and thus adds to the diode losses. Parasitic oscillations may occur when the diode capacitance or any other capacitor resonates with the inductors that were either parasitic or intentionally included in the circuit.

The class-E philosophy in general, states that power dissipation in active components is minimised if [6]:

- The active component is only drawing current without carrying a voltage
- On- and off-switching is done at a moment when the voltage across the active element is zero (to avoid capacitive switch losses) or when the current through the element is zero (to avoid inductive losses)
- The diode-current slope at turn-off is low so that delayed off-switching due to reverse-recovery and the corresponding losses are minimised

There are two basic types of class-E rectifiers: the zero-voltage-switching and the zero-current-switching rectifiers (Figure 2.19). The first type is derived from a conventional half-wave rectifier by shunting the diode with a capacitor $C$ and adding a series-resonance coil $L$. The diode

[6] The class-E requirements are discussed in further detail in Chapter 4.

capacitance is included in $C$ and does not adversely affect the circuit operation. The lead and transformer leak inductances are included in $L$. The shunt capacitor shapes the voltage so that the diode turns on and off at low $dv_D/dt$, reducing the current through the diode junction capacitance at both transitions. Abrupt diode-voltage jumps of $dv_D/dt$ are inhibited by the shunt capacitor.

The zero-current-switching rectifier is the dual version. A coil is put in series with the diode of a conventional rectifier. A resonance capacitor is shunting the diode-coil combination. The diode now switches on and off at low $di_D/dt$ because all $i_D$ jumps are inhibited by the series coil.

The class-E rectifiers feature some additional advantages besides their high conversion efficiency. The conduction angle in conventional rectifiers rarely exceeds 10°. Such narrow current pulses have a low d.c. content and generate substantial amounts of harmonics that have to be filtered out. The diode in a class-E rectifier is conducting for a much larger fraction of the a.c. period with a lower peak current and a reduced diode stress as a result. The lower harmonic content also simplifies the filter that extracts the d.c. part of the rectified signal.

An extensive overview of conventional and class-E rectifiers with all the different voltage and/or current input-output combinations is found in (KAZIMIERCZUK and JÓZWIK, 1990a).

### 2.3.2 SECONDARY COIL AND CAPACITOR TAPPING

Coil-tapping is a second way of creating different equivalent a.c. loads for a given d.c. load, next to playing around with different types of rectifiers (Figure 2.20; KO, 1977). This technique should be applied with care as it causes a second resonance frequency. The coil voltage swings up at the resonance frequency $f_1$ of the secondary coil $L_2$ with the capacitor $C_2$, just like with any other link. But, the coil-split $L_{2B}$ also resonates with $C_2$ at a higher frequency $f_2$, thereby shorting the voltage induced in $L_{2A}$. A simple $R_{load_2}$ -$L_2$-$C_2$-equivalent model cannot represent the occurrence of this voltage dip. It makes thus only sense to define an equivalent a.c. load resistor $R_{load_2}$ below and just slightly above the resonance frequency $f_1$.

Define the following ratios between the number of coil windings:

$$n_{2A} \equiv N.n_2 \text{ and } n_{2B} = (1 - N)n_2, \text{ where } 0 < N < 1.$$

The coil voltages are induced from the primary coil, which makes their amplitude proportional to their number of windings. Hence,

$$V_{L_{2A}} = V_{\sec} N$$

But, this relation is only valid well below the resonance frequency $f_1$. <u>At this $L_2$-$C_2$ resonance</u>, the secondary voltages are rather determined by the large resonance current in the loop $L_2$-$C_2$. The coil voltages are then rather proportional to the coil impedances and hence to the squares of the number of windings.

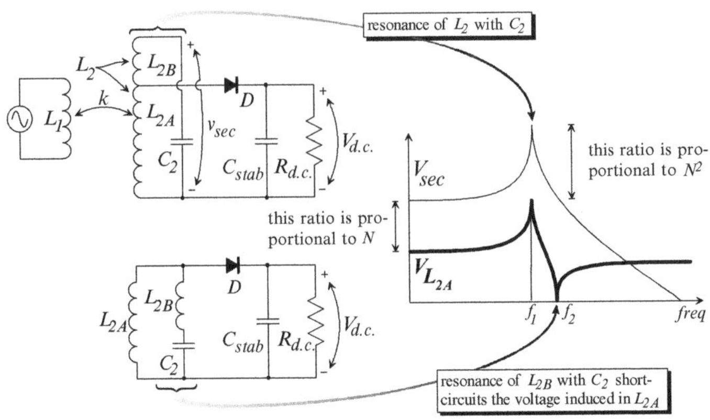


*Secondary coil tapping produces a second resonance effect.* *Figure 2.20.*

This is, however, only valid as long as the induced coil currents can be neglected compared to the resonance current in the loop $L_2$-$C_2$. The quality factor $Q_{sec}$ of the loaded secondary tank should thus be high enough (e.g. > 20) for this assumption to be valid:

$$V_{L_{2A}} \approx V_{sec}\frac{L_{2A}}{L_2} = V_{sec}N^2$$

Equation 2.5. is now recalculated with the coil tap:

$$\Leftrightarrow \begin{cases} P = P_{R_{d.c.}} + P_{diode} = \ldots = \dfrac{V_{d.c.}^2}{R_{d.c.}} + V_{diode}I_{d.c.} = \dfrac{V_{d.c.}}{R_{d.c.}}\left(V_{d.c.} + V_{diode}\right) \\ P = P_{R_{load_2}} = \dfrac{\hat{V}_{sec}^2}{2R_{load_2}} \approx \dfrac{\hat{V}_{L_{2A}}^2}{2R_{load_2}N^4} = \dfrac{\left(V_{d.c.} + V_{diode}\right)^2}{2R_{load_2}N^4} \end{cases}$$

$$\Leftrightarrow R_{load_2} @ L_2 - C_2\ resonance \approx \underbrace{\frac{R_{d.c.}}{2}\left(1 + \frac{V_{diode}}{V_{d.c.}}\right)}_{\text{the equivalent a.c. load for a half-wave rectifier}} \underbrace{\frac{1}{N^4}}_{\text{the correction factor for the coil tap}}$$ *Equation 2.13.*

(Only valid for moderate to high $Q_{sec}$, e.g. > 20)

Capacitor tapping is the dual technique of coil tapping. It does not cause a second resonance frequency. There is only a high-pass roll-off point at a frequency $f_2$, below the actual secondary tank resonance:

$$f_2 = \frac{1}{2\pi R_{a.c.}\left(C_{2A} // C_{2B}\right)}$$

This is because the induced secondary coil voltage is transferred to node ***a*** via the capacitor $C_{2B}$ while that node is connected over $R_{a.c.}$ to the ground (Figure 2.21). The frequency should thus be above $f_2$ for the capacitor impedances to be small enough to drive node ***a***. Equation 2.5. is now recalculated including the tapped capacitors. The expression is valid

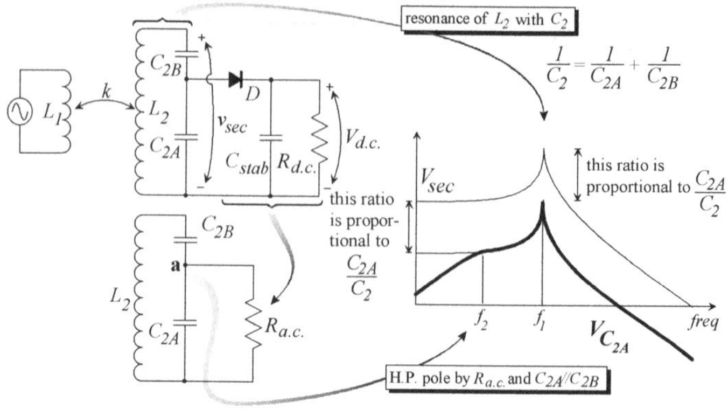


*Figure 2.21.* *Secondary capacitor tapping produces a high-pass roll-off in the voltage transfer.*

for any Q and any $k$ as there are no voltages induced in capacitors, so there is no difference at or away from resonance:

$$V_{C_{2A}} = V_{\text{sec}} \frac{\frac{1}{C_{2A}}}{\frac{1}{C_{2A}}+\frac{1}{C_{2B}}} = V_{\text{sec}} \frac{C_2}{C_{2A}}$$

$$\Leftrightarrow P = P_{R_{load_2}} = \frac{\hat{V}^2_{\text{sec}}}{2R_{load_2}} = \left(\frac{C_2}{C_{2A}}\right)^2 \frac{\hat{V}^2_{C_{2A}}}{2R_{load_2}} = \left(\frac{C_2}{C_{2A}}\right)^2 \frac{\left(V_{d.c.}+V_{diode}\right)^2}{2R_{load_2}}$$

Equation 2.14.

$$\Leftrightarrow R_{load_2} @ f >> f_2 = \underbrace{\frac{R_{d.c.}}{2}\left(1+\frac{V_{diode}}{V_{d.c.}}\right)}_{\text{the equivalent a.c. load for a half-wave rectifier}} \underbrace{\left(\frac{C_2}{C_{2A}}\right)^2}_{\text{the correction factor for the capacitor tap}}$$

### 2.3.3 REGULATORS

The rectified link output voltage or current is most often regulated to make it insensitive to the coil coupling or load variations (DONALDSON, 1983). There are three major regulation mechanisms:

- The linear regulators hold the output on a pre-set value, by continuously controlling a variable shunt or series resistor between the rectifier and the load.
- The switch-mode regulators cut the energy flow from the rectifier to the load when the output exceeds a given threshold and reconnect the rectifier when the output returned to a value below the threshold.
- The switch-mode step-up or step-down regulators first store energy from the input source in a coil or capacitor and then release that energy at the regulator output. The ratio between the time for energy storage and the time of the consecutive release controls the obtained output voltage. Output voltages above the input voltage can be generated, this way.

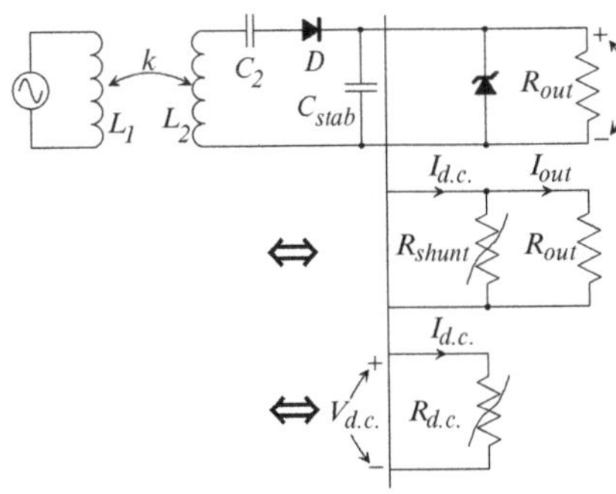


*The Zener diode as a shunt regulator for current-type links.* *Figure 2.22.*

### 2.3.3.1 Linear regulators

***A. The linear shunt regulator***

The shunt regulator is the simplest one in the line of linear regulators. It is a plain Zener-diode shunting both output terminals to clamp the output at the Zener voltage (Figure 2.22). It best fits to a current-type link-output. A voltage-type link would attempt to impose its own voltage and thereby force a high current through the Zener.

The resistance $R_{shunt}$ of the shunt regulator is function of the link output current $I_{d.c.}$ Suppose that $V_{out}$ is the desired output voltage after regulation (of course, $V_{out} = V_{Zener}$). Small currents $I_{d.c.}$ and a small $R_{out}$ keep the regulator input voltage $V_{d.c.}$ below the Zener voltage $V_{Zener}$. As a result, the output is not regulated but follows the input current and voltage:

$$\text{if } I_{d.c.}.R_{out} < V_{out} \text{ then } R_{shunt} = \infty,$$

When a high $I_{d.c.}$ or $R_{out}$ forces $R_{out}.I_{d.c.}$ to surpass $V_{Zener}$, an excess current flows into the shunt to hold the regulator output locked:

$$\text{if } I_{d.c.}.R_{out} > V_{out} \text{ then } R_{shunt} = \frac{V_{out}}{I_{d.c.} - I_{out}}.$$

The equivalent d.c. resistance $R_{d.c.}$ of the loaded regulator becomes:

$$\text{if } I_{d.c.}.R_{out} < V_{out} \text{ then } R_{d.c.} = R_{out},$$
$$\text{if } I_{d.c.}.R_{out} > V_{out} \text{ then } R_{d.c.} \equiv \frac{V_{d.c.}}{I_{d.c.}} = R_{out}\frac{I_{out}}{I_{d.c.}}.$$
*Equation 2.15.*

***The regulator efficiency is defined as the ratio between the useful dissipation in the load $R_{out}$ and the total dissipation in the load and the regulator together.*** Thus,

$$\text{if } I_{d.c.}.R_{out} < V_{out} \text{ then } \eta_{regul} = 1,$$
$$\text{if } I_{d.c.}.R_{out} > V_{out} \text{ then } \eta_{regul} = \frac{I_{out}}{I_{d.c.}}.$$
*Equation 2.16.*

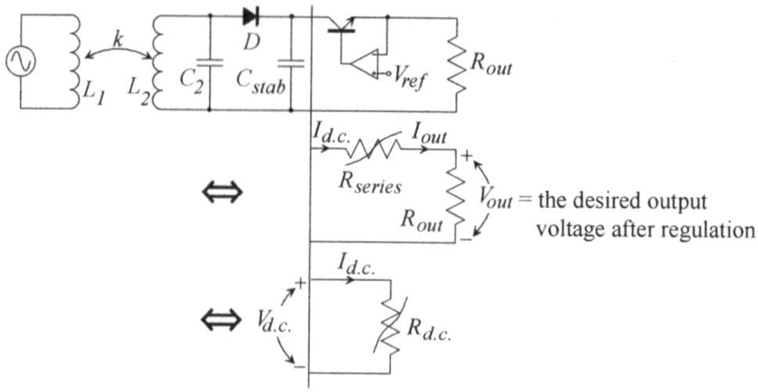


*Figure 2.23.* *The linear series regulator for voltage-type links.*

### *B. The linear series regulator*

The linear series regulator is basically a transistor connected between the rectifier output and the load (Figure 2.23). It is driven by a feedback circuit that compares the output voltage with an internal reference. Series regulators are most suitable for voltage-type link outputs.

The resistance of the series regulator is also function of its input voltage. Suppose again that $V_{out}$ stands for the desired output voltage after regulation. Then,

$$\text{if } V_{d.c.} < V_{out} \text{ then } R_{series} = 0 \text{ and the output is not regulated,}$$

$$\text{if } V_{d.c.} > V_{out} \text{ then } R_{series} = \frac{V_{d.c.} - V_{out}}{I_{out}}.$$

The equivalent resistance $R_{d.c.}$ of the regulator and the load is:

Equation 2.17.

$$\text{if } V_{d.c.} < V_{out} \text{ then } R_{d.c.} = R_{out},$$

$$\text{if } V_{d.c.} > V_{out} \text{ then } R_{d.c.} \equiv \frac{V_{d.c.}}{I_{d.c.}} = R_{out}\frac{V_{d.c.}}{V_{out}}.$$

The regulator efficiency of a series regulator is

Equation 2.18.

$$\text{if } V_{d.c.} < V_{out} \text{ then } \eta_{regul} = 1,$$

$$\text{if } V_{d.c.} > V_{out} \text{ then } \eta_{regul} = \frac{V_{out}}{V_{d.c.}}.$$

### 2.3.3.2 Switch-mode regulators

Linear regulators have the drawback that they dissipate the excess of supply current or voltage they take up. In contrast to this, switch-mode regulators offer theoretical loss-free regulation. Their losses are of second order and occur in the switch-on resistances and in the auxiliary circuits that control the switches. Regulator efficiencies above 85% are well achievable. Several analogue-circuit manufacturers have single-chip step-up and step-down regulators available in the 100 mW–10 W range (Linear Technology, Maxim, National Semiconductors, Texas instruments, etc.). They need only a few external passive components and are well suited for small-sized remote units.

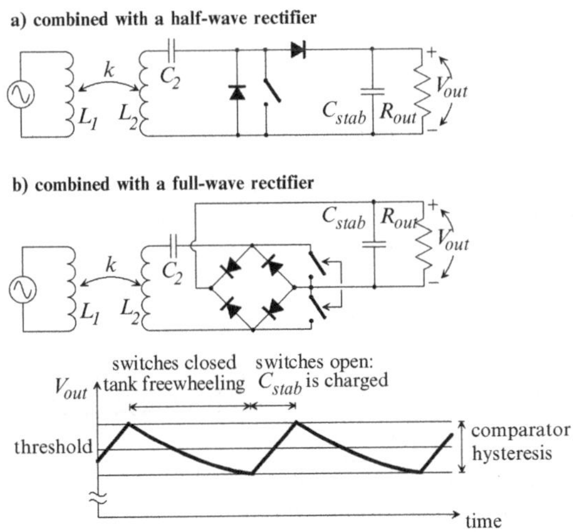


*The switch-mode shunt regulator for current-type links.* *Figure 2.24.*

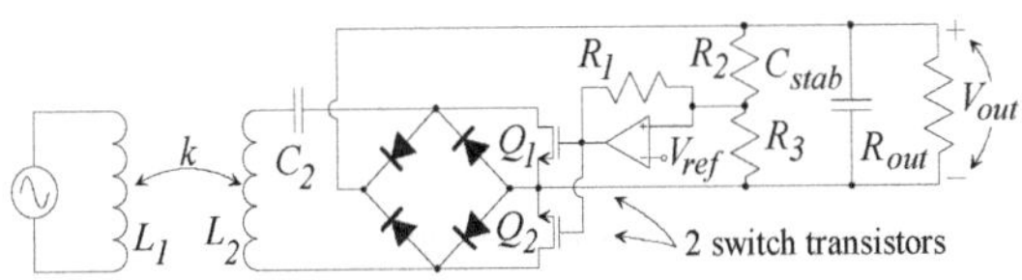


*Circuit of a switching shunt regulator (copied from* SHERMAN *et al., 1985).* *Figure 2.25.*

### *A. The switch-mode shunt regulator*

This regulator switches the power flow to the stabiliser capacitor and the load off when the output voltage $V_{out}$ exceeds a pre-set threshold. This is done by short-circuiting the secondary tank with transistors (Figure 2.24). The output load drains $C_{stab}$ until the output voltage is lowered below the threshold. The short-circuit is then removed, $C_{stab}$ is recharged and the cycle restarts. Theoretically, no power is dissipated as a result of shorting the secondary tank. The tank current is just freewheeling in the lossless secondary loop ($P = I^2R$ with $R = 0$). The only possible dissipation occurs in the secondary coil loss resistor and the switch-on resistor.

The opening and closing of the switches is controlled by comparing the output to a reference voltage. The resistors $R_1$, $R_2$ and $R_3$ are added to create some comparator hysteresis (Figure 2.25; SHERMAN *et al.*, 1985).

Note that the switch-mode shunt regulator should not be combined with a voltage-type link as it has a negative impact on the secondary tank resonance: the switches short the parallel tank, the tank voltage drops to almost zero and it takes several cycles (about equal to the quality factor of the loaded secondary tank) to build-up the tank voltage again to charge $C_{stab}$.

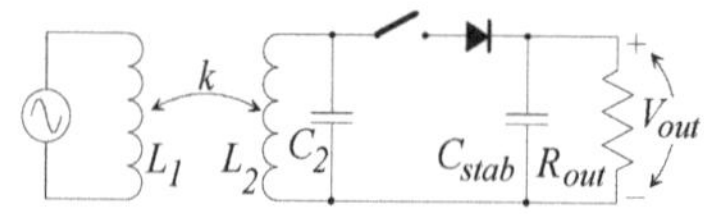


Figure 2.26. *The switch-mode series regulator for voltage-type links.*

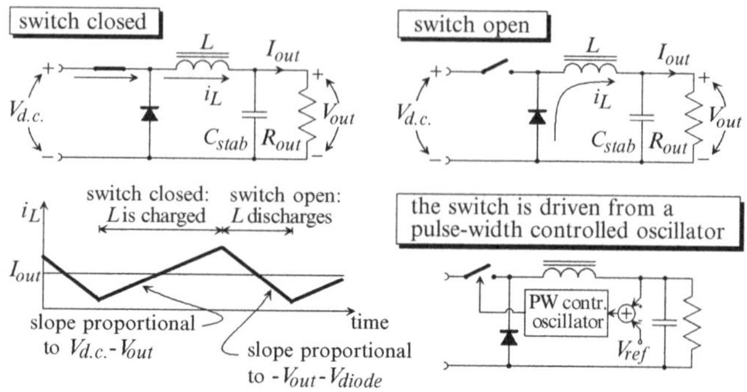


Figure 2.27. *The principle of a voltage-in, voltage-out buck regulator.*

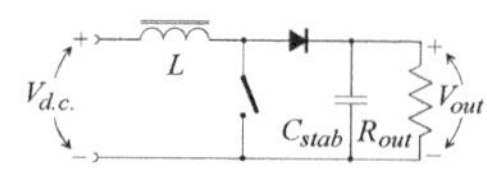


Figure 2.28. *The principle of a voltage-in voltage-out boost regulator.*

***B. The switch-mode series regulator***

This is the reciprocal principle of the shunt regulator (Figure 2.26).

***C. The switch-mode step-down regulator (or buck regulator)***

Buck regulation is based on the temporary storage of energy from the input source (i.e. switch closed) and the successive release of that energy on the output node (i.e. switch open), at a different voltage (Figure 2.27). A coil is the usual storage device.

The switch duty cycle controls the regulator output voltage. A longer time to build-up the current in the storage coil results in a higher regulator output voltage. A feedback circuit compares the output voltage with an internal reference and adapts the switch duty cycle via a pulse-width controlled oscillator running at a constant frequency.

***D. The switch-mode step-up regulator (or boost regulator)***

A simple reconfiguration of the regulator diode, coil and switch creates a boost regulator that allows generating an output voltage higher than the input voltage (Figure 2.28). The output is also regulated by a pulse-width controlled oscillator in a feedback loop.

## 2.3.4 PRIMARY COIL DRIVER

A dedicated amplifier is needed to drive high currents into the primary coil in order to generate the magnetic fields required for the power transfer. The driver requirements were summarised by GUTMANN (1980):

- It is preferably a switch-mode amplifier where the active elements operate as a switch so they only draw currents without carrying a voltage. This is to minimise the dissipation in the active elements.
- The driver output should be a pure sinusoid because only the fundamental component is received at the secondary. The harmonic components do not contribute to the power transfer with a tuned secondary coil, but do cause losses in the primary.
- The primary inductance is tuned with a resonant capacitor.

The latter is to cancel out the large primary leakage inductance that typically occurs with small coupling factors. This leakage inductance causes a large primary coil voltage, given the large coil currents that are required for the inductive powering.

Basically, the primary leakage inductance can be compensated by either a series or a parallel capacitor. Both ways have their benefits and drawbacks. The use of a series-resonant capacitor lowers the amplifier output voltage but maintains the demand for a high output current. The amplifier's output stage needs high-current transistors that require a large base or gate current that also contributes to the driver losses (power MOSFETs have a large gate capacitor due to their large die sizes and take thus large a.c. gate currents). Inductance cancelling with a parallel-resonant capacitor lowers the amplifier output current but maintains the need for a high output voltage. The output stage then contains high-voltage transistors that have large parasitic capacitors. The conclusion is that neither series-tuning nor parallel-tuning enables low-voltage and low-current operation. Luckily, the class-E amplifiers feature a double-tuned circuit. They feature a series-tuned coil with a second capacitor in parallel. This offers an elegant solution as it combines the benefits of low-current operation of a parallel resonance with the low-voltage operation of a series resonance. The class-E drivers are discussed in detail along with their competitors in chapter 4.

There are primarily two approaches for driving a link primary:

- Most links use a small-signal master oscillator (MO) followed by a power amplifier (PA) connected to the primary coil. This MOPA set-up works fine, but at higher coupling, an effect known as ***pole splitting*** occurs (Figure 2.29). If the coil coupling is raised, the apparent primary inductance lowers and the primary resonance frequency increases. This effect is negligible at low coupling, but coupling factors above about 10%, do get primary tanks out of resonance. MILLER J.A. *et al.* (1993) therefore included a feedback loop in their primary coil driver to automatically adapt the driver frequency to the changing tank tuning.

  Pole splitting also occurs for the secondary tank. The equivalent secondary impedance of the driven primary tank dampens the secondary tank and lowers its resonance frequency with increasing coupling. This effect is, however, small.

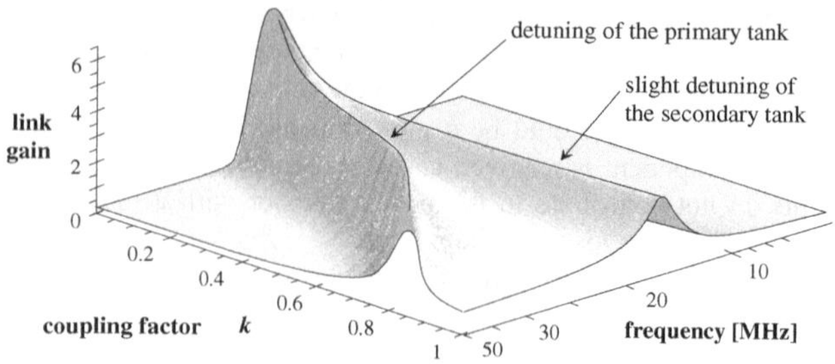


*Figure 2.29.* *Pole-splitting effect simulated on a link with a series-resonant primary and a parallel-resonant secondary, synchronous tuned at 20 MHz.* $L_{S_1} = L_{S_2} = 1.96\mu H, R_{S_1} = R_{S_2} = 5.18\Omega, R_{load_2} = 1k\Omega$

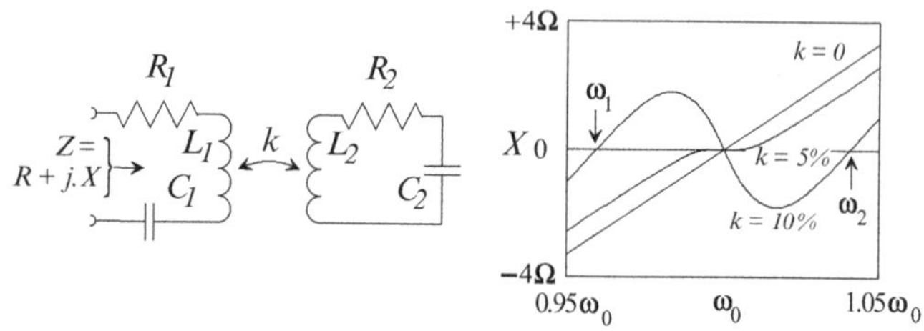


*Figure 2.30.* *Reactive part of the impedance seen looking at the left terminals. Both tanks are synchronous tuned, $L_1 = L_2 = 1\ \mu H$, $C_1 = C_2 = 1\ nF$, $Q_{prim} = 10$ and $Q_{sec} = 20$ (copied from DONALDSON, 1986a).*

- The MOPA approach is straightforward but results in a complicated circuit with two or possibly three circuit tanks to keep in adjustment. A self-oscillating coil driver is simpler – thus less faultprone and cheaper– and it can solve the pole-splitting problem too. It is basically a primary coil driver with its output fed back to its input. This creates a self-oscillating structure where the primary coil and its resonance capacitor, directly determine the driver frequency. The driver frequency is thus always automatically tuned to the driver tank. Despite this, the self-oscillation principle is not problem free. Feeding back the coil voltage, locks the driver at the frequency $\omega_0$ where the reactive part of the primary L–C tank is zero (Figure 2.30). At a coupling above $1/Q_{sec}$ (the quality factor of the loaded, but uncoupled secondary tank), the receiver starts to pull the oscillator. The oscillation at $\omega_0$ becomes unstable (positive $dX/d\omega$), and since two new and stable frequencies $\omega_1$ and $\omega_2$ come up, the oscillator may hop spontaneously between one value and the other (DONALDSON, 1986a).

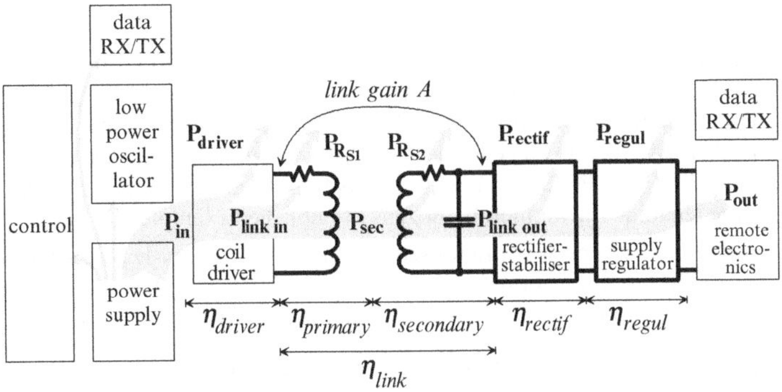


*The power flow in the inductive powering and data transfer system of Figure 2.13.* *Figure 2.31.*

## 2.4 LINK OPTIMISATION

Numerous publications appeared on link design and optimisation since the original paper of SCHUDER *et al.* (1961). Most authors use their own notations and give their own design procedure that fitted their specific application. This makes it for the interested reader difficult to overview the field of available circuits and techniques and to compare the optimisation methods. But, a closer look reveals that all designs, except the two described in sections 2.4.3 and 2.4.4 can be reduced to not more than two basic philosophies:

- Optimisation of the link efficiency $\eta_{link}$, assuming secondary tank resonance, and
- Desensitising the link gain to coupling variations by critical coupling, assuming a resonant primary and a resonant secondary tank

The differences between the design procedures are mostly found in the assumptions and simplifications that were (often implicitly) taken, with a strong dependency on the envisaged application. It is, for instance, common to assume that $k^2 Q_{L_{S_1}} Q_{L_{S_2}}$ is much larger than one. And, this assumption is hard to maintain at low coupling factors.

Note that these methods only focus on finding the set of coils and capacitors that delivers the most optimal link efficiency $\eta_{link}$ (Figure 2.31). The inductive powering system is never considered as a whole, assuming that the driver, rectifier and regulator losses are small compared to the link loss. This assumption is again hard to hold up at small coupling. The weak coupling leaves the primary coil driver more or less freewheeling and makes the driver's losses dominant in the overall system's energy budget. Another important and popular assumption is that the coupling does not influence the tank resonance frequencies. The tanks are calculated as separated circuits with a resonance determined by their own L and C. This becomes a problem at the higher end of the coupling scale, where pole splitting occurs.

### 2.4.1 EFFICIENCY OPTIMISATION

The earliest link designs all started with the computation of the link efficiency $\eta_{link}$. This efficiency was then maximised finding the combination of transfer frequency, primary and secondary inductance that optimally matched with the given equivalent a.c. load (SCHUDER *et al.*, 1961, 1971; KO *et al.*, 1977; FORSTER, 1981). This is by far the most popular optimisation technique, with literally dozens of published examples.

Primary coil tuning was often used to cancel the primary leakage inductance. This was not influencing the link's efficiency or the shifting the operation point for maximal efficiency to other component values.

### 2.4.2 DESENSITISING THE LINK GAIN TO COUPLING VARIATIONS, BY CRITICAL COIL COUPLING

DONALDSON and PERKINS (1983) have found that the voltage gain of a link with a series-resonant primary and a parallel-resonant secondary coil is bell-shaped in function of the coupling factor with a maximum at $k = 1/\sqrt{Q_{prim}Q_{sec}} \equiv k_{crit}$ ($Q_{prim}$ and $Q_{sec}$ are the quality factors of the uncoupled tanks). The rate of change of gain with coupling is zero at this operation point. Small coupling derivations thus hardly change the link output here. This mode is therefore named ***critical coupling.***[7] An identical expression was obtained by HOCHMAIR (1984) for primary and secondary coils that were both in parallel resonance.

There is a minimum value for the coupling factor to allow for critical coupling. The quality factor $Q_{sec}$ of the loaded secondary rarely exceeds 30..40. Higher $Q_{sec}$ would sacrifice on the secondary link efficiency $\eta_{secondary}$. A coil quality factor $Q_{L_{S_2}}$ above 100 is hard to get within a small remote-unit. Hence, producing a $Q_{sec}$ of for example 80 in this case would mean that the useful load has to be smaller than the parasitic coil resistance and that most of the energy induced in the secondary just dissipates in the coil. At the primary side, $Q_{prim}$ is limited to 30..40 by $Q_{L_{S_1}}$ and the tank losses by the driver output. As a result, 2.3% is a minimum value for a critical coupling coefficient.

The penalty for optimised position tolerance is that the theoretical upper limit for the link efficiency $\eta_{link}$ is reduced to 50% compared to the efficiency-optimised links that have an upper limit of 100%.

Critical coupling stabilises the link output to ±8% for coupling variations of ±45%. Additional output regulation can be done by a remote voltage regulator or by sensing the coil coupling and adjusting the driver output accordingly (DONALDSON, 1985).

An alternative mode of critical coupling is described in Chapter 3. It doesn't use primary coil resonance and has a theoretical maximum link efficiency of 100%. However, it needs fairly high coupling factors above about 5%.

---

7 The coupling factor $k = 1/\sqrt{Q_{prim}Q_{sec}}$ should also be smaller than $1/Q_{sec}$, if one wants to run a critically coupled link from a self-oscillating driver and assure a constant transfer frequency and hop-free operation. This makes that $Q_{prim}$ should be larger than $Q_{sec}$ which asks for a coil driver that causes the lowest additional tank loss.

### 2.4.3 LINK STABILISATION THROUGH STAGGER TUNING

Stagger tuning the primary and secondary tanks is a more uncommon technique to desensitise link gain for coupling variations where the resonance frequencies of both tanks are intentionally separated from each other and a constant transfer frequency is chosen right in between them (GALBRAITH *et al.*, 1987). Pole splitting moves the resonance peaks away from the transfer frequency and thereby lowers the gain, with increasing coupling (Figure 2.32). As a coupling increase normally raises the link gain, both effects compensate each other with a properly designed link.

The publications on this technique are rather scarce despite the promising results. This may be due to the complexity associated with the determination of the link components. There are no closed analytic expressions or procedures for the component values that guarantee link operation at maximal efficiency. Moreover, the link design heavily relies on trial and error Spice simulations and prototyping.

There are two important advantages of stagger tuning. The link efficiency is not theoretically limited to 50% as is the case for critical coupling, and the band around the transfer frequency with high gain is considerable broader. There is thus more bandwidth available for data transfer over the link. Stagger tuning is *the* way to get a wide data bandwidth without lowering the coil quality factors and thus sacrificing the link efficiency.

It is also possible to obtain lower link output impedances than with synchronous tuned links.

A stagger tuned link needs a primary coil driver that can efficiently handle a variable load impedance while maintaining a constant driving voltage or current, just because the primary tank is not running at resonance. Only the class-D amplifiers in a MOPA configuration meet these specifications.

Coil drivers that rely on primary tank resonance for lossless operation, such as the (saturating) class-C and class-E amplifiers, cannot be used here.[8] Self-oscillating drivers are also incompatible because they necessarily run at the primary tank resonance.

Stagger tuning has two disadvantages: it needs more coil coupling than critical coupling and it has a lower link gain (Figure 2.32, bottom diagram; DONALDSON, 1988). This means thus a shorter working distance and larger primary coil voltages for a given coil set. Another but less important drawback is that the link gain is a bit more sensitive to coupling variations than with critical coupling. The link gain in function of the coupling is again bell-shaped but with a slightly narrower peak.

[8] Look in Chapter 4 for an overview of all driver types suitable for driving a primary coil.

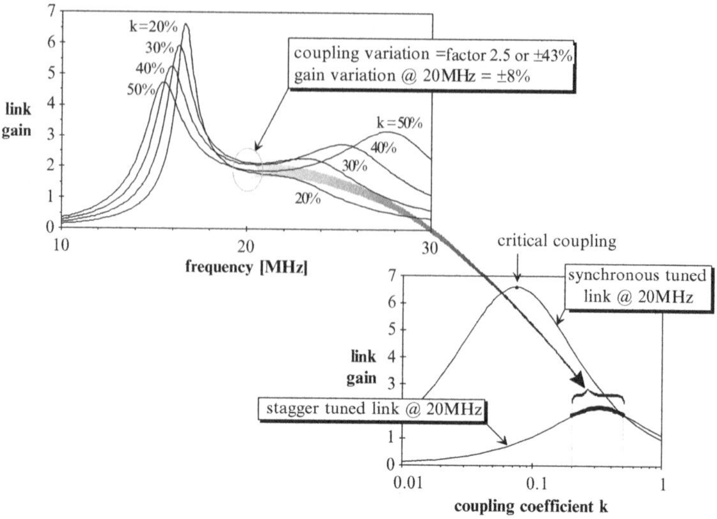


*Figure 2.32.* *The link gain of a stagger tuned link with a parallel-tuned secondary and a series-tuned primary coil. The curves are simulated for the circuit of* GALBRAITH *et al. (1987). The primary tank is tuned at 16.92 MHz and the secondary at 22 MHz.*
$L_{S_1} = L_{S_2} = 1.96\mu H, R_{S_1} = R_{S_2} = 5.18\Omega, R_{load_2} = 1k\Omega$
*The link gain of the same configuration in a critically coupled operation is shown on the bottom diagram. Both tanks were therefore tuned at 20 MHz just like for Figure 2.29.*

## 2.4.4 LINK STABILISATION THROUGH SELF-OSCILLATION

DONALDSON (1987a) found that a series-resonant primary in a self-resonant driver induces a constant tank current in a series-resonant secondary for coupling factors ranging from $1/Q_{sec}$ up to one. The relation between the link gain and the coupling coefficient does not have that typical bell shape like before, but has become an S-curve with a long and constant tail from $1/Q_{sec}$ to one. Both tanks are synchronously tuned and it is the pole-splitting effect occurring at coupling factors above $1/Q_{sec}$ that stabilises the link output (DONALDSON, 1987b; IVAL, 1987).[9]

Special care should be taken as this is also the coupling range for frequency hopping. However, in a practical situation both tanks cannot be tuned exactly synchronous. The driver frequency is then always pulled down if the secondary tank is tuned slightly upwards in frequency and *vice versa*.

---

[9] Donaldson's approach can be seen as <u>a critically coupled link in a self-oscillating set-up where frequency shifts are allowed</u>. The frequency shifts for coupling factors above critical coupling cause the link gain to remain constant instead of declining like the constant frequency case.

ZIERHOFER and HOCHMAIR (1990, 1992) made a similar combination of a stagger-tuned link with a self-oscillating class-E amplifier (EBERT and KAZIMIERCZUK, 1981). They used a series-tuned primary and a parallel-tuned secondary, and attained comparable results of a link gain which remained constant for coupling factors above the point of maximal gain. They needed a minimal coupling factor of 10% because of the stagger tuning.

*THE LINK OPTIMISATION TECHNIQUES CLASSIFIED ACCORDING TO THE COUPLING RANGE.* *TABLE 2.2.*

| | Very low coupling (k < 2%) | Low coupling (2% < k < 3%) | Moderate coupling (3% < k < 10%) | High coupling (k > 10%) |
|---|---|---|---|---|
| Constant frequency | • Classic efficiency optimisation | • Classic efficiency optimisation<br>• Critical coupling | • Classic efficiency optimisation<br>• Critical coupling | • Classic efficiency optimisation<br>• Critical coupling<br>• Stagger tuning |
| Frequency shifting with k | | • Self-oscillating voltage-in, current-out | • Self-oscillating voltage-in, current-out | • Self-oscillating voltage-in, current-out<br>• Stagger tuned link driven by a self-oscillating class E |

The stabilisation technique is only suitable for minimum coupling coefficients above 2.5% to 3% for the very same reasons as with critical coupling. The major difference between both is the non-constant transfer frequency and the restriction to voltage-in, current-out type of links. A current-to-voltage transformer with a low input impedance and a high output impedance, is added at the link output to get a voltage output.

Donaldson *et al.,* did not mention anything on the efficiency of the gain-stabilised links, so it is hard to design them towards optimal efficiency. This aspect is again left to prototyping and simulation. They also predicted a similar stabilisation effect for parallel-resonant secondaries, but could not prove it analytically.

## 2.5 DISCUSSION: OPTIMISATION OF WEAKLY COUPLED LINKS

The previous paragraphs indicate that there is not much choice left in case of poor coil coupling (Table 2.2). The classic efficiency optimisation is possible at any coupling, and critical coupling for coupling factors down to 2% or 3%. But, all other methods require coupling factors way beyond the specifications of the type of applications for which this study was originally carried out (k < 1%).

Conventional optimisation methods design the coils and capacitors separately and then connected them to a high-efficiency r.f. amplifier, hoping for optimal system operation. However, due to the weak coupling, the coil driver remains virtually unloaded causing poor driver efficiency. It is not uncommon for a weakly coupled link, properly designed according to one of the previously reviewed methods, that the equivalent load formed by the primary coil losses and the secondary power consumption, is much smaller than the parasitic on-resistance of the driver transistor. The total system dissipation then largely depends on the driver losses and the link efficiency almost plays a secondary role. The rectifier and regulator selection, and the optimisation of the link coils and coil driver should be carried out together, in a single procedure, to obtain optimal system efficiency. ZIERHOFER and HOCHMAIR (1990) were the first to recognise this problem.

**Link optimisation for weakly coupled applications:**
the rectifier and regulator selection, and the optimisation of the link coils and coil driver should be carried out together, in a single procedure, to obtain optimal system efficiency.

The above consideration doesn't imply that inductive powering is restricted to well-coupled applications. HEETDERKS (1988) first demonstrated that useful amounts of power can be coupled to millimetre and sub-millimetre coils for powering untethered sensor and stimulator implants. He derived useful expressions for the link output voltage and its output impedance. He calculated component values for some potential applications and obtained link efficiencies between 0.0015%, for a *k* of 0.02%, and 1.25% for a *k* of 0.5%. There was no further efficiency optimisation carried out.

Heetderks' results were later put in practise in some brilliant work on ultra-miniaturised, inductively powered muscle stimulators for paraplegic patients (LOEB *et al.*, 1991; STROJNIK *et al.*, 1992; CAMERON *et al.*, 1993; TROYK *et al.*, 1996; NARDIN *et al.*, 1996). Their design approach was exactly opposite to the classic link optimisation. They performed no link optimisation at all, but used the highest-Q coils that fit in their implants and generated a magnetic powering field as efficiently as possible (TROYK and SCHWAN, 1992a, b). The field amplitude was increased until they got the implants running. The driver consumption was just a matter of maintaining the powering field since the presence of several dozens of implants was not noticed at the driver due to the extremely low efficiency. Clearly, this approach may be fine for secondary power consumptions in the microwatt range, but would be rather wasteful if one targets the milliwatt range like in the present work.

The goal of the present study is to craft efficient inductive links with an output capacity of several milliwatts for coupling factors below the mythical 1% level, by a careful optimisation of the driven inductive link (cf. Chapter 5) and by the on-line adjustment of the driver output power to the actual implant needs (cf. Chapter 6).

### 2.5.1 OPTIMISATION OF THE DRIVEN INDUCTIVE LINK

Although the vast majority of the optimisation procedures available in literature can be reduced to either maximising link efficiency, or maximising link gain through critical coupling, they are described in a fairly application-specific manner; e.g. they treat only the case of series-resonant secondary and not the parallel-resonant one, or the assumptions taken are not valid for other applications. There is no general procedure that covers a whole family at once. DONALDSON and PERKINS stated in 1983: "Magnetic coupling between coils at radio frequency is often used for transmitting power, data or both to implanted devices. No general treatment of the design of such coils has been published, perhaps because the specifications imposed are so divers that entirely different designs result."[10] The

[10] The MRC Neuroprosthesis Unit therefore developed the Polychrest set-up (DONALDSON, 1986b). It is a general breadboard where all link parameters can be adjusted over a wide range for the experimental *ad hoc* generation of the link data for a given application. It is however felt that experimental generation from scratch, assumes a comprehensive understanding of the inductive link problem and the influence of each of the many parameters involved. An *a priori* calculation of the link components remains mandatory before any meaningful experiments can

present study contests the above statement by proposing a general optimisation flow that includes driver optimisation. It is complemented with an exhaustive list of exact design formulae for all possible primary and secondary configurations, and for operation at maximal efficiency or at critical coupling. Output stabilisation through stagger tuning or by using a self-resonant driver is not included because it is incompatibility with low coil coupling.

### 2.5.2 ON-LINE ADJUSTMENT OF THE DRIVER OUTPUT POWER

Inductive links should be designed to provide adequate power in worst case coupling conditions, which means that the driver output power is too high with actual coupling conditions that are by definition better than worst case. It is not atypical that a link that is tuned to accommodate worst-case situations produces a remote supply voltage that is two to three times too high with nominal coupling. Tuning the driver output down to match the actual coupling level could then reduce its power consumption by a factor of 4–9. The voltage at the remote coil is linear proportional to the primary coil voltage and to the coupling coefficient, as shown in Chapter 3. The primary coil voltage can thus be halved if the actual coupling factor is twice its worst-case value. And as the driver consumption is largely proportional to its output power, and hence to the square of the coil voltage, one potentially saves a factor of four.[11]

> On-line optimisation of the driver output power reduces overhead losses in case of unpredictable coupling.

So, in addition to the optimised efficiency of a carefully designed link, substantial power savings can be realised by on-line adjusting driver output power to the actual needs, especially with remote units that can move around and show a rather unpredictable coil coupling. This works for all coupling values, no matter how small they are. It is best implemented by feeding the value of the received power back to the driver and continuously adjusting the driver output to maintain a preset power level at the secondary.

### 2.5.3 ADDITIONAL SYSTEM OPTIMISATION

The feedback of the received-power amplitude is also useful to improve the user ergonomics of an inductive-powering system particularly with non-technical users. Acoustic or visual indication of the received signal strength allows positioning the primary coil for the best coupling and may confirm that the link is functioning properly.

Ideally, an inductive powering system should automatically adjust its output power, primary resonance characteristics and transfer frequency. A primary coil that is made flexible is more comfortable for a patient as it interferes less with body movements. Flexible coils are easily taped onto the patient's body for temporary measurements or incorporated in a shirtsleeve or a trouser leg, for permanent powering a muscle stimulator,

---

be carried out. The indisputable value of the Polychrest set-up lies in its flexibility to change link parameters to rapidly verify a calculated solution and to check its sensitivity to tolerances.

[11] A constant secondary load $R_{load_2}$ is assumed here. The situation is not as straightforward with a remote supply regulator as $R_{load_2}$ varies with the received voltage. The main line of thought remains valid, though.

for example. But geometrically deforming the primary coil or holding it close to metal objects changes its resonance behaviour. The switch-mode driver should be able to withstand this or adjust itself to the changing resonance characteristics.

Frequency tuning is needed when a single coil driver is combined with several remote units. For instance, a clinician would have only one external system to monitor the telemetric implants of all his patients. It would be ridiculous to need an individual driver for each of them. But, the secondary resonance frequency is different from implant to implant because of the component tolerances. It is possible to select coil-capacitor pairs with identical resonance frequencies, but that may be prohibitive in volume production.[12] It is more effective to provide a primary coil-driver that performs a frequency sweep of ±10% and automatically locks itself at the most optimal frequency based on the feedback information of the received-power level.

## 2.6 CONCLUSIONS

This review of link topologies and design procedures identified the following for weakly coupled applications:

- Efficiency-optimised or critically coupled links in either a MOPA or a self-oscillating configuration are the candidates for weakly coupled systems.
- The approximate design formulae found in literature are not valid at low coupling.
- At weak coil coupling, an inductive powering system cannot be designed component by component as commonly shown in literature; it has to be optimised as one big system. The rectifier and regulator selection, and the optimisation of the link coils and coil driver should be carried out all at once.
- On-line adjustment of the driver output power can substantially reduce overhead losses in case of unpredictable coupling.

Auto tuning of the link parameters will get the system self-supporting, robust and user-friendly for the technically challenged.

## 2.7 REFERENCES

**Bîrca-Galateanu S.** and J.-L. Cocquerelle, "Class E half-wave low dv/dt rectifier operating in a range of frequencies around resonance," *IEEE Trans. Circuits Syst. Part I*, vol. 42, pp. 83–94, 1995.

**Cameron T.,** G.E. Loeb, J.R. Richmond, R.A. Peck, J.H. Schulman, P. Strojnik and P.R. Troyk, "Micromodular electronic devices to activate paralyzed muscles and limbs," *Proc. 15th. Annual Int. Conf. IEEE-EMBS*, San Diego, CA, pp. 1242–1243, 1993.

***Donaldson N. de N.**, "Voltage regulators for implants powered by coupled coils," *Med. Biol. Eng. Comput.*, vol. 21, pp. 756–761, 1983.

**Donaldson N. de N.**, "Use of feedback with voltage regulators for implants powered by coupled coils," *Med. Biol. Eng. Comput.*, vol. 23, pp. 291, 1985.

[12] The remote unit is often too small to accept an adjustable capacitor. Moreover, secondary resonance capacitors are often in the nanofarad range where there are no adjustable versions available.

***Donaldson N. de N.**, "Comments on "efficient transdermal links with coupling-insensitive gain"," *IEEE Trans. Biomed. Eng.*, vol. BME-35, pp. 280–280, 1988.

***Donaldson N. de N.** and T.A. Perkins, "Analysis of resonant coupled coils in the radio frequency transcutaneous links," *Med. Biol. Eng. Comput.*, vol. 21, pp. 612–627, 1983.

**Donaldson P.E.K.**, "Frequency-hopping in r.f. energy-transfer links," *Electron. Wireless World*, pp. 24–26, Aug. 1986a.

**Donaldson P.E.K**, "Polychrest: a development system for radio-frequency-coupled implants," *J. Biomed. Eng.*, vol. 8, pp. 137–142, 1986b.

***Donaldson P.E.K.**, "Power for neurological prostheses: a simple inductive r.f. link with improved performance," *J. Biomed. Eng.*, vol. 9, pp. 194–197, 1987a.

***Donaldson P.E.K.**, "Three separation-insensitive radio frequency inductive links," *J. Med. Eng. Technol.*, vol. 11, pp. 23–29, 1987b.

***Donaldson P.E.K**, "Twenty years of neurological prosthesis-making," *J. Biomed. Eng.*, vol. 9, pp. 291–298, 1987c.

***Ebert J.** and M.K. Kazimierczuk, "Class E high-efficiency tuned power oscillator," *IEEE J. Solid-State Circuits*, vol. SC-16, pp. 62–66, 1981.

**Forster I.C.**, "Theoretical design and implementation of a transcutaneous, multichannel stimulator for neural prosthesis applications," *J. Biomed. Engng.*, vol. 3, pp. 107–120, 1981.

**Forster I.C.,** "Corrections on the paper: theoretical design and implementation of a transcutaneous, multichannel stimulator for neural prosthesis applications," *J. Biomed. Engng.*, vol. 5, pp. 176, 1983.

***Galbraith D.C.**, M. Soma and R.L. White, "A wide-band efficient inductive transdermal power and data link with coupling insensitive gain," *IEEE Trans. Biomed. Eng.*, vol. BME-34, pp. 265–275, 1987.

**Gutmann R.J.**, "Application of RF circuit design principles to distributed power converters," *IEEE Trans. Ind. Electron. Contr. Instrum.*, vol. IECI-27, pp. 156–164, 1980.

***Heetderks W.J.**, "RF powering of millimeter- and submillimeter-sized neural prosthetic implants," *IEEE Trans. Biomed. Eng.*, vol. BME-35, pp. 323–327, 1988.

**Hochmair E.S.**, "System optimization for improved accuracy in transcutaneous signal and power transmission," *IEEE Trans. Biomed. Eng.*, vol. BME-31, pp. 177–186, 1984.

**Ival T.**, "Does your coupling coefficient matter?," *Electron. Wireless World*, vol. 93, no. 6, pp. 577–579, 1987.

**Kazimierczuk M.K.**, "Analysis of class E zero-voltage-switching rectifier," *IEEE Trans. Circuits Syst.*, vol. CAS-37, pp. 747–755, 1990.

**Kazimierczuk M.K.** and J. Józwik, "Class E zero-voltage-switching rectifier with a series capacitor," *IEEE Trans. Circuits Syst.*, vol. CAS-36, pp. 926–928, 1989a.

**Kazimierczuk M.K.** and J. Józwik, "Class $E^2$ narrow-band resonant DC/DC converters," *IEEE Trans. Instr. Meas.*, vol. 38, pp. 1064–1068, 1989b.

**Kazimierczuk M.K.** and J. Józwik, "Class E zero-voltage-switching and zero-current-switching rectifiers," *IEEE Trans. Circuits Syst.*, vol. CAS-37, pp. 436–444, 1990a.

**Kazimierczuk M.K.** and J. Józwik, "Analysis and design of class E zero-current-switching rectifier," *IEEE Trans. Circuits Syst.*, vol. CAS-37, pp. 1000–1009, 1990b.

***Ko W.H.**, S.P. Liang and C.D.F. Fung, "Design of radio-frequency powered coils for implant instruments," *Med. Biol. Eng. Comput.*, vol. 15, pp. 634–640, 1977.

***Loeb G.E.**, C.J. Zamin, J.H. Schulman and P.R. Troyk, "Injectable microstimulator for functional electrical stimulation," *Med. Biol. Eng. Comput.*, vol. 29, pp. NS13-NS19, 1991.

**Miller J.A.,** G. Bélanger and T. Mussivand, "Development of an autotuned transcutaneous energy transfer system," *ASAIO J.*, vol. 39, pp. M706–M710, 1993.

**Mitamura Y.**, E. Okamoto, A. Hirano and T. Mikami, "Development of an implantable motor-driven assist pump system," *IEEE Trans. Biomed. Eng.*, vol. BME-37, pp. 146–156, 1990.

**Mueller J.S.** and H.T. Nagle, "Feasibility of inductive powering of miniature low-power biotelemetry for use with microfabricated biomedical sensors," in: *Biotelemetry XIII (Proc. 13th Int. Symposium on Biotelemetry)*, Eds. C. Cristalli, C.J. Amlaner Jr. and M.R. Neuman, Williamsburg, VI, pp. 372–377, 1996.

***Nardin M.**, B. Ziaie, J. Von Arx, A. Coghlan, M. Dokmeci and K. Najafi, "An inductively powered microstimulator for functional neuromuscular stimulation," in: *Biotelemetry XIII (Proc. 13th Int. Symposium on Biotelemetry)*, Eds. C. Cristalli, C.J. Amlaner Jr. and M.R. Neuman, Williamsburg, VI, pp. 99–104, 1996.

***Schuder J.C.**, H.E. Stephenson and J.F. Townsend, "High-level electromagnetic energy transfer through a closed chest wall," *Inst. Radio Engrs. Int. Conv. Record*, vol. 9, pp. 119–126, 1961.

**Schuder J.C.**, J.H. Gold and H.E. Stephenson, Jr., "An inductively coupled RF system for the transmission of 1kW of power through the skin," *IEEE Trans. Biomed. Eng.*, vol. BME-18, pp. 265–273, 1971.

**Sherman C.W.**, W.C. Clay, K.A. Dasse and B.D.T. Daly, "A transcutaneous energy transmission (TET) system for high-power prosthetics," *Proc. IEEE 7th Annu. Conf. EMBS*, Chicago, IL, Sept. 27–30, pp. 804–808, 1985.

**Strojnik P.**, J.H. Schulman, G.E. Loeb, and P.R. Troyk, "Multichannel FES system with distributed microstimulators," *Proc. 14th. Annu. Int. Conf. IEEE-EMBS*, Paris, France, pp. 1352, 1992.

**Troyk P.R.** and M.A.K. Schwan, "Class E driver for transcutaneous power and data link for implanted electronic devices," *Med. Biol. Eng. Comput.*, vol. 30, pp. 69–75, 1992a.

***Troyk P.R.** and M.A.K. Schwan, "Closed-loop class E transcutaneous power and data link for MicroImplants," *IEEE Trans. Biomed. Eng.*, vol. BME-39, pp. 589–599, 1992b.

**Troyk P.R.**, M.A.K. Schwan, G.E. Loeb and J. Schulman, "Micromodules for implanted telemetry and stimulation," in: *Biotelemetry XIII (Proc. 13th Int. Symposium on Biotelemetry)*, Eds. C. Cristalli, C.J. Amlaner Jr. and M.R. Neuman, Williamsburg, VI, pp. 105–111, 1996.

***Zierhofer C.M.** and E.S. Hochmair, "High-efficiency coupling-insensitive transcutaneous power and data transmission via an inductive link," *IEEE Trans. Biomed. Eng.*, vol. BME-37, pp. 716–722, 1990.

**Zierhofer C.M.** and E.S. Hochmair, "The class-E concept for efficient wide-band coupling-insensitive transdermal power and data transfer," *Proc. 14th. Annu. Int. Conf. IEEE-EMBS*, Paris, France, pp. 382–383, 1992.

# 3 Exact link formulae

Weak coupling renders the approximate link formulae invalid. This chapter therefore derives a set of exact formulae for efficiency-optimised and for critically coupled coil sets. The formulae are exact over the entire coupling range. The only assumption is that the link is running at the phase-resonance frequency of the secondary tank. This means that the procedure covers all possible link configurations for low coupled applications but that gain stabilisation by stagger tuning or self-oscillating drivers, is not included as they are impossible at the coupling coefficients envisaged for the applications for which this study was originally carried out.



K. Van Schuylenbergh and R. Puers, *Inductive Powering: Basic Theory and Application to Biomedical Systems*, Analog Circuits and Signal Processing, 77–101. 

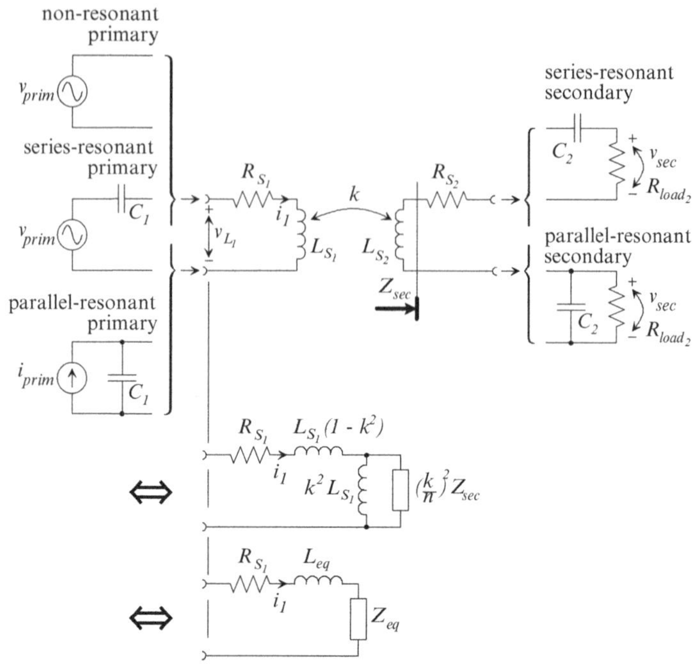


*Figure 3.1.* *All inductive link combinations and the two equivalent representations that are used to reduce the impedance $Z_{sec}$ to an equivalent impedance at the primary side. The impedance $Z_{sec}$ has been transformed to the primary according to Figure 2.6 for mutually coupled coils.*
*The primary sources are assumed to be ideal and 100 % sinusoidal. Actual circuit signals are often not sinusoidal. In that case, the above symbols point to the first harmonic component, since neither the d.c. component nor the higher harmonics contribute to the power transfer when secondary resonance is applied.*

The six combinations of a parallel- or series-resonant secondary and a parallel-, series- or non-resonant primary are studied (Figure 3.1). Two kinds of link optimisations are discussed for each of these six link combinations: optimisation of the system efficiency and optimisation of the displacement tolerance by critical coupling.

The equations for the link components (i.e., coils and capacitors) and performance are calculated in the following order. The equivalent impedance $Z_{eq}$ of the secondary reflected to the primary is first computed to find an operation mode where the coupling and load variations have a minimal effect on the tuning of primary circuits. The general expressions for the link efficiency and gain are then calculated. This is followed by the derivation of the operation modes for maximal link efficiency or critical coupling. This order is followed twice: once in section 3.1 for the parallel-resonant secondary and once in section 3.2 for the series-resonant secondary.

All expressions needed to perform an actual link design are highlighted in a frame.

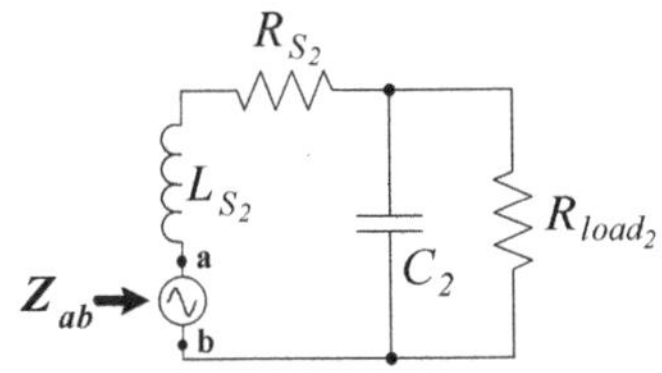


$$Z_{ab} = R_{S_2} + j\omega L_{S_2} + \frac{1}{j\omega C_2 + \frac{1}{R_{load_2}}}$$

$$\omega_{resP} \equiv \omega \;@\mathrm{Im}\{Z_{ab}\} = 0$$

$$\omega_{resP} = \sqrt{\frac{1}{L_{S_2}C_2} - \frac{1}{R_{load_2}^2 C_2^2}}$$

*The phase-resonance frequency $\omega_{resP}$ of a parallel-tuned secondary. It is the frequency at which the impedance $Z_{ab}$ seen by a voltage source in series with $L_{S2}$, is real. The source represents, for instance, a magnetically induced voltage in $L_{S2}$.* *Figure 3.2.*

# 3.1 INDUCTIVE LINKS WITH A PARALLEL-RESONANT SECONDARY

## 3.1.1 EQUIVALENT SECONDARY IMPEDANCE

The secondary circuit is reduced to an equivalent secondary impedance $Z_{eq}$ in series with the primary coil, according to Figure 2.6. The two equivalent link representations in Figure 3.1 indicate that

$$j\omega L_{S_1}\left(1-k^2\right) + \frac{1}{\frac{1}{j\omega k^2 L_{S_1}} + \frac{1}{\left(\frac{k}{n}\right)^2 Z_{\sec}}} = j\omega L_{S_1} + Z_{eq}$$

$$\text{with } Z_{\sec} = R_{S_2} + \frac{1}{j\omega C_2 + \frac{1}{R_{load_2}}}$$

Hence,

$$Z_{eq} = \frac{1}{\frac{1}{j\omega k^2 L_{S_1}} + \frac{1}{\left(\frac{k}{n}\right)^2 Z_{\sec}}} - j\omega k^2 L_{S_1}$$

Or,

$$Z_{eq} = \omega^2 k^2 L_{S_1} L_{S_2} * \frac{R_{load_2} + R_{S_2} + \omega^2 C_2^2 R_{load_2}^2 R_{S_2} - j\omega\left(L_{S_2} - C_2 R_{load_2}^2 + \omega^2 C_2^2 L_{S_2} R_{load_2}^2\right)}{\left(R_{load_2} + R_{S_2}\right)^2 + \omega^2\left[L_{S_2}^2 + R_{load_2}^2 C_2\left(R_{S_2}^2 C_2 - 2L_{S_2}\right)\right] + \omega^4 L_{S_2}^2 C_2^2 R_{load_2}^2}$$

This equivalent secondary impedance $Z_{eq}$ is real at the frequency

$$\omega = \sqrt{\frac{1}{L_{S_2}C_2} - \frac{1}{R_{load_2}^2 C_2^2}} = \omega_{resP}$$ *Equation 3.1.*

which is the phase-resonance frequency $\omega_{resP}$ of the loaded secondary tank circuit if it was not coupled to the primary coil (Figure 3.2).

> From here on, the driver frequency $\omega$ is always assumed to be equal to the secondary phase-resonance frequency $\omega_{resP}$.

Running an inductive link at the frequency $\omega_{resP}$ has the major advantage that $Z_{eq}$ remains purely resistive regardless variations in the coil coupling or the secondary power consumption. The primary coil reactance is thus not changing. Only the damping of the primary circuits is affected. A transfer frequency equal to $\omega_{resP}$ makes inductive links compatible with switch-mode amplifiers.

The substitution of $\omega$ by $\omega_{resP}$ in the expression of $Z_{eq}$, results in:

$$R_{eq} \equiv Z_{eq} @ \omega_{resP} = \frac{k^2 L_{S_1}\left(C_2 R_{load_2}^2 - L_{S_2}\right)}{R_{load_2} C_2 \left(R_{load_2} C_2 R_{S_2} + L_{S_2}\right)}$$

Or,

*Equation 3.2.*

$$R_{eq} = k^2 L_{S_1} \omega_{resP} \frac{\alpha . Q_{L_{S_2}}}{\alpha + Q_{L_{S_2}}}$$

*Equation 3.3.*

$$\text{where } \alpha \equiv \omega_{resP} C_2 R_{load_2}$$

***The parameter $\alpha$ is*** has no dimensions and corresponds to ***the ratio of the a.c. load $R_{load_2}$ to the reactance of the capacitor $C_2$*** [(1)].

## 3.1.2 LINK EFFICIENCIES

### 3.1.2.1 Primary link efficiency

***The primary link efficiency is*** defined as ***the ratio of the power $P_{sec}$ that reaches the secondary circuit to the power $P_{link\ in}$ put into the inductive link.*** It is calculated as the ratio of the power dissipated in $R_{eq}$ to the total power dissipated in both $R_{eq}$ and $R_{S_1}$ (Figure 3.3).

Hence,

$$\eta_{primary} \equiv \frac{P_{sec}}{P_{link\ in}} = \frac{R_{eq}}{R_{eq} + R_{S_1}}$$

$$\eta_{primary} = \frac{M^2\left(C_2 R_{load_2}^2 - L_{S_2}\right)}{R_{S_1} R_{load_2} C_2 L_{S_2}\left(R_{S_2} R_{load_2} C_2 + L_{S_2}\right) + M^2\left(C_2 R_{load_2}^2 - L_{S_2}\right)}$$

---

[1] The parameter $\alpha$ is similar to the load factor $a$ defined by FORSTER (1991; 1983). The parameter $\alpha$ is, however, more general since it describes all kind of links with either series- or parallel-resonant secondary and series-, parallel-, or non-resonant primary, and this without any assumption on the coil coupling or quality factors.

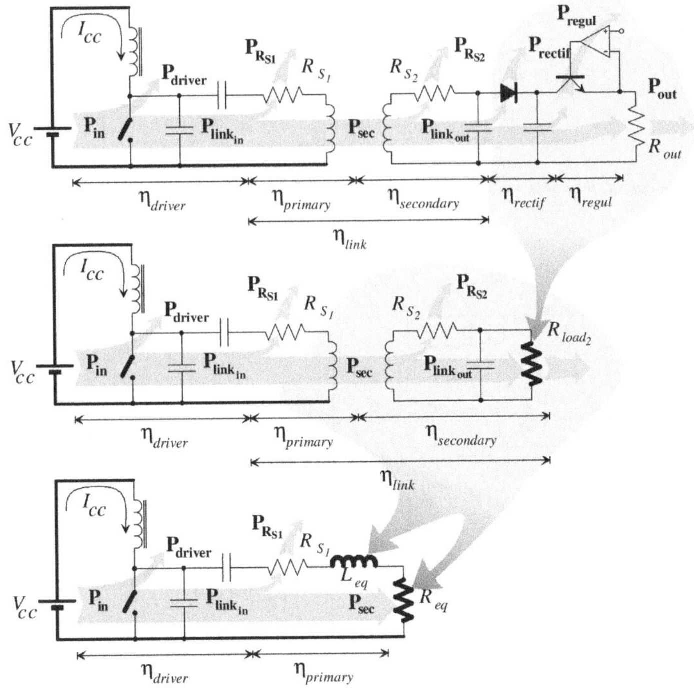

*The power distribution in an inductive link. The concept is illustrated here on a class-E driven link with a half-wave rectifier and a series regulator, but remains valid for any other combination.* *Figure 3.3.*

Or,

$$\eta_{primary} = \frac{k^2 Q_{L_{S_1}} Q_{L_{S_2}}}{1 + \frac{Q_{L_{S_2}}}{\alpha} + k^2 Q_{L_{S_1}} Q_{L_{S_2}}}$$ *Equation 3.4.*

This expression is not changed by primary coil tuning.

### 3.1.2.2 Secondary link efficiency

***The secondary link efficiency is*** defined as ***the ratio of the useful power $P_{link\ out}$ dissipated in the a.c. load to the power $P_{sec}$ that reaches the secondary circuit.*** It is calculated from the power distribution between $R_{S_2}$ and $R_{load_2}$. But, because $C_2$ shunts $R_{load_2}$, the voltages across both resistors are not in phase and $\eta_{secondary}$ is no simple resistance ratio (Figure 3.4). The computation is somewhat elaborate, but the result is surprisingly simple:

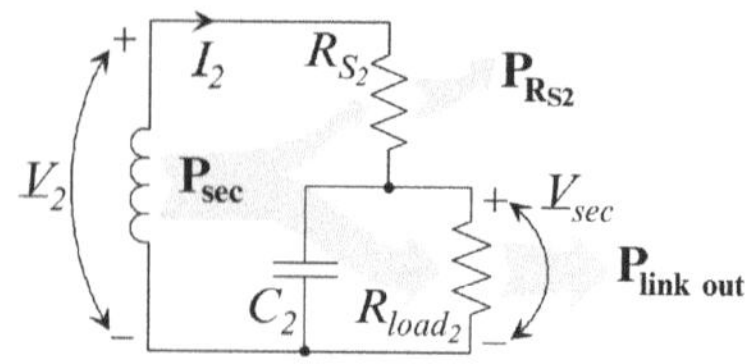


*Figure 3.4. The calculation of the secondary link efficiency for a parallel-resonant secondary.*

$$\eta_{secondary} \equiv \frac{P_{link\ out}}{P_{sec}} = \frac{|\underline{V}_{sec}|^2 / R_{load_2}}{\text{Re}\{\underline{V}_2 \cdot \underline{I}_2\}}$$

$$= \frac{\left| \frac{\underline{V}_2 \left( \frac{1}{R_{load_2}} + j\omega_{resP} C_2 \right)^{-1}}{R_{S_2} + \left( \frac{1}{R_{load_2}} + j\omega_{resP} C_2 \right)^{-1}} \right|^2}{R_{load_2}} \text{Re}\left\{ \frac{\underline{V}_2 \cdot \underline{V}_2}{R_{S_2} + \left( \frac{1}{R_{load_2}} + j\omega_{resP} C_2 \right)^{-1}} \right\}^{-1}$$

$$= \ldots$$

$$= \frac{L_{S_2}}{R_{load_2} C_2 R_{S_2} + L_{S_2}}$$

Or,

Equation 3.5.
$$\eta_{secondary} = \frac{Q_{L_{S_2}}}{\alpha + Q_{L_{S_2}}}$$

### 3.1.2.3 Total link efficiency

The total link efficiency is the product of the primary and the secondary link efficiencies. The expression is valid for all links with a parallel-resonant secondary, regardless whether the primary coil is tuned or not.

Equation 3.6.
$$\eta_{link} = \frac{k^2 Q_{L_{S_1}} Q_{L_{S_2}}^2}{\left(1 + \frac{Q_{L_{S_2}}}{\alpha} + k^2 Q_{L_{S_1}} Q_{L_{S_2}}\right)\left(\alpha + Q_{L_{S_2}}\right)}$$

## 3.1.3 LINK GAIN

***The link gain is*** defined as ***the ration of the a.c. voltage $V_{sec}$ across the a.c. load $R_{load_2}$ to the a.c. link input voltage $V_{prim}$ across the primary coil and its occasional resonance capacitor*** (Figure 3.1).

### 3.1.3.1 Non-resonant primary coil

It is impossible to find manageable expressions for the link gain from the equivalent schemes on Figure 3.1. It is however not necessary to know the entire complex transfer function $\underline{V}_{sec}(\omega)/\underline{V}_{prim}(\omega)$ because one is only interested in the amplitude gain at one particular frequency $\omega_{resP}$, and not in the phase difference. The link gain can therefore better be computed via the link efficiency:

$$A \equiv \frac{V_{\sec}}{V_{prim}} = \frac{\sqrt{P_{\text{link out}} R_{load_2}}}{V_{prim}} = \frac{\sqrt{P_{\text{link in}} \eta_{link} R_{load_2}}}{V_{prim}}$$

The link input power $P_{link\ in}$ for a primary voltage $V_{prim}$ is found from Figure 3.1:

$$\begin{aligned} P_{link\ in} &= I_1^2 \left(R_{eq} + R_{S_1}\right) \\ &= \frac{V_{prim}^2}{\left|j\omega_{resP} L_{S_1} + R_{eq} + R_{S_1}\right|^2}\left(R_{eq} + R_{S_1}\right) \\ &= \frac{V_{prim}^2}{\omega_{resP}^2 L_{S_1}^2 + \left(R_{eq} + R_{S_1}\right)^2}\left(R_{eq} + R_{S_1}\right) \end{aligned}$$

This expression and the expression for $\eta_{link}$, are then substituted in the above formula for $A$ to yield:

$$A = \sqrt{\frac{R_{eq} Q_{L_{S_2}} R_{load_2}}{\left[\omega_{resP}^2 L_{S_1}^2 + \left(R_{eq} + R_{S_1}\right)^2\right]\left(\alpha + Q_{L_{S_2}}\right)}}$$

The following three parameters are now substituted in this expression:

- The a.c. load $R_{load_2}$ (via the definition of $\alpha$): $R_{load_2} = \dfrac{\alpha}{\omega_{resP} C_2}$
- The tank capacitor $C_2$ (via the definition of $\omega_{resP}$):

$$\omega_{resP} = \sqrt{\frac{1}{L_{S_2} C_2} - \frac{\omega_{resP}^2}{R_{load_2}^2 C_2^2 \omega_{resP}^2}}$$

$$\Rightarrow \omega_{resP}^2 = \frac{1}{L_{S_2} C_2} - \frac{\omega_{resP}^2}{\alpha^2}$$

$$\Rightarrow \quad C_2 = \frac{\alpha^2}{L_{S_2} \omega_{resP}^2 \left(1 + \alpha^2\right)} \qquad \textit{Equation 3.7.}$$

- The equivalent resistor $R_{eq}$ (via Equation 3.2.)

This finally results in the following expression for the link gain:

$$A = k.n.Q_{L_{S_1}} Q_{L_{S_2}} \sqrt{\frac{\alpha^2 + 1}{Q_{L_{S_1}}^2 \left(\alpha + Q_{L_{S_2}}\right)^2 + \left(\alpha k^2 Q_{L_{S_1}} Q_{L_{S_2}} + \alpha + Q_{L_{S_2}}\right)^2}} \qquad \textit{Equation 3.8.}$$

### 3.1.3.2 Series-resonant primary coil

$$A_{stp} \equiv \frac{V_{sec}}{V_{prim}} = \frac{\sqrt{P_{\text{link out}} R_{load_2}}}{V_{prim}} = \frac{\sqrt{P_{\text{link in}} \eta_{link} R_{load_2}}}{V_{prim}}$$

The link input power $P_{link\ in}$ for a primary voltage $V_{prim}$ is given by:

$$P_{link\ in} = I_1^2 \left(R_{eq} + R_{S_1}\right)$$

$$= \frac{V_{prim}^2}{\left|\frac{1}{j\omega_{resP} C_1} + j\omega_{resP} L_{S_1} + R_{eq} + R_{S_1}\right|^2} \left(R_{eq} + R_{S_1}\right)$$

The primary coil is at resonance with the series capacitor $C_1$. Hence,

$$\omega_{resP} = \frac{1}{\sqrt{L_{S_1} C_1}} \Rightarrow P_{\text{link in}} = \frac{V_{prim}^2}{R_{eq} + R_{S_1}}$$

This expression and the expressions for $\eta_{link}$, $R_{load_2}$, $C_2$ and $R_{eq}$ are then substituted in the above formula for $A_{stp}$ to yield:

*Equation 3.9.*

$$A_{stp} = \frac{k.n.Q_{L_{S_1}} Q_{L_{S_2}} \sqrt{\alpha^2 + 1}}{\alpha k^2 Q_{L_{S_1}} Q_{L_{S_2}} + \alpha + Q_{L_{S_2}}}$$

### 3.1.3.3 Parallel-resonant primary coil

Parallel-resonant primaries are fed by current-type drivers. The link gain is thus better expressed as a ***transimpedance*** $r_{m_{ptp}}$ :

$$r_{m_{ptp}} \equiv \frac{V_{sec}}{I_{prim}} = \frac{\sqrt{P_{\text{link out}} R_{load_2}}}{I_{prim}} = \frac{\sqrt{P_{\text{link in}} \eta_{link} R_{load_2}}}{I_{prim}}$$

The link input power $P_{link\ in}$ for a primary current $I_{prim}$ is found as:

$$P_{link\ in} = I_1^2 \left(R_{eq} + R_{S_1}\right)$$

$$= I_{prim}^2 \left| \frac{\frac{1}{j\omega_{resP} C_1}}{\frac{1}{j\omega_{resP} C_1} + j\omega_{resP} L_{S_1} + R_{eq} + R_{S_1}} \right|^2 \left(R_{eq} + R_{S_1}\right)$$

The primary coil is in resonance with the parallel capacitor $C_1$, thus,

$$\omega_{resP} = \frac{1}{\sqrt{L_{S_1} C_1}} \Rightarrow P_{link\ in} = \frac{I_{prim}^2}{\omega_{resP}^2 C_1^2 \left(R_{eq} + R_{S_1}\right)} = \frac{I_{prim}^2 \omega_{resP}^2 L_{S_1}^2}{R_{eq} + R_{S_1}}$$

The transimpedance $r_{m_{ptp}}$ then finally becomes:

*Equation 3.10.*

$$r_{m_{ptp}} = \omega_{resP} L_{S_1} A_{stp} = \omega_{resP} L_{S_1} \frac{k.n.Q_{L_{S_1}} Q_{L_{S_2}} \sqrt{\alpha^2 + 1}}{\alpha k^2 Q_{L_{S_1}} Q_{L_{S_2}} + \alpha + Q_{L_{S_2}}}$$

### 3.1.4 LINK OPTIMISATION

**Rule of thumb no. 1**: Optimising $\alpha$ for a given a.c. load $R_{load2}$ and transfer frequency $\omega_{resP}$, corresponds to matching $C_2$ and $L_{S2}$ to the load $R_{load2}$.

The link optimisation involves the derivation of an optimal $\alpha$ to yield either maximal link efficiency or critical coupling.

The matching $C_{2_{opt}}$ follows directly from the definition of $\alpha$:

$$C_{2_{opt}} = \frac{\alpha_{opt}}{\omega_{resP} R_{load_2}} \qquad \text{Equation 3.11.}$$

The matching $L_{S_2opt}$ is found by substituting Equation 3.7. in the definition of $\alpha$:

$$\alpha_{opt} = \omega_{resP} R_{load_2} C_{2_{opt}} \Rightarrow \alpha_{opt} = \omega_{resP} R_{load_2} \frac{\alpha_{opt}^2}{L_{S_2opt}\,\omega_{resP}^2 \left(1+\alpha_{opt}^2\right)}$$

Thus,

$$L_{S_2opt} = \frac{R_{load_2}}{\omega_{resP}} \left( \frac{\alpha_{opt}}{\alpha_{opt}^2 + 1} \right) \qquad \text{Equation 3.12.}$$

The optimal $\alpha$-values that correspond to maximal link efficiency or to critical coil coupling, are studied now in detail.

#### 3.1.4.1 Solution at maximal link efficiency

The maximal link efficiency occurs at an $\alpha$-value that results from

$$\frac{\partial \eta_{link}}{\partial \alpha} = 0$$

The solution of this equation is

$$\alpha_{\eta_{link_{max}}} = \frac{Q_{L_{S_2}}}{\sqrt{1+X}} \qquad \text{Equation 3.13.}$$

$$\text{where } X \equiv k^2 Q_{L_{S_1}} Q_{L_{S_2}} \qquad \text{Equation 3.14.}$$

The maximal link efficiency follows from substituting this $\alpha$-value in the expression for $\eta_{link}$:

$$\eta_{link_{max}} = \frac{X}{\left(1+\sqrt{1+X}\right)^2} \qquad \text{Equation 3.15.}^{2}$$

[2] This expression is identical to the expression (6) in (KO *et al.*, 1977) that was found with approximate formulae, assuming a high quality factor for the loaded secondary.

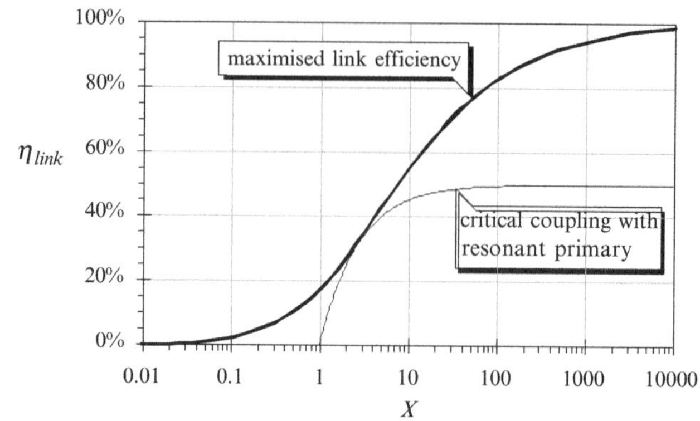


*Figure 3.5. The maximal achievable link efficiency for a parallel-resonant secondary as a function of the only variable X (in bold). The efficiency at critical coupling of a link with a resonant primary and a parallel-resonant secondary is drawn in hairline for comparison. This operation mode is discussed on page 89.*
*It will be shown in paragraph 3.2, that the expressions for the link efficiencies of a series-resonant secondary are identical to those of a parallel-resonant secondary, which makes this plot valid for both cases.*

> **Rule of thumb no. 2:** optimising a coil set involves maximisation of the coil coupling and the coil quality factors.

The maximal link efficiency is a monotonous increasing function in the single variable $X$ which proofs our rule no. 2 (Figure 3.5)[3].

The link gains at maximal link efficiency result from the substitution of the $\alpha$-value from Equation 3.13. in Equation 3.8., Equation 3.9. and Equation 3.10. respectively:

*Equation 3.16.*
$$A @ \eta_{link_{max}} = \frac{k.n.Q_{L_{S_1}}}{1+\sqrt{1+X}} \sqrt{\frac{1+X+Q^2_{L_{S_2}}}{1+X+Q^2_{L_{S_1}}}}$$

*Equation 3.17.*
$$A_{stp} @ \eta_{link_{max}} = \frac{k.n.Q_{L_{S_1}}}{1+\sqrt{1+X}} \sqrt{\frac{1+X+Q^2_{L_{S_2}}}{1+X}}$$

*Equation 3.18.*
$$r_{m_{ptp}} @ \eta_{link_{max}} = \omega_{resP} L_{S_1} A_{stp} @ \eta_{link_{max}}$$

[3] The maximum link efficiency was found by searching the extremes of the function $\eta_{link}(\alpha)$. The second derivative of $\eta_{link}(\alpha)$ in $\alpha_{\eta_{link_{max}}}$ is negative and proves that this is a maximum and not a minimum:

$$\left.\frac{\partial^2 \eta_{link}}{\partial \alpha^2}\right|_{\alpha=\alpha_{\eta_{link_{max}}}} = -\frac{2X\sqrt{(1+X)^3}\left(2+X+2\sqrt{1+X}\right)}{Q^2_{L_{S_2}}\left(1+\sqrt{1+X}\right)^6}$$

The equivalent secondary resistor at maximal link efficiency is found by substituting the $\alpha$-value from Equation 3.13. in Equation 3.2.:

$$R_{eq}@\eta_{link_{max}} = \frac{\omega_{resP}L_{S_1}}{Q_{L_{S_1}}}\frac{X}{1+\sqrt{1+X}} = R_{S_1}\frac{X}{1+\sqrt{1+X}}$$

*Equation 3.19.*

### 3.1.4.2 Solution at critical coupling

An inductive link is said to work at ***critical coupling*** if its ***output voltage is insensitive to small coupling variations***. This corresponds, assuming that the driver output is independent of the coil coupling, to an operation point at maximal link gain in function of the coupling factor.

#### *A. Non-resonant primary coil*

This situation corresponds to a primary coil driven by an amplifier with a constant output voltage independent of the coupling. It does not matter whether this driver contains tuning capacitors or not, as long as its output is constant. The critical coupling factor is found from $\frac{\partial A}{\partial k} = 0$. This equation has four mathematical solutions:

$$\Rightarrow k_{crit}^2 = \pm\frac{\left(\alpha + Q_{L_{S_2}}\right)\sqrt{1+Q_{L_{S_1}}^2}}{\alpha Q_{L_{S_1}} Q_{L_{S_2}}}$$

However, the link-design problem is normally stated just the other way around: "given a coupling factor *k*, find the $\alpha$-value $\alpha_{crit}$ to run the link at critical coupling." The expression is therefore reversed to:

$$\alpha_{crit} = \frac{Q_{L_{S_2}}}{\pm\frac{k^2 Q_{L_{S_1}} Q_{L_{S_2}}}{\sqrt{1+Q_{L_{S_1}}^2}} - 1}$$

The minus sign in the denominator is dropped because the parameter $\alpha$, being the ratio of the a.c. load to the reactance of $C_2$, is always positive:

$$\alpha_{crit} = \frac{Q_{L_{S_2}}}{\frac{k^2 Q_{L_{S_1}} Q_{L_{S_2}}}{\sqrt{1+Q_{L_{S_1}}^2}} - 1}$$

*Equation 3.20.* [4]

[4] It is easily proven that $\alpha_{crit}$ corresponds to a gain maximum and not a minimum, since the second derivative of the gain at $\alpha_{crit}$ is always negative:

$$\left.\frac{\partial^2 A}{\partial k^2}\right|_{\alpha=\alpha_{crit}} = -\frac{n}{Q_{L_{S_2}} k^3} * \sqrt{\frac{2\sqrt{1+Q_{L_{S_1}}^2}\left[Q_{L_{S_2}}^2\left(1+Q_{L_{S_1}}^2\right)+\left(k^2 Q_{L_{S_1}} Q_{L_{S2}} - \sqrt{1+Q_{L_{S_1}}^2}\right)^2\right]}{\left(1+\sqrt{1+Q_{L_{S_1}}^2}\right)^3}}$$

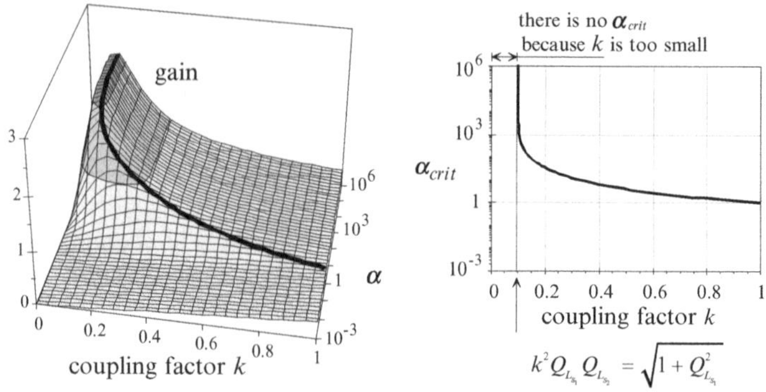


*Figure 3.6.* *The gain of an inductive link with a non-resonant primary and a parallel-resonant secondary, as a function of the coupling factor and the parameter α (left). It has a maximum for moderate to high coupling factors, but below a k-value defined by Equation 3.21., there is no maximum found anymore.*
*The right diagram indicates where the gain maxima occur.*
*Both diagrams are for* $L_{S_1} = 200\mu H, L_{S_2} = 20\mu H$ *and* $Q_{L_{S_1}} = Q_{L_{S_2}} = 100.$

Note that the value for $\alpha_{crit}$ can still become negative although this has no physical meaning. There is thus an extra condition to be met by the coupling factor $k$ and the coil quality factors $Q_{L_{S_1}}$ and $Q_{L_{S_2}}$ to have a link gain that shows a maximum in function of $k$ (Figure 3.6):

*Equation 3.21.*

$$k^2 Q_{L_{S_1}} Q_{L_{S_2}} \geq \sqrt{1+Q_{L_{S_1}}^2}$$

This expression reduces to a condition for $Q_{L_{S_2}}$ at higher quality factors:

$$Q_{L_{S_2}} \geq \frac{1}{k^2}$$

**Rule of thumb no. 3**: Critical coupling with a non-resonant primary coil is only possible at moderate to high coil-coupling factors (k > 10 %).

The lower the coupling factor, the higher the quality factors have to be to produce a peak in the gain function for the critical coupling. A coupling factor of 1 % would need an unrealistic secondary coil quality factor of 10,000!

The link efficiency, the link gain and the equivalent resistance at critical coupling are found by substituting the $\alpha_{crit}$ value into Equation 3.6., Equation 3.8. and Equation 3.2. respectively.

*Equation 3.22.*

$$\eta_{link} @ \alpha_{crit} = \frac{\sqrt{1+Q_{L_{S_1}}^2} \cdot \left(k^2 Q_{L_{S_1}} Q_{L_{S_2}} - \sqrt{1+Q_{L_{S_1}}^2}\right)}{k^2 Q_{L_{S_1}} Q_{L_{S_2}} \left(1+\sqrt{1+Q_{L_{S_1}}^2}\right)}$$

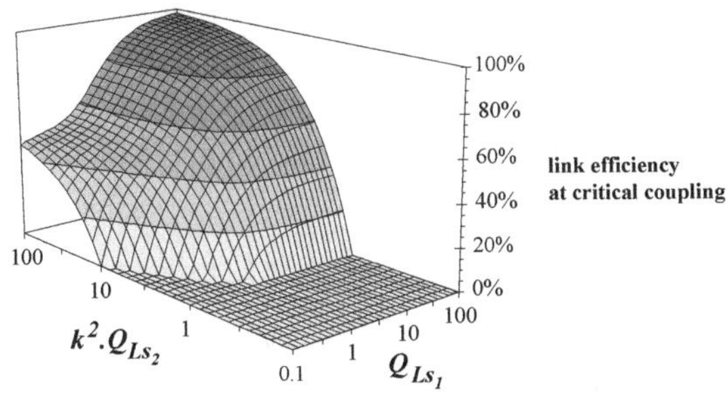


Figure 3.7. *The efficiency at critical coupling of a link with a non-resonant primary as function of $k^2Q_{L_{S_2}}$ and the primary coil quality factor $Q_{L_1}$. The area with zero efficiency is for k and $Q_L$ combinations without a peak in the gain function so that critical coupling is not possible. As links with a parallel-resonant or a series-resonant secondary have identical expressions for the link efficiency, this plot is valid for both.*

$$A@\alpha_{crit} = \frac{n}{kQ_{L_{S_2}}} \cdot \sqrt{\frac{Q^2_{L_{S_2}}\left(1+Q^2_{L_{S_1}}\right) + \left(k^2 Q_{L_{S_1}} Q_{L_{S_2}} - \sqrt{1+Q^2_{L_{S_1}}}\right)^2}{2\left(1+Q^2_{L_{S_1}} + \sqrt{1+Q^2_{L_S}}\right)}}$$

Equation 3.23.

$$R_{eq}@\alpha_{crit} = \frac{\omega_{resP} L_{S_1}}{Q_{L_{S_1}}} \sqrt{1+Q^2_{L_{S_1}}} = R_{S_1}\sqrt{1+Q^2_{L_{S_1}}}$$

Equation 3.24.

The k, $Q_{L_1}$ and $Q_{L_2}$ values are again preferably as large as possible to attain the highest link efficiency (Figure 3.7).

***B. Series- or parallel-resonant primary coil***

The critical coupling factor is found from $\frac{\partial A_{stp}}{\partial k} = 0$ or $\frac{\partial r_{m_{ptp}}}{\partial k} = 0$, which have identical solutions. Both cases are thus discussed together.
Two mathematical solutions are found:

$$\Rightarrow k_{crit} = \pm\sqrt{\frac{\alpha + Q_{L_{S_2}}}{\alpha Q_{L_{S_1}} Q_{L_{S_2}}}}$$

The reversed expression becomes:

$$\alpha_{crit} = \frac{Q_{L_{S_2}}}{X-1}$$

Equation 3.25. [5]

[5] $\alpha_{crit}$ corresponds to a gain maximum and not a minimum because the second derivative of the gain at $\alpha_{crit}$, is always negative:

$$\left.\frac{\partial^2 A_{stp}}{\partial k^2}\right|_{\alpha=\alpha_{crit}} = -\frac{n}{2Q_{L_{S_2}} k^3}\sqrt{(1-X)^2 + Q^2_{L_{S_2}}}$$

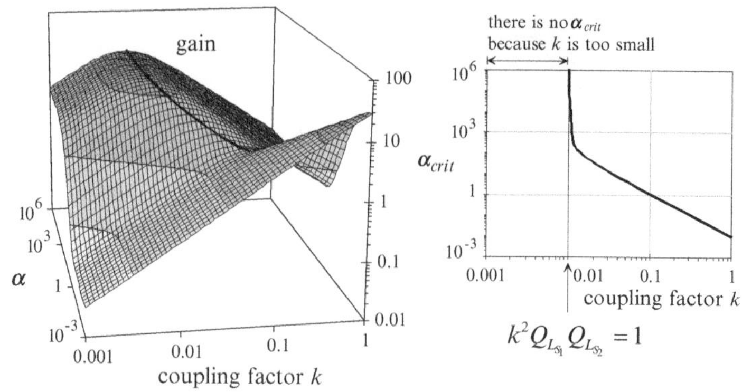


*Figure 3.8.* *The gain of an inductive link with a series-resonant primary and a parallel-resonant secondary, as a function of the coupling factor and the parameter $\alpha$ (left). The gain maxima are indicated in bold. The right diagram indicates where these gain maxima occur. Note that logarithmic scales are used as opposed to Figure 3.6.*
*Both diagrams are for $L_{S_1} = 200\mu H, L_{S_2} = 20\mu H$ and $Q_{L_{S_1}} = Q_{L_{S_2}} = 100$.*

The condition to be met by the coupling factor $k$ and the coil quality factors $Q_{L_{S_1}}$ and $Q_{L_{S_2}}$ to have a positive $\alpha_{crit}$ is (Figure 3.8):

*Equation 3.26.*

$$k^2 Q_{L_{S_1}} Q_{L_{S_2}} \geq 1$$

**Rule of thumb no. 4**: Critical coupling with a resonant primary coil is possible at low, but not at very low coil coupling (k > 1 %).

Again, the lower the coupling factor, the higher the quality factors must be to find a gain peak for the critical coupling. But, the condition here is less stringent than the one for non-resonant primaries. A coupling factor of 1 % requires coil quality factors of 100, which is achievable although it is about the limit. Higher values are difficult to realise in small telemetry units.

The link efficiency, the equivalent resistance and the link gains $A$ and $r_m$ at critical coupling are found by substituting $\alpha_{crit}$ into Equation 3.6., Equation 3.2., Equation 3.9. and Equation 3.10. respectively.

*Equation 3.27.*

$$\eta_{link} @ \alpha_{crit} = \frac{X-1}{2X}$$

*Equation 3.28.*

$$R_{eq} @ \alpha_{crit} = \frac{\omega_{resP} L_{S_1}}{Q_{L_{S_1}}} = R_{S_1}$$

For a series-resonant primary:

*Equation 3.29.*

$$A_{stp} @ \alpha_{crit} = \frac{n}{2kQ_{L_{S_2}}} \cdot \sqrt{Q_{L_{S_2}}^2 + (1-X)^2}$$

For a parallel-resonant primary:

*Equation 3.30.*

$$r_{m_{ptp}} @ \alpha_{crit} = \omega_{resP} L_{S_1} A_{stp} @ \alpha_{crit}$$

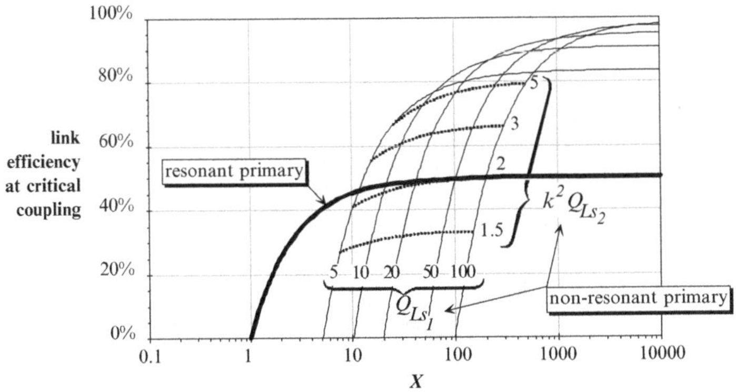


*A comparison for critical coupling, between links with or without primary resonance. The link efficiency is expressed as a function of X. This plot is valid for links with either a parallel- or a series-resonant secondary.* *Figure 3.9.*

> **Rule of thumb no. 5**: Primary coil resonance should be omitted whenever possible (i.e., $k > 10$ %) to yield a higher link efficiency.

The link efficiency at critical coupling is again only function of X. This product should once more be taken as large as possible for the best link efficiency. Primary coil resonance makes it possible to obtain critical coupling with less-coupled coils, but there is a price to pay. The link efficiency with resonant primaries can never exceed 50 %. The non-resonant links are more efficient at larger X values (Figure 3.9).

### *C. Coupling bandwidth of critically coupled links*

Operation at maximal link gain was introduced to desensitise inductive links from coupling variations. However, nothing has been said yet about the link's sensitivity for coupling variations around this maximum. A link gain with a broad maximum is less sensitive than a gain with a sharp and peaky maximum.

In order to have an idea about the coupling bandwidth, the relative gain reduction $dA$ when the coupling factor is changed by a relative amount $dk$ is calculated, for a link running at critical coupling for a given $k$. This expression in case of a non-resonant primary is rather complex:

$$dA@dk \equiv 1 - \left.\frac{A\big(k(1+dk)\big)}{A_{crit}(k)}\right|_{\alpha=\alpha_{crit}(k)}$$

$$= 1 - \left(1 + \frac{dk^2\left(4 + 4dk + dk^2\right)\sqrt{1+Q_{LS_1}^2}}{2(1+dk)^2\left(1+\sqrt{1+Q_{LS_1}^2}\right)}\right)^{-1/2} \qquad \textit{Equation 3.31.}$$

The normal design problem is the inverse one: "what is the coupling range if one allows relative gain variations of $dA$?" It is, though, impossible to revert Equation 3.31. and find a closed expression for $\pm dk$ as a function of $dA$. Nonetheless, the expression indicates that the coupling range is only function of $dA$ and the primary coil quality factor $Q_{L_{S_1}}$.

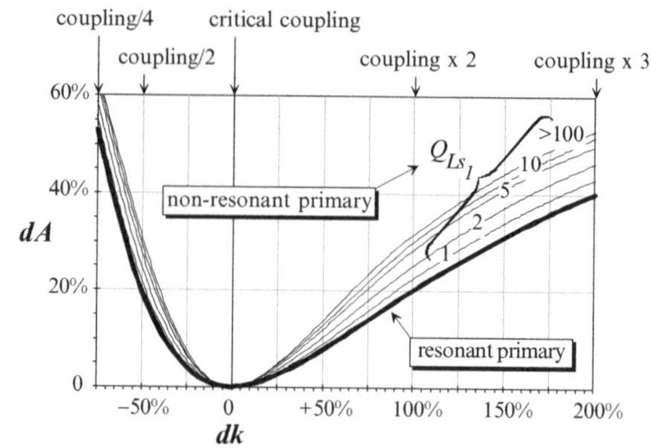


*Figure 3.10.* *The relative gain decrease dA of critically coupled links, as a function of the relative deviation dk in the coupling factor. This diagram is valid for links with a parallel- or a series-resonant secondary.*

The expression for a primary-resonant link is much simpler:

*Equation 3.32.*

$$dA@dk \equiv 1 - \left.\frac{A(k(1+dk))}{A_{crit}(k)}\right|_{\alpha=\alpha_{crit}(k)} = \frac{dk^2}{dk^2 + 2dk + 2}$$

The coupling bandwidth corresponding to a relative gain change $dA$, is not function of coil quality factors, but only of $dA$:

*Equation 3.33.*

$$\text{relative coupling range} = 2\frac{\sqrt{2dA - dA^2}}{1 - dA}$$

The above expressions are difficult to compare as such. They are therefore plotted in Figure 3.10. This shows that both link types are very comparable. The links without primary coil resonance are slightly more sensitive to coupling variations.

## 3.2 INDUCTIVE LINKS WITH A SERIES-RESONANT SECONDARY

This paragraph is reciprocal to the previous one and is therefore kept somewhat condensed. All the calculations are similar, apart from the secondary tank capacitor $C_2$ that is now in series with the secondary inductor $L_{S_2}$, its loss resistor $R_{S_2}$ and the a.c. load $R_{load_2}$ (Figure 3.1).

> The parasitic secondary coil capacitor is assumed to be negligible compared to $C_2$, to get valid results from the expressions for a series-resonant secondary.

The parallel tuning of the secondary allowed incorporating the parasitic capacitor of the secondary coil into $C_2$. This merit is, however, disposed with the series-resonant secondary.

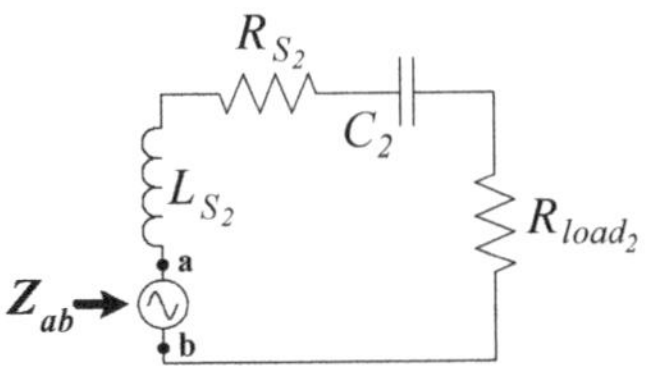


$$\begin{cases} Z_{ab} = R_{S_2} + j\omega L_{S_2} + \dfrac{1}{j\omega C_2} + R_{load_2} \\ \omega_{resP} \equiv \omega \quad @\text{Im}\{Z_{ab}\} = 0 \end{cases}$$

$$\Rightarrow \omega_{resP} = \sqrt{\frac{1}{L_{S_2} C_2}}$$

*The phase-resonance frequency $\omega_{resP}$ for a series-tuned secondary. It is the frequency at which the impedance $Z_{ab}$ seen by a voltage source in series to $L_{S2}$, is real. The source represents for instance a voltage induced in $L_{S2}$.* *Figure 3.11*

### 3.2.1 EQUIVALENT SECONDARY IMPEDANCE

Suppose that the tank capacitor $C_2$ in Figure 3.1 is put in series with the coil. The impedance $Z_{eq}$ is again calculated as

$$Z_{eq} = \frac{1}{\dfrac{1}{j\omega k^2 L_{S_1}} + \dfrac{1}{\left(\dfrac{k}{n}\right)^2 Z_{sec}}} - j\omega k^2 L_{S_1},$$

but with a different $Z_{sec}$:

$$Z_{sec} = R_{S_2} + \frac{1}{j\omega C_2} + R_{load_2}.$$

$$\Leftrightarrow Z_{eq} = \frac{\omega^3 k^2 L_{S_1} L_{S_2} C_2 \left[\omega C_2 \left(R_{load_2} + R_{S_2}\right) + j\left(1 - \omega^2 L_{S_2} C_2\right)\right]}{\omega^2 C_2^2 \left(R_{load_2} + R_{S_2}\right)^2 + \left(1 - \omega^2 L_{S_2} C_2\right)^2}$$

This equivalent secondary impedance $Z_{eq}$ is real at the frequency

$$\boxed{\omega = \sqrt{\frac{1}{L_{S_2} C_2}} = \omega_{resP} = \omega_{resA} \equiv \omega_{res}}$$ *Equation 3.34.*

This frequency corresponds to the phase-resonance frequency $\omega_{resP}$ and the amplitude-resonance frequency $\omega_{resA}$ of the loaded secondary tank circuit if it was not coupled to the primary coil (Figure 3.11). Both resonance frequencies coincide for series-resonant links and are therefore named $\omega_{res}$ from here on.

> The driver frequency $\omega$ is assumed to be equal to the secondary resonance frequency $\omega_{res}$.

The substitution of $\omega$ by $\omega_{res}$ in the expression of $Z_{eq}$, results in:

$$R_{eq} \equiv Z_{eq} @\omega_{res} = \frac{k^2 L_{S_1}}{C_2 \left(R_{load_2} + R_{S_2}\right)}$$

Or,

Equation 3.35.

$$R_{eq} = \frac{k^2 L_{S_1} \omega_{res}}{\alpha + \frac{1}{Q_{L_{S2}}}}$$

Equation 3.3.

$$\text{where } \alpha \equiv \omega_{res} C_2 R_{load_2}$$

## 3.2.2 LINK EFFICIENCIES

### 3.2.2.1 Primary link efficiency

$$\eta_{primary} \equiv \frac{P_{sec}}{P_{link\ in}} = \frac{R_{eq}}{R_{eq} + R_{S_1}} = \frac{k^2 L_{S_1}}{k^2 L_{S_1} + R_{S_1} C_2 \left(R_{S_2} + R_{load_2}\right)}$$

Or,

Equation 3.36.

$$\eta_{primary} = \frac{k^2 Q_{L_{S_1}}}{\alpha + k^2 Q_{L_{S_1}} + \frac{1}{Q_{L_{S_2}}}}$$

Primary coil tuning does not influence this expression.

### 3.2.2.2 Secondary link efficiency

As the tank capacitor is not shunting either $R_{S_2}$ or $R_{load_2}$, the secondary link efficiency now becomes a simple ratio of resistances:

$$\eta_{secondary} \equiv \frac{P_{link\ out}}{P_{sec}} = \frac{R_{load_2}}{R_{S_2} + R_{load_2}}$$

Or,

Equation 3.37.

$$\eta_{secondary} = \frac{\alpha}{\alpha + \frac{1}{Q_{L_{S_2}}}}$$

### 3.2.2.3 Total link efficiency

The total link efficiency is the product of the primary and the secondary efficiencies. This expression is also independent of primary coil tuning.

Equation 3.38.

$$\eta_{link} = \frac{k^2 Q_{L_{S_1}} \alpha}{\left(\alpha + k^2 Q_{L_{S_1}} + \frac{1}{Q_{L_{S_2}}}\right)\left(\alpha + \frac{1}{Q_{L_{S_2}}}\right)}$$

## 3.2.3 LINK GAIN

### 3.2.3.1 Non-resonant primary coil

The link gain is again computed via the link efficiency:

$$A \equiv \frac{V_{sec}}{V_{prim}} = \frac{\sqrt{P_{\text{link out}} R_{load_2}}}{V_{prim}} = \frac{\sqrt{P_{\text{link in}} \eta_{link} R_{load_2}}}{V_{prim}}$$

The link input power $P_{link\ in}$ for a given primary coil voltage $V_{prim}$ is also found from Figure 3.1:

$$P_{\text{link in}} = I_1^2\left(R_{eq} + R_{S_1}\right) = \ \dots\ = \frac{V_{prim}^2\left(R_{eq} + R_{S_1}\right)}{\omega_{res}^2 L_{S_1}^2 + \left(R_{eq} + R_{S_1}\right)^2}$$

This expression and the expression for $\eta_{link}$, are then substituted in the above formula for $A$ to yield:

$$A = \sqrt{\frac{\alpha R_{eq} R_{load_2}}{\left[\omega_{res}^2 L_{S_1}^2 + \left(R_{eq} + R_{S_1}\right)^2\right]\left(\alpha + \frac{1}{Q_{L_{S_2}}}\right)}}$$

Three parameters are now substituted in this expression:

- The a.c. load $R_{load_2}$ (via the definition of $\alpha$): $R_{load_2} = \dfrac{\alpha}{\omega_{res} C_2}$
- The tank capacitor $C_2$ (via the definition of $\omega_{res}$): $C_2 = \dfrac{1}{L_{S_2}\omega_{res}^2}$ *Equation 3.39.*
- The equivalent resistor $R_{eq}$ (via Equation 3.35.)

The result is:

$$A = k.n.Q_{L_{S_1}} Q_{L_{S_2}} \frac{\alpha}{\sqrt{Q_{L_{S_1}}^2\left(1 + \alpha Q_{L_{S_2}}\right)^2 + \left(\alpha Q_{L_{S_2}} + 1 + k^2 Q_{L_{S_1}} Q_{L_{S_2}}\right)^2}}$$

*Equation 3.40.*

#### 3.2.3.2 Series-resonant primary coil

$$A_{\text{stp}} \equiv \frac{V_{\text{sec}}}{V_{prim}} = \frac{\sqrt{P_{\text{link out}} R_{load_2}}}{V_{prim}} = \frac{\sqrt{P_{\text{link in}} \eta_{link} R_{load_2}}}{V_{prim}}$$

The link input power $P_{link\ in}$ for a given primary voltage $V_{prim}$ and a primary coil that resonates with the series capacitor $C_1$ is found as:

$$\left.\begin{aligned} P_{\text{link in}} &= I_1^2\left(R_{eq} + R_{S_1}\right) \\ \omega_{res} &= \frac{1}{\sqrt{L_{S_1} C_1}} \end{aligned}\right\} \Rightarrow \dots \Rightarrow P_{\text{link in}} = \frac{V_{prim}^2}{R_{eq} + R_{S_1}}$$

This expression and the expressions for $\eta_{link}$, $R_{load_2}$, $C_2$ and $R_{eq}$ are substituted in the above formula for $A_{stp}$ to produce:

$$A_{stp} = \frac{k.n.Q_{L_{S_1}} Q_{L_{S_2}} \alpha}{k^2 Q_{L_{S_1}} Q_{L_{S_2}} + \alpha Q_{L_{S_2}} + 1}$$

*Equation 3.41.*

#### 3.2.3.3 Parallel-resonant primary coil

$$r_{m_{ptp}} \equiv \frac{V_{\text{sec}}}{I_{prim}} = \frac{\sqrt{P_{\text{link out}} R_{load_2}}}{I_{prim}} = \frac{\sqrt{P_{\text{link in}} \eta_{link} R_{load_2}}}{I_{prim}}$$

The link input power $P_{link\ in}$ for a primary current $I_{prim}$ for a primary coil that resonates with the parallel capacitor $C_1$ is given by:

$$\left.\begin{aligned} P_{\text{link in}} &= I_1^2\left(R_{eq}+R_{S_1}\right) \\ \omega_{res} &= \frac{1}{\sqrt{L_{S_1}C_1}} \end{aligned}\right\} \Rightarrow \ldots \Rightarrow P_{\text{link in}} = \frac{I_{prim}^2\omega_{res}^2L_{S_1}^2}{R_{eq}+R_{S_1}}$$

The transimpedance $r_{m_{ptp}}$ is again equal to the product of the primary coil impedance with the voltage gain of the link:

Equation 3.42.
$$r_{m_{ptp}} = \omega_{res}L_{S_1}A_{stp} = \omega_{res}L_{S_1}\frac{k.n.Q_{L_{S_1}}Q_{L_{S_2}}\alpha}{k^2Q_{L_{S_1}}Q_{L_{S_2}}+\alpha Q_{L_{S_2}}+1}$$

### 3.2.4 LINK OPTIMISATION

The $C_{2_{opt}}$ for an optimal $\alpha$, follows directly from the definition of $\alpha$:

Equation 3.11.
$$C_{2_{opt}} = \frac{\alpha_{opt}}{\omega_{res}R_{load_2}}$$

$L_{S_2opt}$ results from substituting Equation 3.39. in the definition of $\alpha$:

Equation 3.43.
$$L_{S_2opt} = \frac{R_{load_2}}{\alpha_{opt}\omega_{res}}$$

#### 3.2.4.1 Solution at maximal link efficiency

Solving the equation $\frac{\partial \eta_{link}}{\partial \alpha} = 0$ yields

Equation 3.44.
$$\alpha_{\eta_{link_{\max}}} = \frac{\sqrt{1+X}}{Q_{L_{S_2}}},$$

Equation 3.14.
with the same definition for $X$ as before: $X \equiv k^2Q_{L_{S_1}}Q_{L_{S_2}}$.

This $\alpha$-value substituted in the expression for $\eta_{link}$ produces the expression for the maximal link efficiency:

Equation 3.15.[6]
$$\eta_{link_{\max}} = \frac{X}{\left(1+\sqrt{1+X}\right)^2}$$

This expression is identical to the one for a parallel-resonant secondary.

[6] The second derivative of $\eta_{link}(\alpha)$ in $\alpha_{\eta_{link_{\max}}}$ is again negative and proves that $\eta_{link_{\max}}$ is a maximum and not a minimum:

$$\left.\frac{\partial^2\eta_{link}}{\partial\alpha^2}\right|_{\alpha=\alpha_{\eta_{link_{\max}}}} = -\frac{2XQ_{L_{S_2}}^2\left[2+X+2\sqrt{1+X}\right]}{\sqrt{1+X}\left(1+\sqrt{1+X}\right)^6}$$

Replacing the $\alpha$-value from Equation 3.44. in Equation 3.40., Equation 3.41. and Equation 3.42. yields the link gains at maximal link efficiency. Note the remarkable similarity between Equation 3.16. and Equation 3.45. and between Equation 3.17. and Equation 3.46. The only difference is the $Q^2_{L_{S_2}}$ term that is each time missing in the numerator.

$$A@\eta_{link_{max}} = \frac{k.n.Q_{L_{S_1}}}{1+\sqrt{1+X}}\sqrt{\frac{1+X}{1+X+Q^2_{L_{S_1}}}}$$ *Equation 3.45.*

$$A_{stp}@\eta_{link_{max}} = \frac{k.n.Q_{L_{S_1}}}{1+\sqrt{1+X}}$$ *Equation 3.46.*

$$r_{m_{ptp}}@\eta_{link_{max}} = \omega_{res}L_{S_1}A_{stp}@\eta_{link_{max}}$$ *Equation 3.47.*

The equivalent secondary resistor at maximal link efficiency is found by substituting the $\alpha$-value from Equation 3.44. in Equation 3.35.:

$$R_{eq}@\eta_{link_{max}} = \frac{\omega_{res}L_{S_1}}{Q_{L_{S_1}}}\frac{X}{1+\sqrt{1+X}} = R_{S_1}\frac{X}{1+\sqrt{1+X}}$$ *Equation 3.19.*

This expression is also identical to the one for a parallel-resonant secondary.

### 3.2.4.2 Solution at critical coupling

*A. Non-resonant primary coil*

The critical coupling factor is again described by the equation $\frac{\partial A}{\partial k} = 0$ that has four mathematical solutions:

$$\Rightarrow k^2_{crit} = \pm\frac{\left(1+\alpha Q_{L_{S_2}}\right)\sqrt{1+Q^2_{L_{S_1}}}}{Q_{L_{S_1}}Q_{L_{S_2}}}$$

This expression reverses to:

$$\alpha_{crit} = \pm\frac{k^2 Q_{L_{S_1}}}{\sqrt{1+Q^2_{L_{S_1}}}} - \frac{1}{Q_{L_{S_2}}}$$

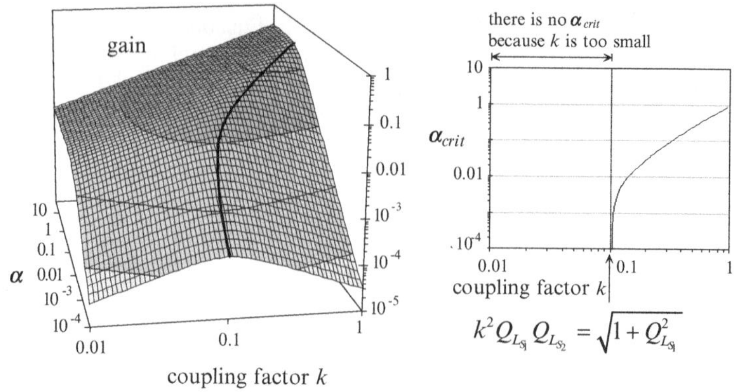


*Figure 3.12. The gain of an inductive link with a non-resonant primary and a series-resonant secondary, as a function of the coupling factor and the parameter $\alpha$ (left). The gain maxima are indicated in bold. The right diagram indicates where these gain maxima occur. Note that the plots are on logarithmic scales.*
*Both diagrams are for $L_{S_1} = 200\mu H, L_{S_2} = 20\mu H$ and $Q_{L_{S_1}} = Q_{L_{S2}} = 100$.*

The minus sign in the numerator is omitted to get a positive $\alpha$:

Equation 3.48. [7]

$$\alpha_{crit} = \frac{k^2 Q_{L_{S_1}}}{\sqrt{1+Q_{L_{S_1}}^2}} - \frac{1}{Q_{L_{S_2}}}$$

The condition for the coupling factor $k$ and the coil quality factors $Q_{L_{S_1}}$ and $Q_{L_{S_2}}$ to have a positive $\alpha_{crit}$, is the very same one as for the case of a parallel-resonant secondary (Figure 3.12):

Equation 3.21.

$$k^2 Q_{L_{S_1}} Q_{L_{S_2}} \geq \sqrt{1+Q_{L_{S_1}}^2}$$

Finally, the link efficiency, the link gain and the equivalent resistance at critical coupling are found by substituting the $\alpha_{crit}$ value into Equation 3.38., Equation 3.40. and Equation 3.35. respectively. The formulae for the link efficiency and the equivalent secondary resistor are again identical to their counterparts of a parallel-resonant secondary.

[7] There is still to prove that $\alpha_{crit}$ corresponds to a gain maximum and not a minimum. The second derivative of the gain at $\alpha_{crit}$ is given by

$$\left.\frac{\partial^2 A}{\partial k^2}\right|_{\alpha=\alpha_{crit}} = \frac{n}{Q_{L_{S_2}} k^3}\left(\sqrt{1+Q_{L_{S_1}}^2} - k^2 Q_{L_{S_1}} Q_{L_{S2}}\right)\sqrt{\frac{2\sqrt{1+Q_{L_{S_1}}^2}}{\left(1+\sqrt{1+Q_{L_{S_1}}^2}\right)^3}}$$

The first and the third factor are always positive so only the second factor $\left(\sqrt{1+Q_{L_{S_1}}^2} - k^2 Q_{L_{S_1}} Q_{L_{S_2}}\right)$ determines the sign of the second derivative. This factor is negative for $k$ and $Q_{L_S}$ values that meet Equation 3.21., which proofs that it is a maximum.

$$\eta_{link}@\alpha_{crit}=\frac{\sqrt{1+Q_{L_{S_1}}^2}.\left(k^2Q_{L_{S_1}}Q_{L_{S_2}}-\sqrt{1+Q_{L_{S_1}}^2}\right)}{k^2Q_{L_{S_1}}Q_{L_{S_2}}\left(1+\sqrt{1+Q_{L_{S_1}}^2}\right)}$$ *Equation 3.22.*

$$A@\alpha_{crit}=\frac{n}{kQ_{L_{S_2}}}.\frac{k^2Q_{L_{S_1}}Q_{L_{S_2}}-\sqrt{1+Q_{L_{S_1}}^2}}{\sqrt{2\left(1+Q_{L_{S_1}}^2+\sqrt{1+Q_{L_s}^2}\right)}}$$ *Equation 3.49.*

$$R_{eq}@\alpha_{crit}=\frac{\omega_{res}L_{S_1}}{Q_{L_{S_1}}}\sqrt{1+Q_{L_{S_1}}^2}=R_{S_1}\sqrt{1+Q_{L_{S_1}}^2}$$ *Equation 3.24.*

### B. Series- or parallel-resonant primary coil

The critical coupling factor is again found from $\frac{\partial A_{stp}}{\partial k}=0$ or $\frac{\partial r_{m_{ptp}}}{\partial k}=0$.

Both equations have two mathematical solutions:

$$\Rightarrow k_{crit}=\pm\sqrt{\frac{\alpha Q_{L_{S_2}}+1}{Q_{L_{S_1}}Q_{L_{S_2}}}}$$

The reversed expression is:

$$\alpha_{crit}=\frac{X-1}{Q_{L_{S_2}}}$$ *Equation 3.50.*[8]

The condition to be met by the coupling factor $k$ and the coil quality factors $Q_{L_{S_1}}$ and $Q_{L_{S_2}}$ to have a positive $\alpha_{crit}$ is identical to the one of a link with a parallel-resonant secondary (Figure 3.13):

$$k^2Q_{L_{S_1}}Q_{L_{S_2}}\geq 1$$ *Equation 3.26.*

The link efficiency, the equivalent resistance and the link gains at critical coupling, are found by substituting the $\alpha_{crit}$ value into Equation 3.38., Equation 3.35., Equation 3.41. and Equation 3.42. respectively. The link efficiency and the equivalent secondary resistor are again equal to their equivalents for a parallel-resonant secondary.

---

8 The second derivative of $A_{stp}$ in $\alpha_{crit}$ is negative and proves that $\alpha_{crit}$ corresponds to a maximum and not a minimum:

$$\left.\frac{\partial^2 A_{stp}}{\partial k^2}\right|_{\alpha=\alpha_{crit}}=-n\frac{X-1}{2Q_{L_{S_2}}k^3}$$

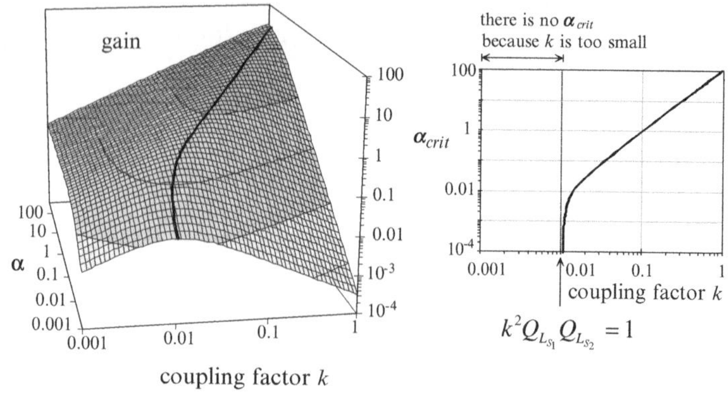


*Figure 3.13.* *The gain of an inductive link with a series-resonant primary and secondary, as a function of the coupling factor and the parameter α (left). The gain maxima are indicated in bold. The right diagram indicates where these gain maxima occur. Both diagrams are for* $L_{S_1} = 200\mu H, L_{S_2} = 20\mu H$ *and* $Q_{L_{S_1}} = Q_{L_{S_2}} = 100$.

*Equation 3.27.*

$$\eta_{link} @\alpha_{crit} = \frac{X-1}{2X}$$

*Equation 3.28.*

$$R_{eq} @\alpha_{crit} = \frac{\omega_{resF} L_{S_1}}{Q_{L_{S_1}}} = R_{S_1}$$

For a series-resonant primary:

*Equation 3.51.*

$$A_{stp} @\alpha_{crit} = n\frac{X-1}{2kQ_{L_{S_2}}}$$

For a parallel-resonant primary:

*Equation 3.52.*

$$r_{m_{ptp}} @\alpha_{crit} = \omega_{resF} L_{S_1} A_{stp} @\alpha_{crit}$$

### *C. Coupling range of critically coupled links*

The links with a resonant primary coil are again compared to those without primary coil resonance in terms of coupling sensitivity at critical coupling. The relative gain decrease *dA* if the coupling factor is changed with a relative value *dk*, is identical to what was found for the links with a parallel-resonant secondary. The expressions are therefore not repeated.

## 3.3 CONCLUSIONS

It was shown that the widespread and commonly used approximate design formulae are unsatisfactory when dealing with poor coil coupling. Therefore, a new set of exact formulae was derived. The use of dimensionless variables –the coupling factor *k*, the coil quality factors $Q_{L_{S_1}}$ and $Q_{L_{S_2}}$, the parameter $\alpha$ and *n,* the square root of the inductance ratio– dramatically simplified the expressions and rendered them workable. This analysis also indicated that the link efficiency was essentially independent of the primary inductance regardless primary tuning. The voltage gains equaled $k.n$ multiplied by a factor that is independent of the primary inductance.

Three optimised operation modes were then studied: maximal link efficiency, and maximal link gain with and without primary coil resonance (also known as critical coupling). The maximal link efficiency and the link efficiency for critical coupling with a resonant primary were monotonous increasing functions in only one parameter, i.e. the product $k^2 Q_{L_{S_1}} Q_{L_{S_2}}$. This product should thus be maximised for any kind of application.

There were two operation modes identified –one using primary coil resonance and one without– where the link gain as a function of the coil coupling attained a maximum. Operating an inductive link in such a "critically coupled" mode desensitised its gain to coil movements. However, there was a lower limit found for the coupling factor below which the gain lost its peaked shape and became a monotonous function. The price for a desensitised gain was that the link efficiency could never be made better than 50%, at least in case of a resonant primary. The mode without primary resonance did not show this efficiency limit and was surprisingly never published before. It required however more coupling ($k^2 \geq 1/Q_{L_{s2}}$ for high Q-values, instead of $1/\left(Q_{L_{s1}} Q_{L_{s2}}\right)$). The conclusion was thus that primary resonance was best omitted when possible with the given coil coupling.

There was a remarkable similarity observed between the link formulae for parallel- and series-resonant secondaries. There were identical expressions found for:

- the link efficiency and the equivalent secondary resistor in the three optimised modes (i.e. maximal link efficiency and critical coupling with or without primary coil resonance);
- the conditions for the coupling factor and the coil quality factors to produce a gain peak and make critical coupling possible (with and without primary resonance); and
- the relative gain reduction $dA$ if the coupling is changed around its critical value by a relative amount $dk$ (with and without primary resonance).

This similarity means that the choice between series- or parallel-tuning of the secondary is not guided by efficiency since there is no difference, but by the amplitude of the equivalent a.c. load $R_{load_2}$ (Figure 2.12).

## 3.4 REFERENCES

**Forster I.C.**, "Theoretical design and implementation of a transcutaneous, multichannel stimulator for neural prosthesis applications," *J. Biomed. Engng.*, vol. 3, pp. 107–120, 1981.

**Forster I.C.**, "Corrections on the paper: Theoretical design and implementation of a transcutaneous, multichannel stimulator for neural prosthesis applications," *J. Biomed. Engng.*, vol. 5, pp. 176, 1983.

***Ko W.H.**, S.P. Liang and C.D.F. Fung, "Design of radio-frequency powered coils for implant instruments," *Med. & Biol. Eng. & Comput.*, vol. 15, pp. 634-640, 1977.

# 4 Primary coil drivers

Because we recognised in chapter 2 that the optimisation of weakly coupled inductive powering systems should include the coil driver, the different amplifiers that are applied for this purpose are reviewed first. Two specific features characterise inductive link drivers: the driver load is essentially inductive and the drivers generate large a.c. currents to produce the magnetic field for the inductive powering. An inductive powering set-up is also intended to operate at a specific transfer frequency. This makes it possible to tune out the large primary leakage impedance with a resonance capacitor and have a more effective use of the available a.c. voltage. It is



K. Van Schuylenbergh and R. Puers, *Inductive Powering: Basic Theory and Application to Biomedical Systems*, Analog Circuits and Signal Processing, 103–143. 

perfectly possible to drive the primary coil with a class-A or -B amplifier (HANSEN *et al.*, 1982), but coil tuning opens a range of more efficient alternatives, from the class-C, in saturating mode or not, over the class-D, to the most efficient class-E topology. Their common concept is that they all have the primary tank freely resonating and use the power supply and the driver transistor just to compensate for the tank losses. These amplifiers, except the non-saturating class C, are switch-mode circuits where the transistors operate as switches controlled by clock-like input signals.

Primary coil drivers are not linear amplifiers since their output signal is not an amplified representation of their input signal. The coil drivers merely act as d.c.-to-a.c. power inverters. They take up d.c. energy from the supply source and convert that as efficient as possible into a.c. energy to power the primary coil. There is no linear temporal relation between input and output. The output is a sinusoidal voltage across or a sinusoidal current through the primary coil, whereas the input signals are pulses.

The amplifier design equations developed in this chapter, are not the typical textbook versions that describe the powering of a given constant load (often 50Ω). The coils and capacitors in a conventional driver design have no other function than to shape the driver signals in such a way that an optimal d.c.-to-a.c. conversion is obtained for powering the driver load. The capacitor and inductor values are set during the amplifier design and the tank quality factors are mostly taken rather low, in the range of 5 to 15, for a high driver efficiency. But, the situation is different with inductive powering. The values for the primary coil and its resonance capacitor result from the link design, which means that they are not available anymore for optimising the driver circuit. A low tank quality factor is also not realistic and even undesired, especially in case of poor coil coupling. The primary coil resistor is made as small as possible for minimal coil loss and the equivalent resistor of the secondary power consumption is small due to the poor coupling. The resistive load is also not some constant typically 50Ω resistor, but is often much smaller and varies with the secondary consumption and the coil coupling.

The poor coupling not only changes the way to look at driver design, it also interferes with the design of the coil set. The customary design procedures for moderate to well-coupled links, determine the values for the primary coil and its resonance capacitor during the link design. The present study identified this as the major cause of problematic driver performance at weak coupling, because the primary coils optimised for link performance, do not match the driver requirements very well. This leads to absurd designs where the driver would need a 0.3 V supply and draw 20 A or 100 V and 2 mA. Moreover, the driver load resistor is often much smaller than the switch-on resistance, and the supply voltage is often comparable to the switch-saturation voltage. Hence, the high driver efficiency intrinsic to switch-mode topologies is completely compromised. The novel approach presented in this text, therefore combines link design with driver design, to get both right in one single procedure.

In a strict mathematical sense, the optimisation of the driven inductive link corresponds to the minimisation of some cost function (that is efficiency$^{-1}$) with boundary conditions, like operation at critical coupling. Such problems can be solved numerically. But, a large number of parameters determine the link and driver design: physical coil layout, transfer frequency, coil inductances and quality factors, resonance capacitors, the rectifier, regulator and driver types, spectrum regulation. This already imposes an important computational problem, and –what is maybe more important– the insight of how parameters interact is lost with a one-step and pure mathematical approach.

It is preferred to break the overall optimisation down into smaller parts with minimal mutual interaction and to iterate once or twice when a present sub-optimisation step requires the revision of a previous one. A maximum insight is also obtained if these sub-optimisations are allocated to physical sub-blocks of the driven inductive link.

The optimisation procedure is fully elaborated in chapter 5. Before this puzzle is put together, all parts are laid out first. This already began in chapter 2, with the discussion of rectifiers and regulators. The exact link equations were then derived in chapter 3. This fourth chapter now reviews the driver types that are suitable for inductive links. The puzzle is then completed in the chapter 5 where the optimisation procedure is explained. Nonetheless, the <u>optimisation concept</u> is already introduced here to better situate the bits and pieces that follow.

1. The design starts with the optimisation of the <u>magnetic</u> link properties.
2. It is then followed by the <u>electronic</u> optimisation of the <u>secondary</u> circuit, and
3. The <u>electronic</u> optimisation of the <u>primary</u> circuit with the coil driver. First-order-simplified expressions are used for the driver calculations. This is allowed as well designed link drivers run at a high efficiency.
4. Spice control simulations of the driven inductive link, complete the design, to check for errors and to verify if it was justified to use first-order-simplified driver models.

The interaction between the first and the second part is minimal. Magnetic parameters purely relate to coil geometries and materials and do not impact the electronic link parameters. Once they are optimised, they need no further revision in the remainder of the design. If the procedure proposed in the next chapter is followed, the secondary components are independently selected from the primary ones. The selection of the best suited driver type and the matching of the coil set to that driver, is only carried out at a primary level and does not affect secondary component values. It was observed and it is proven further on, that the interaction between coil set and driver is controlled by one single parameter: the primary coil inductance. This parameter has no influence on the link efficiency and is, at the same time, a determining factor with respect to driver performance. All primary link components are thus expressed in function of this parameter and the value for optimal system efficiency is selected, taking the supply constraints and maximum transistor ratings into account.

The four driver classes suitable for inductive powering –class C, saturating or not, class D and E– are now discussed in detail. Note that the class-B amplifiers used in some very early link designs are quite suboptimal for inductive powering and thus further ignored. The modelling of switch transistors and some considerations on the importance of proper supply decoupling, are also included in this chapter. This review then closes off with a discussion of the specific problem of the high tank Q when driving weakly coupled links. A circuit modification is proposed to lower the tank Q of a class-E driver to restore its high efficiency, while maintaining a high coil Q for maximal link efficiency.

Note that the driver efficiencies mentioned in this chapter are collector efficiencies, meaning that the power required to drive the transistor base or gate is not included. This contribution, however, cannot just be neglected and must be considered when comparing amplifier designs. For example, driving a 1 nF gate of a power MOSFET at 1 MHz already takes 10 mA for a gate-voltage swing of 10 V. Furthermore, the rapid gate-voltage transitions needed in switch-mode drivers, require dedicated buffer drivers that provide the large current peaks. Switching the 1 nF power MOSFET gate from 0 to 10 V in 50 ns demands 200 mA and a source resistance below 10 Ω ($\tau < 10$ ns).

Driver efficiency is not only a matter of saving power. Moving the driver efficiency from 90 % to 97.5 % represents a fairly small efficiency improvement (only about 8.5 %) but means that the driver dissipation is cut by a factor of four. A smaller switch transistor can thus be used and rapid switch transitions are more readily achieved with simpler circuits.

## 4.1 CLASS C

A class-C amplifier is built around a freely oscillating parallel-LC tank. The tank dissipation is compensated by a short current pulse from an active element, each time when the voltage across this element is minimal. The shape of current pulse is the positive segment of a negative-biased sine wave. The fact that the active element carries only currents during short intervals and at minimal voltage results in a lower amplifier dissipation compared to the classes A or B where the active element is conducting for respectively 100 or 50 % of the time.

A class-C primary coil driver is obtained by replacing the amplifier tank with the parallel-tuned primary link coil (Figure 4.1). The tank coil is then formed by the equivalent inductance of the primary, coupled to the secondary. The tank capacitor is the primary coil tuning capacitor. The tank resistor constitutes of the coil losses and the dissipation in the secondary.

Figure 4.1 pictures a MOSFET as the active element, but class-C drivers can be built with any active element that possesses a current output characteristic: bipolar transistors, vacuum tubes etc.

The tank impedance $\underline{Z}_{tank}$ seen at the transistor terminals is

$$\underline{Z}_{tank} = \frac{1}{j\omega C_1} \, // \left(j\omega L_{eq} + R_{S_1} + R_{eq}\right)$$

Class-C drivers run at the phase resonance of the driver tank. Hence,

$$\omega_{tank} \equiv \omega \;@\left(\mathrm{Im}\left\{\underline{Z}_{tank}\right\} = 0\right) \Leftrightarrow \omega_{tank} = \sqrt{\frac{1}{L_{eq}C_1} - \frac{\left(R_{S_1} + R_{eq}\right)^2}{L_{eq}^2}}$$

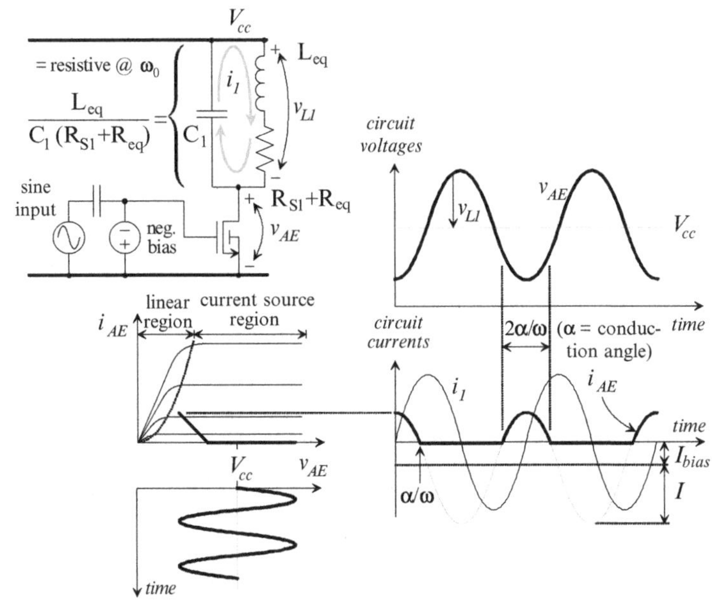


*A class-C primary coil driver.* *Figure 4.1.*

At this frequency, the driver-tank impedance is purely resistive and the transistor voltage and current are perfectly in phase:

$$R_{tank} \equiv \underline{Z}_{tank} @\omega_{tank} = \frac{L_{eq}}{\left(R_{S_1} + R_{eq}\right)C_1} = \frac{\omega^2 L_{eq}^2 + \left(R_{S_1} + R_{eq}\right)^2}{R_{S_1} + R_{eq}}$$

The last term is found by substituting $C_1$ from the frequency $\omega_{tank}$.

The transistor current is the positive segment of a negative-biased sine wave. It is zero for the remainder of the period. The ***conduction angle α*** is defined to indicate ***the time interval –α/ω, +α/ω where the active element is conducting***.

$$\begin{cases} i_{AE} = I\cos(\omega t) - I_{bias} = I\left[\cos(\omega t) - \cos\alpha\right] & \text{for} -\frac{\alpha}{\omega} < t \le \frac{\alpha}{\omega} \\ i_{AE} = 0 & \text{for} \frac{\alpha}{\omega} < t \le T - \frac{\alpha}{\omega} \end{cases}$$

The useful a.c. driver output power is determined by $R_{tank}$ and the a.c. voltage across this resistor:

$$P_{driver\ out} = \frac{\hat{V}_{AE_\omega}^2}{2R_{tank}}$$

The symbol $\hat{V}_{AE_\omega}$ stands for the amplitude of the fundamental harmonic component at frequency $\omega$, of the switch voltage $v_{AE}$. Its RMS value is denoted as $V_{AE_\omega}$.

The a.c. output power corresponds to the link input power $P_{link\ in}$ (Figure 3.3). Higher order power terms are not transferred over the link due to the secondary resonance.

The voltage $\hat{V}_{AE_\omega}$ across $R_{tank}$ is found as the first harmonic $\hat{I}_{AE_\omega}$ of the transistor current[1] times the resistance $R_{tank}$. Hence,

$$P_{link\ in} = \frac{\hat{I}^2_{AE_\omega} R_{tank}}{2} = \frac{I^2}{8\pi^2}\left[2\alpha - \sin(2\alpha)\right]^2 R_{tank}$$

Note that $\hat{I}_{AE_\omega}$ corresponds to $\hat{I}_{prim}$ in Figure 3.1.

The driver consumption is calculated from the d.c. component $I_{AE_{d.c.}}$ of the transistor current. $I_{AE_{d.c.}}$ equals the net supply current $I_{cc}$. Thus,

$$P_{in} = V_{cc} I_{AE_{d.c.}} = \frac{V_{cc} I}{\pi}(\sin\alpha - \alpha\cos\alpha)$$

The class-C driver efficiency is the ratio of last two expressions:

$$\eta_{driver} \equiv \frac{P_{link\ in}}{P_{in}} = \frac{\left[2\alpha - \sin(2\alpha)\right]^2}{(\sin\alpha - \alpha\cos\alpha)} \frac{IR_{tank}}{8\pi V_{cc}}$$

The transistor voltage must remain large enough throughout the complete conduction cycle, to maintain a current-source characteristic and avoid transistor saturation. Hence, $V_{cc} = v_{AE_{min}} + \hat{I}_{prim} R_{tank}$ and

$$\eta_{driver} = \frac{\left[2\alpha - sin(2\alpha)\right]^2}{(sin\alpha - \alpha cos\alpha)} \frac{IR_{tank}}{8\pi\left(v_{AE_{min}} + \hat{I}_{prim} R_{tank}\right)}$$

These equations allow calculating the circuit components of a class-C coil driver. All driver parameters are now expressed in terms of the primary coil inductance, to fit in the optimisation procedure of chapter 5. The resulting class-C design process is summarised in the frame on the right.

A driver transistor is subjected to three types of device stresses: dissipation, voltage and current. The driver efficiency describes the dissipation stress: the maximal driver output for a given transistor with a dissipation limit *DL* equals $DL/(1-\eta_{driver})$. There is no similar parameter defined yet to describe the power limitations due to voltage and current stresses. The current and voltage limits are considered together as they can be exchanged for each other: increasing the coil inductance $L_{S1}$ and reducing the capacitor $C_1$ scales down the circuit currents and scales up the circuit voltages accordingly. The ***output power capability p*** is defined as ***the driver output divided by the product of the maximal active element current with the maximal active element voltage.***

---

[1] The current through the active-element is an even function. Its Fourier development contains only cosine terms and is given by:

$$i_{AE}(t) = I_{AE_{d.c.}} + \hat{I}_{AE_\omega} cos(\omega t) + \hat{I}_{AE_{2\omega}} cos(2\omega t) + \hat{I}_{AE_{3\omega}} cos(3\omega t) + \ldots$$

$$\text{where}\quad \begin{cases} I_{AE_{d.c.}} = \dfrac{I}{T}\displaystyle\int_{-\alpha/\omega}^{\alpha/\omega}\left[cos(\omega t) - cos\,\alpha\right]dt = \dfrac{I}{\pi}(sin\,\alpha - \alpha\,cos\,\alpha) \\ \hat{I}_{AE_\omega} = \dfrac{2I}{T}\displaystyle\int_{-\alpha/\omega}^{\alpha/\omega}\left[cos(\omega t) - cos\,\alpha\right]cos(\omega t)dt = \dfrac{I}{2\pi}\left[2\alpha - sin(2\alpha)\right] \end{cases}$$

**THE CLASS-C DESIGN FLOW**

The link output power and output voltage are given.

The secondary components, the required link output voltage $V_{sec}$, and the transfer frequency $\omega$ result from the secondary optimisation.

Choose $L_{S_1}$ and conduction angle $\alpha$.

Compute $L_{eq}$, $R_{eq}$ and the link gain $r_{m_{ptp}}$ with the equations of chapter 3.

$$\hat{I}_{prim} = \hat{V}_{sec} / r_{m_{ptp}}$$

$$C_1 = \frac{L_{eq}}{\omega^2 L_{eq}^2 + \left(R_{S_1} + R_{eq}\right)^2}$$

$$R_{tank} = \frac{\omega^2 L_{eq}^2 + \left(R_{S_1} + R_{eq}\right)^2}{R_{S_1} + R_{eq}}$$

$$I = \frac{2\pi}{2\alpha - \sin(2\alpha)} \hat{I}_{prim}$$

$$I_{cc} = I(\sin\alpha - \alpha\cos\alpha)/\pi$$

$$i_{AE_{max}} = I(1 - \cos\alpha)$$

The active element needs a minimal voltage $v_{AE_{min}}$ to maintain a current-source characteristic up to this maximum current $i_{AE_{max}}$.

$$V_{cc} = v_{AE_{min}} + \hat{I}_{prim} R_{tank}$$

$$\eta_{driver} = \frac{2\alpha - sin(2\alpha)}{4(sin\alpha - \alpha cos\alpha)} \left( \frac{\hat{I}_{prim} R_{tank}}{v_{AE_{min}} + \hat{I}_{prim} R_{tank}} \right)$$

$$v_{AE_{max}} = V_{cc} + \hat{I}_{prim} R_{tank}$$

The maximal driver output of a circuit that fully exploits the transistor voltage and current limitations $v_{AE_{max}}$ and $i_{AE_{max}}$ –the load coil and capacitor are thus selected such that the transistor reaches its current and its voltage limits– is equal to $p.v_{AE_{max}} i_{AE_{max}}$.

The output power capability of the class-C topology is:

$$p \equiv \frac{P_{link\ in}}{v_{AE_{max}} i_{AE_{max}}} = \frac{2\alpha - \sin(2\alpha)}{4\pi(1 - \cos\alpha)} \left( \frac{\hat{I}_{prim} R_{tank}}{v_{AE_{min}} + 2\hat{I}_{prim} R_{tank}} \right)$$

This expression is plotted in Figure 4.2 together with the class-C driver efficiency. The left diagram indicates that loss-free operation is theoretically achieved for a zero conduction angle, and a zero minimal voltage for the current source. But, supplying a finite amount of power in a zero conduction angle demands an infinite transistor current.

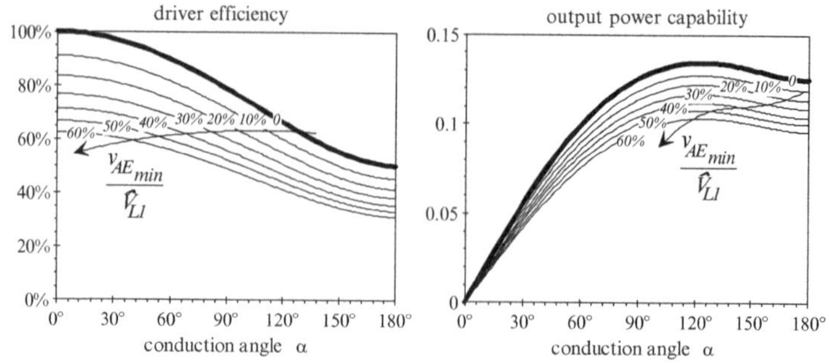


*Figure 4.2.* *The class-C efficiency and output power capability in function of the conduction angle $\alpha$ and the ratio between the minimal active element voltage and the amplitude of the coil voltage $v_{L1}$ (peak value, not RMS!).*

Or otherwise stated, the maximal output power is zero for a real transistor since its current bound is finite. This is also expressed by the zero output power capability in the right diagram of Figure 4.2.

Operating the active element as a current source is not good for the efficiency, since there is a minimal voltage $v_{AE_{min}}$ needed during conduction. What is more, class-C operation is never loss-free for non-zero conduction angles, even for a zero $v_{AE_{min}}$. Since the transistor voltage has a sinusoidal shape, it is non-zero before and after its minimum (Figure 4.2, left). For both reasons, it is better to leave the current-source operation mode for the active element, and run it in a switch mode. Ideal switches are lossfree since they conduct without voltage across their terminals.

The current-mode operation is also problematic for reliability reasons. It is difficult to guarantee a prescribed transistor current, because of the temperature dependency of current gains, the large variability between transistors from different batches or manufacturers, and the non-ideal transistor dynamics that come into play at higher frequencies. If the transistors are driven hard enough to saturate, they cease to be current sources and operate as switches. Transistor control is then reduced to abrupt on-off switching instead of the subtle control of device currents. Using a transistor as a switch and not as a current source, makes circuit operation less sensitive to transistor characteristics.

The conclusion is thus that coil drivers must their transistors as a switch to obtain a good reliability and a high efficiency.

## 4.2 MODELLING OF SWITCH TRANSISTORS

A standard approach in circuit simulation is to replace non-linear devices, by their equivalent models (cf. the level 1, 2, 3 MOSFET models and the bipolar Ebers-Moll models used in Spice simulations).

However, most device models either are small-signal models or are too complicated for hand calculations. Fortunately, switch transistors work in a narrow region of their operation range. The transistor is used as a switch, meaning that its gate or base terminal is driven to such a level that the low source-drain or collector-emitter voltage is guaranteed during conduction.

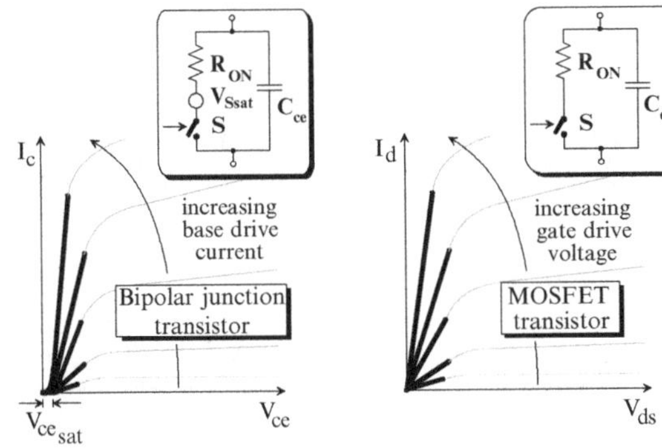


*The operation range of a transistor functioning as a switch and the corresponding transistor models suitable for switch-mode signal calculations.* *Figure 4.3.*

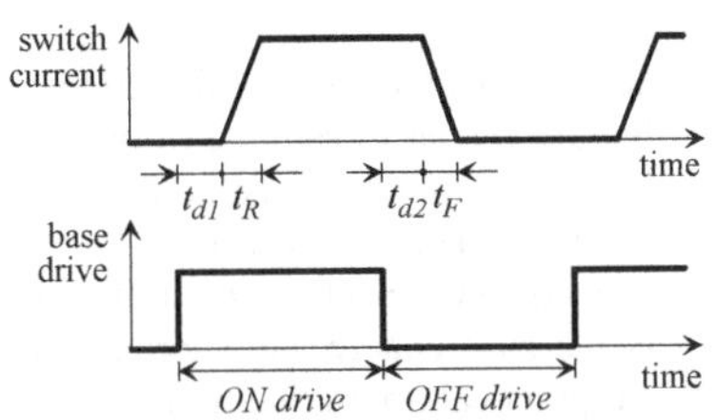


*Modelling transistor dynamics* *Figure 4.4.*

This justifies using an equivalent transistor model that consist of an ideal switch in series with an ***on-resistance*** $R_{ON}$ and a ***saturation voltage*** $V_{Ssat}$ (Figure 4.3). The saturation voltage is the residual switch voltage when the switch current approaches zero and has nothing to do with the saturation of a bipolar transistor that looses its current-source characteristic at small voltages. MOSFET's are well modelled by the switch and the on-resistance only. Bipolar transistors however, need all three parameters. Because of the unequal base-emitter and base-collector junction area, a voltage drop $V_{CEsat}$ arises during conduction.

It is more common with switch-mode circuits, to talk about the switch duty cycle instead of its conduction angle. ***The duty cycle is the ratio between the switch-on time and the total time of a switch on-off cycle.***

Transistor switching speeds are limited (Figure 4.4.). The non-ideal switch dynamics are expressed by the rise and fall times $t_R$ and $t_F$, and the delay times $t_{d1}$ and $t_{d2}$. The ***current rise time*** $t_R$ is ***the time it takes for the drain or collector current to build up and reach its steady state value, after the gate or base drive has been set***. The ***current rise time*** $t_R$ is defined in a similar way, as ***the time the current declines to zero, after the transistor is switched off***. The ***switch delay times*** are defined as ***the reaction times between the reversal of the switch drive and the beginning of the corresponding current change***.

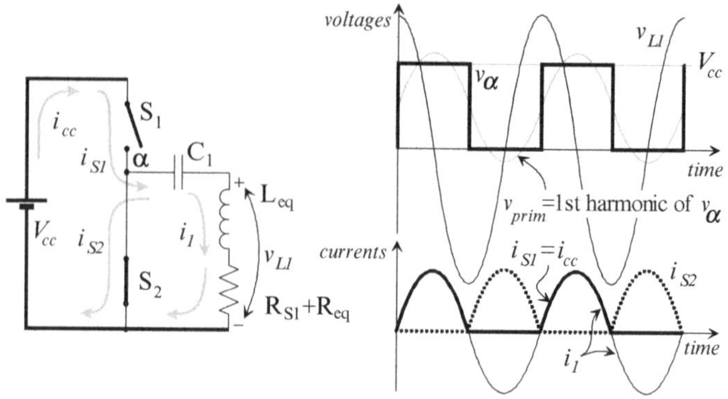


*Figure 4.5.* *An ideal class-D primary coil driver.*
*The switches $S_1$ and $S_2$ are complementary driven at a 50 % duty cycle.*

MOS switches have pretty constant delay times as listed in the component data sheets. Bipolar transistors that are forced to saturate to operate as switches accumulate charge in their base region that must be depleted during $t_{d2}$, before the switch actually turns off. This turn-off delay is better known as the storage time $t_S$ and depends on the actual situation: e.g. a negative base drive actively extracts the base charge and shortens the storage time. SOKAL (1976) published a practical approximation for $t_S$ at high switch frequencies where the transistor-on time $t_{ON}$ becomes smaller than the lifetime $\tau_S$ of excess charge carriers in the base region.

The delay times are not explicitly included in the switch model since they only change the duty cycle or shift the on-time somewhat.

## 4.3 CLASS D

Class-D drivers are characterised by a push-pull configuration of complementary driven switches that connect the output node $\alpha$ in Figure 4.5 to either the ground or the positive supply rail. The amplifier has a voltage-type output and is thus most suitable for links with a series-resonant primary. The switches run at a duty cycle of 50 %[2], and define the output voltage without regard of the load network. This is in contrast with the class-C driver, where the amplifier tank shaped the sinusoidal output signals and the transistor only supplied the current pulses to keep up the tank oscillation. Class-D amplifiers can also drive any kind of load,

[2] Class-D drivers are switching push-pull amplifiers with a band-pass filtered output, while class-S drivers have a low-pass filtered output. The class-S switches are not driven by a 50 % duty cycle but by a pulse-width-modulated signal. Class S is often used to generate high-power audio-frequency signals with a better efficiency than class A or B.

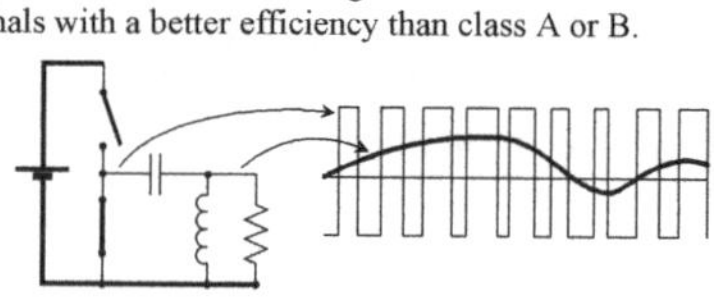

The definitions of class D and S are sometimes unclear: some authors use “class D” and “class S” interchangeably, while others reverse the definitions mentioned here.

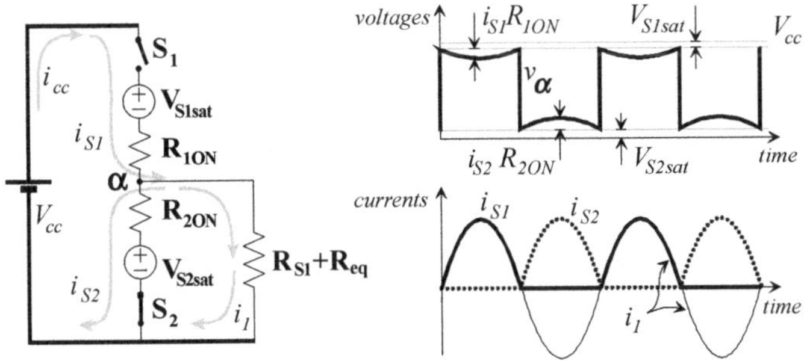


*A class-D driver with non-ideal switches running at the resonance frequency of the load tank. $C_1$ and $L_{eq}$ are not drawn as they compensate each other.* *Figure 4.6.*

be it resistive, inductive or capacitive. Tank resonance is not imperative for driver operation but maximises the coil current for a given supply and is therefore supposed here. Hence,

$$\omega = \omega_{tank} \equiv \sqrt{\frac{1}{C_1 L_{eq}}}.$$

At quality factors of the loaded primary coil above 10 to 20, the output current $i_1$ is virtually sinusoidal. It is solely determined by the loaded-coil resistance $R_{S_1} + R_{eq}$ and the first harmonic $\hat{V}_{\alpha_\omega}$ of the output voltage $v_\alpha$[3]. The d.c. term $V_{\alpha_{d.c.}}$ is blocked by the tank capacitor and the higher order harmonics are suppressed by the tank resonance. The first harmonic $\hat{V}_{\alpha_\omega}$ corresponds to $\hat{V}_{prim}$ in Figure 3.1.

---

[3] The first harmonic $\hat{V}_{\alpha_\omega}$ is found as follows. Figure 4.6 shows the output voltage of a class D stage with a perfectly tuned load tank but non-ideal switches. The pulse signal flips between $V_{S2_{sat}}$ and $V_{cc} - V_{S1_{sat}}$, and has bumpy top and bottom plateaus because of the voltage drops across the switch ON-resistors. It is split in three sub-parts to find its first harmonic:

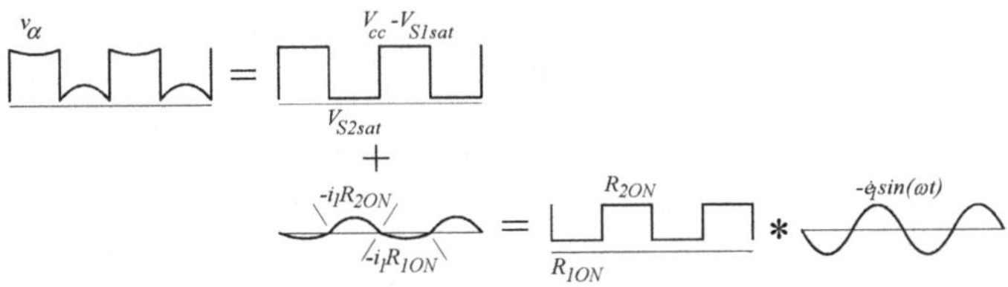


The first harmonic of the upper pulse signal between $V_{S2_{sat}}$ and $V_{cc} - V_{S1_{sat}}$ is:

$$2\frac{V_{cc} - V_{S1_{sat}} - V_{S2_{sat}}}{\pi} sin(\omega t)$$

The voltage drop across the switch-on resistors, depicted on the lower trace, is the product of a pulse train between $R_{2_{ON}}$ and $R_{1_{ON}}$, and a sine wave 180° out of phase with the driver output signal. The Fourier development of this composite signal is the convolution of both components. Its first harmonic term is the d.c. part of the pulse signal times the sine wave:

$$-\frac{R_{1_{ON}} + R_{2_{ON}}}{2} \hat{I}_1 \, sin(\omega t)$$

Its amplitude equals:

$$\hat{V}_{\alpha_\omega} = 2\frac{V_{cc} - V_{S1_{sat}} - V_{S2_{sat}}}{\pi} - \frac{R_{1_{ON}} + R_{2_{ON}}}{2}\hat{I}_1$$

$$\Leftrightarrow \hat{I}_1 = \frac{\hat{V}_{\alpha_\omega}}{R_{S_1} + R_{eq}} = \frac{2}{\pi}\frac{V_{cc} - V_{S1_{sat}} - V_{S2_{sat}}}{R_{S_1} + R_{eq} + \frac{R_{1_{ON}} + R_{2_{ON}}}{2}}$$

The power supplied to the resistive load $R_{eq}$ plus $R_{S_1}$, is found as

$$P_{link\,in} = \left(R_{S_1} + R_{eq}\right)\frac{\hat{I}_1^2}{2} = \frac{2\left(R_{S_1} + R_{eq}\right)}{\pi^2}\left(\frac{V_{cc} - V_{S1_{sat}} - V_{S2_{sat}}}{R_{S_1} + R_{eq} + \frac{R_{1_{ON}} + R_{2_{ON}}}{2}}\right)^2$$

The average current through the switch $S_1$ determines the d.c. driver consumption:

$$P_{in} = V_{cc}\,I_{cc} = V_{cc}I_{S1_{d.c.}} = V_{cc}\left(\frac{1}{T}\int_0^{T/2}\hat{I}_1\sin(\omega t)dt\right) = \frac{V_{cc}\hat{I}_1}{\pi}$$

$$= \frac{2V_{cc}}{\pi^2}\frac{V_{cc} - V_{S1_{sat}} - V_{S2_{sat}}}{R_{S_1} + R_{eq} + \frac{R_{1_{ON}} + R_{2_{ON}}}{2}}$$

The driver efficiency is the ratio between both previous expressions:

*Equation 4.1.*

$$\eta_{driver} \equiv \frac{P_{link\,in}}{P_{in}} = \frac{1 - \frac{V_{S1_{sat}} + V_{S2_{sat}}}{V_{cc}}}{1 + \frac{R_{1_{ON}} + R_{2_{ON}}}{2\left(R_{S_1} + R_{eq}\right)}}$$

This is an upper bound for the driver efficiency as the following important loss factors are not included:

- A large current will flow trough both transistors straightway from supply to ground, should they conduct simultaneously.
- The parasitic capacitors between the switch terminals give rise to switching losses. At the moment when $S_2$ switches on, the parasitic output capacitor is charged to $V_{cc}$ - $V_{S1sat}$. This charge is not transferred to the load but immediately dissipated in the switch $S_2$.
- This parasitic capacitor makes that the output-node voltage cannot change instantaneously. As a result, some of the switch current is already flowing while the switch voltage has not returned to zero yet.

Switching losses and retarded output transients become problematic for frequencies above 2 MHz or so.

Simultaneous conduction of both transistors is particularly dangerous because of the high associated switch dissipation. A conservative approach to avoid this, would be to create a short time gap between the opening of one transistor and the closing of the other one. However, this would suddenly interrupt the inductor current and cause large voltage transients.

**THE CLASS-D DESIGN FLOW ($Q_{tank} > 15$ and low $\omega$ assumed)**

The link output power and output voltage are given.

The secondary components, the required link output voltage $V_{sec}$, and the transfer frequency $\omega$ result from the secondary optimisation..

Choose $L_{S_1}$.

Compute $L_{eq}$, $R_{eq}$ and the link gain $A$ with the equations of chapter 3.

$$\hat{V}_{prim} = \frac{\hat{V}_{sec}}{A}$$

$$C_1 = \frac{1}{\omega^2 L_{eq}}$$

$$V_{cc} = \frac{\pi \hat{V}_{prim}}{2}\left[1 + \frac{R_{1_{ON}} + R_{2_{ON}}}{2\left(R_{S_1} + R_{eq}\right)}\right] + V_{S1_{sat}} + V_{S2_{sat}}$$

Equation 4.2.

$$I_{cc} = \frac{\hat{V}_{prim}}{\pi\left(R_{S_1} + R_{eq}\right)}$$

$$\eta_{driver} = \frac{1 - \dfrac{V_{S1_{sat}} + V_{S2_{sat}}}{V_{cc}}}{1 + \dfrac{R_{1_{ON}} + R_{2_{ON}}}{2\left(R_{S_1} + R_{eq}\right)}}$$

$$i_{S1_{max}} = i_{S2_{max}} = \frac{\hat{V}_{prim}}{R_{S_1} + R_{eq}}$$

$$v_{S1_{max}} = v_{S2_{max}} = V_{cc}$$

There is thus always some switch overlap needed for a smooth driver output, and switch timing remains critical in most situations to keep dissipation under control.

The design flow for class-D driven links is given in the above frame. The expressions are only valid for the lower frequencies. There was no further attempt made here to improve the analysis and count for the limited switch dynamics at higher frequencies because of the inevitable drawbacks cited here.

However, the class-D amplifiers still feature the important benefit that output-tank resonance is not critical for high-efficiency operation (Figure 4.7). For this reason, for instance, they are the only choice for driving stagger-tuned links. Shifting the frequency away from resonance does not dramatically change the switch dissipation. It only induces negative switch currents and steep current jumps. The power supply must be able to take up these negative currents, efficiently store the associated charge to release it back when the supply current becomes positive again. Otherwise, the high driver efficiency is compromised. This aspect deserves some extra attention and is therefore discussed in the next section 4.4.

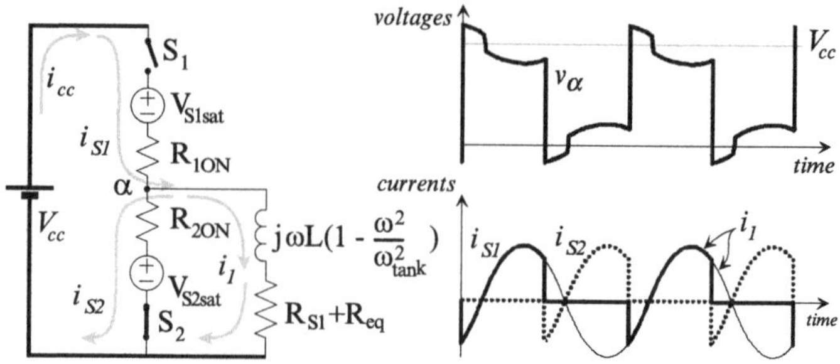


*Figure 4.7. A class-D driver with non-ideal switches driven slightly above tank resonance. The load is inductive and the tank current is somewhat lagging behind the output voltage. Note how the polarity of the switch saturation voltages reverse with the switch current.*

Note that the driver switches must also be able to pass negative currents and endure the negative voltages. This is inherent to MOS devices, but bipolar transistors that are unidirectional devices need additional diodes in parallel.

The expressions in the above frame learn that the driver efficiency increases with the primary coil inductance in two different ways: first, via the resistors $R_{eq}$ and $R_{S_1}$, and second via $V_{cc}$. The equivalent secondary resistor $R_{eq}$ is linear proportional to $L_{S_1}$ according to the Equations 3.2 and 3.35. A similar linear relation holds for the parasitic coil resistor $R_{S_1}$ that equals $\omega L_{S_1}$ divided by the coil quality factor. This quality factor is pretty independent of the exact number of turns, and thus the specific inductance value, but is mainly function of the frequency, the coil materials, its surroundings and the assembly method.

> Increasing the primary coil inductance improves the driver efficiency

The link gain is proportional to $\sqrt{L_{S_2}/L_{S_1}}$ (Equations 3.8, 3.9, 3.40 and 3.41). Increasing the primary coil inductance reduces this transformer ratio, so that a larger $V_{prim}$ is needed to obtain the same $V_{sec}$. This asks for a larger $V_{cc}$ and again increases $\eta_{driver}$.

The output power capability of a class-D driver is given by:

$$p \equiv \frac{P_{link\,in}}{v_{S1_{max}} i_{S1_{max}}} = \frac{\hat{V}_{prim}}{2V_{cc}} = \frac{1}{\pi\left[1+\dfrac{R_{1_{ON}}+R_{2_{ON}}}{2\left(R_{S_1}+R_{eq}\right)}\right]+2\dfrac{V_{S1_{sat}}+V_{S2_{sat}}}{\hat{V}_{prim}}}$$

The value for perfect switches is 0.318, which is more than twice the class-C value. The class-D topology is thus less stressing for the transistors.

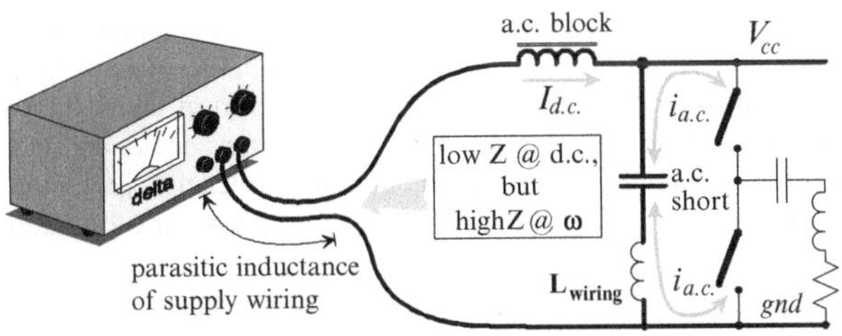


*A.c. decoupling of the supply impedance from a class-D driver. This creates a power supply with a low impedance at the switch frequency and which is able to take up negative current and efficiently release that energy lateron when the supply current turns positive anew.* *Figure 4.8.*

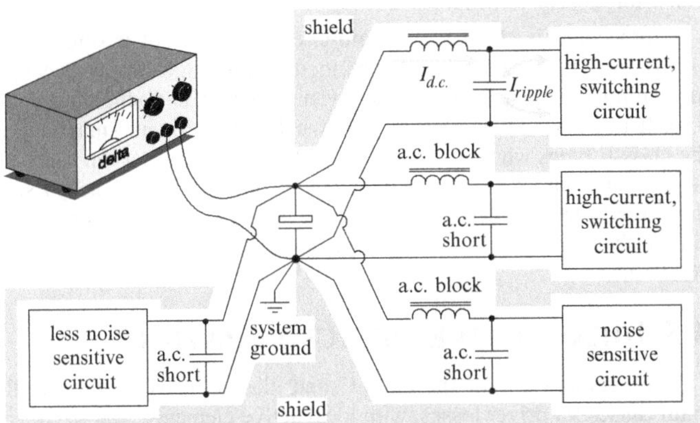


*Star-wise supply distribution, thorough a.c. decoupling and shielding.* *Figure 4.9.*

## 4.4 THE IMPORTANCE OF SUPPLY DECOUPLING

Supply decoupling is often reduced to a big electrolyte capacitor at the power supply terminals and some decoupling capacitors here and there on the circuit board. But, this is definitely insufficient once there are switch currents of several 100 mA and frequencies of some 10 kHz involved. Supply lines are implicitly supposed to have low impedance, also at the operation frequency. The reality is, however, largely different if no special precautions are taken.

Adequate supply decoupling of a class-D driver is shown in Figure 4.8. The idea behind this and the next figure is to create small a.c. islands for each high-current circuit, and a.c. isolate each island from the power supply and from the rest of the electronics. The power supply provides the d.c. current. All a.c. and ripple currents are drawn from the local decoupling capacitor. It is important that the parasitic resonances of the choke coil and the decoupling capacitor amply exceed the driver frequency.

For the latter reason, the parasitic inductance $L_{wiring}$ of the loop formed by the decoupling capacitor, the tank circuit and the switch transistor, must be as small as possible. The supply tracks on the printed circuit board, are best laid out in parallel and right next to each other. The decoupling capacitor must be a ceramic type with just enough capacitance to remove the ripple, but small enough for a high in-circuit resonance frequency. Dedicated low-Q decoupling capacitors are preferred above high-Q capacitors to avoid supply ringing. A high-frequency choke must be used for the same reason that is just large enough to filter the ripple, but small enough to keep its in-circuit resonance way above the operation frequency.

Electrolytic or tantalum capacitors are useless for decoupling above 5 or 10 kHz, as they already resonate there. They are useful in the d.c. path between the power supply and the choke coil, but high-frequency capacitors are really compulsory at the opposite side, where the rippling currents are produced. Soldering some 100 nF SMD capacitors between the supply tracks on the printed circuit board, right underneath the switch transistor, is very effective.

This procedure is standard in radio-circuit design, but is often ignored elsewhere. The best results are obtained by thorough supply decoupling in combination with starwise supply distribution and shielding (Figure 4.9). It is worth considering for all switch-mode drivers –maybe except the class-E types where the d.c. feed coil already serves as an a.c. block– as it cleans output signals, reduces parasitic ringing and keeps supply lines free from high-current glitches. The result is a greatly improved circuit performance and less high-frequency noise induced in the rest of the electronics.

## 4.5 IDEAL ACTIVE-DEVICE BEHAVIOUR

The foregoing study of the class-C and class-D drivers revealed three main causes for driver losses, with ideal active elements:

1. The non-zero active-element voltage during conduction, when using the active device as a current source (solution: operate the active elements as switches),
2. The large current that flows from the supply, through both transistors, to the ground, when they conduct simultaneously (solution: avoid two-pole switch structures), and
3. The switching losses of parasitic capacitors (solutions: switch on at zero switch voltage and transfer the charge stored in the parasitic switch capacitor, to the load and do not dissipate it in the switch).

Non-ideal active elements add two more loss factors:

1. The non-zero on-resistance of the switch, and
2. The non-zero saturation voltage of the switch.

Ideal active-device behaviour for maximal amplifier efficiency, has been summarised by the Sokals in the following eight conditions (Figure 4.10; SOKAL and SOKAL, 1975a):

1. The voltage $v_S$ across the active device, is minimal while conducting a current (the "on state")
2. The current $i_S$ through the active device is minimal when there a voltage across its terminals (the "off state")

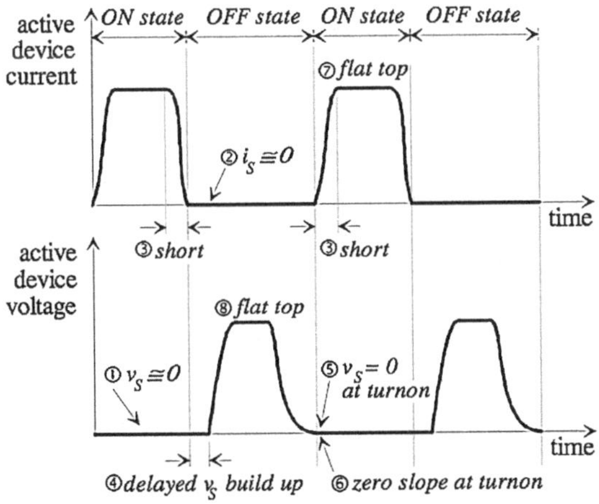


*Ideal active-device behaviour for maximal amplifier efficiency.* *Figure 4.10.*

3. The active device switches fast from the on to the off state and *vice versa*.
4. The $v_S$ build-up is delayed somewhat after turn-off.
5. $v_S$ has already returned to zero at the moment of turn-on.
6. $v_S$ has a zero slope at turn-on, to allow small timing errors or limited switch-on speeds, without additional losses.
7. The $v_S$ transient is flat topped.
8. And finally, the $i_S$ transient is also flat topped.

It has been shown, however, that it is theoretically impossible to meet all requirements within one practical circuit (MOLNÁR, 1984; KAZIMIERCZUK, 1986a; HERMAN and ZULINSKI, 1990). For that reason, practical circuit classes always select subsets of the above list.

**Saturating-class-C** is defined in this study as an operation mode that meets two specifications[4]:

1. The amplifier transistor operates as a **switch**:
   - ⇒ Low-impedant on state (to meet condition 1 in the above list).
   - ⇒ High-impedant off state (condition 2).
   - ⇒ Fast on-to-off and off-to-on transitions (condition 3).
2. The **switch voltage $v_S$** is **zero at turn-on** (condition 5).

A third specification is added to define the **class-E** mode:

3. The amplifier transistor operates as a **switch** (conditions 1,2 and 3).
4. The **switch voltage $v_S$** is **zero at turn-on** (condition 5).
5. The **switch voltage $v_S$** has a **zero slope at turn-on** (condition 6).

[4] Class definitions differ from author to author. Classes are often used to categorise circuit topologies, while in this and many other studies, circuit classes are seen as operation modes. A given circuit topology can then run in different classes depending on how it is tuned.

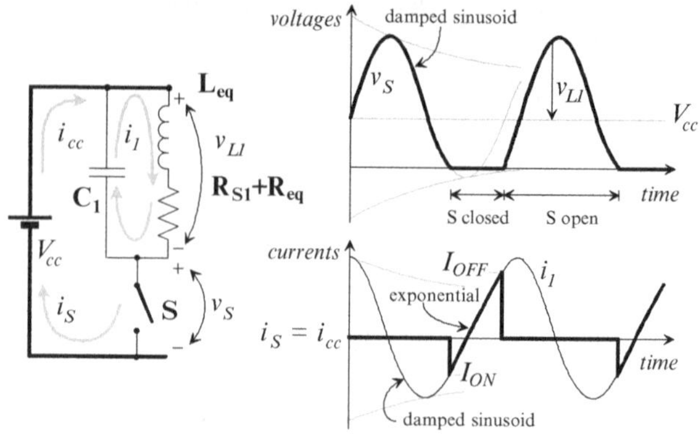


*Figure 4.11. The saturating-class-C primary coil driver.*

## 4.6 SATURATING CLASS C[5]

The saturating-class-C amplifier is built around a parallel L-C tank that normally consists of the resistive load plus an additional coil and capacitor (Figure 4.11). But in an inductive link, the tank is formed by the tuned primary coil, coupled to the secondary. The tank resistor is no separate component like in the conventional amplifier, but represents the parasitic coil loss and the secondary consumption.

Circuit operation is somewhat similar to the non-saturating class C, except that the tank dissipation is not compensated by a pulsating current source, but by repetitively switching on a power supply. The driver tank is left to oscillate freely when the amplifier switch is open. The tank current describes an exponentially damped oscillation at the tank-resonance frequency $\omega_{tank}$. The switch voltage oscillates around the supply $V_{cc}$.

$$\begin{cases} i_{L_{OFF}}(t') = I_{OFF} \cdot \dfrac{\cos(\omega_{tank} t' - \theta)}{\cos\theta} \exp\left(-\dfrac{t'}{\alpha'}\right) \\ v_{S_{OFF}}(t') = V_{cc} \cdot \left[1 + \dfrac{\sin(\omega_{tank} t' - \tau)}{\sin\tau} \exp\left(-\dfrac{t'}{\alpha'}\right)\right] \end{cases}$$

---

[5] The term "saturating-class-C amplifier" originates from the "saturation" of bipolar transistors that lose their current-source characteristic for small collector-emitter voltages. Although the corresponding region of a MOSFET is called the "linear region", the amplifier name was kept when the MOSFET switches were introduced afterwards, leaving a rather confusing terminology.

The first-order design equations of saturating-class-C amplifiers are developed in appendix A3. The ***first-order approximation*** is valid for small switch losses and assumes that the switch on-resistor, its saturation voltage and the switch fall and rise times, barely influence the circuit voltages and currents. ***The circuit signals are first calculated for ideal loss-free switches. The switch dissipation is then computed considering the individual loss components one by one and assuming that the simplified circuit signals remain unchanged.*** E.g. the losses caused by the switch on-resistor are computed with zero saturation voltage and ideal rise and fall times. ***Finally, the individual loss contributions are summed.*** It is clear that a first-order approximation becomes meaningless when the driver efficiency drops below 90 % or so. It remains though useful in such case, to decide if a faster switch or a switch with a lower on-resistance is needed to improve the driver performance.

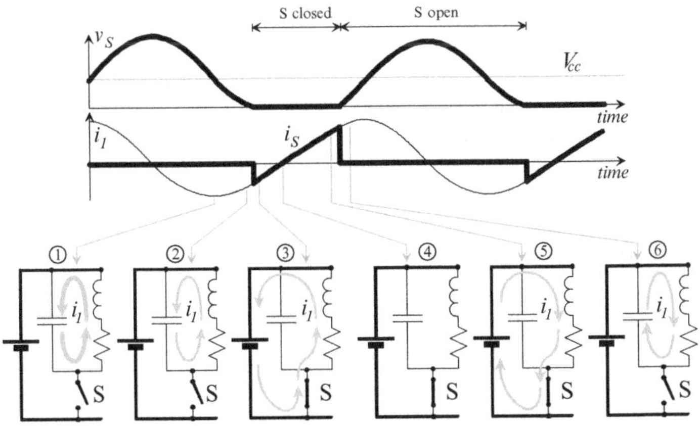


*The switch and coil currents in a saturating-class-C driver.* *Figure 4.12.*

$I_{OFF}$ is the coil current at the moment of switch-off. The tank ringing would normally die after approximately $Q^*_{L_{eq}}$ cycles because of the tank dissipation (the tank quality factor $Q^*_{L_{eq}} \equiv \omega L_{eq} / (R_{S_1} + R_{eq})$; the asterisk indicates that the primary is coupled to the secondary). The switch $S$ therefore recharges the tank each cycle[6]. This switch holds the capacitor voltage at $V_{cc}$ during the on-time, and charges the coil to compensate the tank dissipation. The coil current then starts at a negative $I_{ON}$ and rises exponentially to a positive and larger $I_{OFF}$.

$$\begin{cases} i_{L_{ON}}(t) = i_S(t) = \dfrac{V_{cc}}{R_{eq} + R_{S_1}} + \left( I_{ON} - \dfrac{V_{cc}}{R_{eq} + R_{S_1}} \right) . \exp\left( -\dfrac{t}{\alpha} \right) \\ v_{S_{ON}}(t) = 0 \end{cases}$$

As the capacitor is charged to $V_{cc}$ and the coil carries a non-zero current $I_{OFF}$ at the moment of switch-off, the driver tank is amply energised so that the amplitude of the successive ringing exceeds the supply voltage. The switch voltage thus crosses the horizontal axis after about half a period, and turns negative. Closing the switch to recharge the tank, at this moment of zero crossing, ensures a virtual lossfree switch conduction and avoids that the parasitic capacitor is discharged in the switch. This alternation of switch opening and closing is repeated at the driver frequency $\omega$. This switch frequency $\omega$ is lower than the momentary frequency $\omega_{tank}$ of the tank ringing. The driver is thus not running at tank resonance.

The switch of a properly tuned saturating-class-C driver conducts both negative as positive currents and should be adapted for this (Figure 4.12). The key point for high efficiency is again that the power supply should be able to take up returning energy and release it efficiently. Saturating-class-C

---

[6] Closing the switch after each two or three tank oscillations turns the driver into a frequency doubler or tripler as the switch frequency $\omega$ is about two or three times smaller than the tank frequency $\omega_{tank}$. The design equations in appendix A3 contain the parameters $m$ or $n$ to include also these modes.

drivers cannot just be hooked up to any kind of supply line. They do not run properly without supply decoupling with a high-frequency capacitor and a choke coil.

Note also how parasitic switch and wiring capacitors appear in parallel to the tank capacitor and become functional circuit components. Saturating-class-C drivers are thus very suitable for high-frequency applications.

The saturating-class-C design equations are often simplified assuming low tank damping:

$$\text{high } Q^*_{L_{eq}} \Rightarrow \begin{cases} i_{L_{ON}}(t) \approx I_{ON} + \dfrac{V_{cc}}{L_{eq}} t \\ v_{S_{ON}}(t) = 0 \end{cases}$$

$$\text{high } Q^*_{L_{eq}} \Rightarrow \begin{cases} i_{L_{OFF}}(t') \approx I_{OFF} \cdot \dfrac{\cos(\omega_{tank} t' - \theta)}{\cos\theta} \\ v_{S_{OFF}}(t') \approx V_{cc} \cdot \left[1 + \dfrac{\sin(\omega_{tank} t' - \theta)}{\sin\theta}\right] \end{cases} \quad \text{where } \omega_{tank} \equiv \sqrt{\frac{1}{L_{eq} C_1}}$$

These expressions are more practical and still accurate enough for inductive-link design. A complete set of design equations valid for any tank quality factor is developed in appendix A3 for the interested reader. It can be of use for other applications or to check the validity of the high-Q simplification in case an exceptionally high coil coupling reduces the tank quality factor.

High-efficiency operation with a zero switch voltage at turn-on, requires a precise combination of switch frequency $\omega$, tank frequency $\omega_{tank}$ and switch duty cycle $D$. This condition is derived in Appendix A3 in terms of the duty cycle and $W$ which is ***the tank frequency $\omega_{tank}$ normalised to the driver frequency $\omega$***:

*Equation 4.3.*

$$W \equiv \frac{\omega_{tank}}{\omega} = \frac{1}{1-D}\left[\frac{1}{2} + \frac{1}{\pi}\arctan\left(\frac{1}{\pi D W}\right)\right]$$

This equation in $W$ has no analytical solutions and is solved by numerical itteration. Some values are given in Table 4.1. and Figure 4.13.

The signal parameters for saturating-class-C operation with zero turn-on voltage, are

$$\begin{cases} \theta = \arctan\left(\dfrac{1}{x}\right) \\ x = \pi D W \\ I_{ON} = -\dfrac{V_{cc}}{2L_{eq}} t_{ON} \\ I_{OFF} = -I_{ON} \end{cases}$$

The output power of a perfectly tuned driver equals:

$$P_R = \frac{V_{cc}^2}{6\omega L_{eq} Q^*_{L_{eq}}}\left[\pi^2 D^2 (3 - D) + \frac{3}{W^2}\right]$$

*NUMERICAL SOLUTIONS FOR W AND THE POWER OUTPUT CAPABILITY* **p** *FOR SATURATING-CLASS-C OPERATION. THESE ARE FIRST-ORDER-SIMPLIFIED SOLUTIONS FOR HIGH* $Q^*_{L_{eq}}$. *TABLE 4.1.*

| *D* | *W* |
|---|---|
| 0 | 1 |
| 0.1 | 1.00314 |
| 0.2 | 1.02274 |
| 0.3 | 1.06960 |
| 0.4 | 1.15368 |
| 0.5 | 1.29155 |
| 0.6 | 1.51761 |
| 0.65 | 1.68570 |
| 0.7 | 1.91414 |
| 0.75 | 2.23856 |
| 0.8 | 2.73031 |
| 0.85 | 3.55599 |
| 0.9 | 5.21552 |
| 0.95 | 10.2089 |

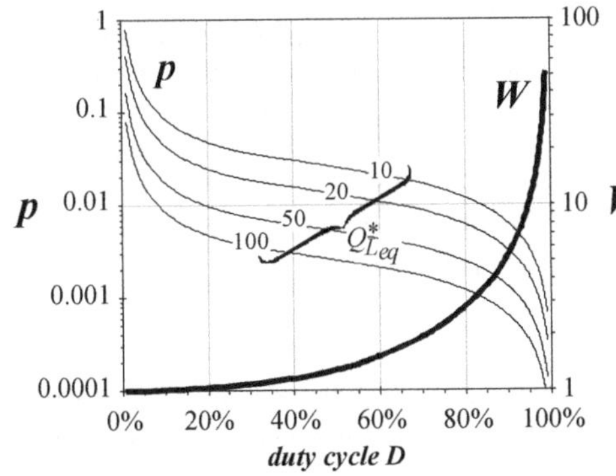


*The frequency ratio W and the power output capability p, for a saturating-class-C driver. These are first-order-simplified solutions for high* $Q^*_{L_{eq}}$. *Figure 4.13.*

Improper tuning occurs when the switch voltage is not exactly zero at turn-on. A short and intense current peak through the switch rapidly charges the tank capacitor to nullify the switch voltage before the normal cycle takes over. A term $f\,CV_{ON}^2/2$ adds to the switch dissipation and the driver efficiency is reduced accordingly. Although the averaged additional dissipation may fall well within the transistor specifications, improper tuning should be avoided. The abrupt current peaks largely exceed the regular switch currents and represent a high momentary dissipation that is harmful for the switch.

The losses in a properly tuned driver are due to the switch-on resistor $R_{ON}$, its saturation voltage $V_{S_{sat}}$ and the non-zero switch times $t_R$ and $t_F$:

$$\eta_{sat.\,class\,C} = \frac{p_R}{1 + p_{R_{ON}} + p_{V_{S_{sat}}} + p_{t_F} + p_{t_R}}$$ *Equation 4.4.*

$$\text{where}\begin{cases} p_R \equiv \dfrac{P_R}{P_{V_{cc}}} = 1 \\[2ex] p_{R_{ON}} \equiv \dfrac{P_{R_{ON}}}{P_{V_{cc}}} = A\dfrac{R_{ON}}{\left(R_{eq}+R_{S_1}\right)}2\pi D \\[2ex] p_{V_{S_{sat}}} \equiv \dfrac{P_{V_{S_{sat}}}}{P_{V_{cc}}} = A\dfrac{V_{S_{sat}}}{V_{cc}}\dfrac{3\omega L_{eq}}{\left(R_{eq}+R_{S_1}\right)} = 3A\,Q^*_{L_{eq}}\dfrac{V_{S_{sat}}}{V_{cc}} \\[2ex] p_{t_F} \equiv \dfrac{P_{t_F}}{P_{V_{cc}}} = A\dfrac{t_F^2\omega}{8C\left(R_{eq}+R_{S_1}\right)} = \dfrac{A}{8}\omega^2W^2t_F^2Q^*_{L_{eq}} \\[2ex] p_{t_R} \equiv \dfrac{P_{t_R}}{P_{V_{cc}}} = A\dfrac{t_R^2\omega}{8C\left(R_{eq}+R_{S_1}\right)} = \dfrac{A}{8}\omega^2W^2t_R^2Q^*_{L_{eq}} \end{cases}$$

(this formula continues overleaf)

The symbol $A$ is a shortcut for $\dfrac{\pi D^2 W^2}{3+(3-D)\pi^2 D^2 W^2}$

The driver is fed by a low-impedance supply $V_{cc}$. The supply voltage and the duty cycle set the circuit voltages, almost independent of the switch parasitics. The circuit currents are determined by the circuit voltages and the impedance of the loaded coil. The supply current $I_{cc}$ found for the first-order simplified circuit doesn't take the switch losses into account. A "real" circuit therefore draws a somewhat higher current $I^*_{cc}$ from the supply to compensate for the switch losses:

$$I^*_{cc} = \frac{I_{cc}}{\eta_{sat.\,class\,C}} = \frac{P_R}{V_{cc}\eta_{sat.\,class\,C}}$$

The output power capability $p$ is found as the ratio between $P_R$ and the product of the switch peak voltage with the maximal switch current:

$$p \equiv \frac{P_R}{v_{S_{\max}} i_{S_{\max}}} = \frac{P_R}{V_{cc}\left(1+\dfrac{1}{\sin\theta}\right) I_{OFF}} = \frac{W^2\pi^2 D^2 (3-D)+3}{6Q^*_{L_{eq}} W^2 \pi D\left(1+\sqrt{1+W^2\pi^2 D^2}\right)}$$

The output power capability $p$ is only function of $Q^*_{L_{eq}}$ and the duty cycle $D$, since $D$ and the frequency ratio $W$ are related to each other via Equation 4.3. Figure 4.13 shows the $p$ data for some $Q^*_{L_{eq}}$ values.

The driver output voltage $v_L$ across the tank coil and resistor, is not sinusoidal because of the flat segment during switch-on. However, only its first harmonic component $\hat{V}_{L_\omega}$ actually contributes to the power transfer. This component is the $\hat{V}_{prim}$ in Figure 3.1 and is given by:

*Equation 4.5.*

$$\hat{V}_{L_\omega} = \frac{2V_{cc}W^2\left|\pi D\cos(\pi D)-\sin(\pi D)\right|}{\pi\left(W^2-1\right)} = \hat{V}_{prim}$$

The tank dissipated caused by this voltage $\hat{V}_{prim}$, is also the power $P_{link\ in}$ delivered to the link. Note again that this tank resistor represents the primary coil losses plus the dissipation in the secondary (Figure 3.3).

$$P_{link\ in} = V_{L_\omega} I_{L_\omega} \cos\varphi_Z = \frac{\hat{V}^2_{L_\omega}\cos\varphi_Z}{2\sqrt{\left(R_{S_1}+R_{eq}\right)^2+\omega^2 L^2_{eq}}}$$

$$= \frac{\hat{V}^2_{L_\omega}\left(R_{S_1}+R_{eq}\right)}{2\left[\left(R_{S_1}+R_{eq}\right)^2+\omega^2 L^2_{eq}\right]} = \frac{2V^2_{cc}W^4\left[\pi D\cos(\pi D)-\sin(\pi D)\right]^2}{\pi^2\left(W^2-1\right)^2\left(R_{S_1}+R_{eq}\right)\left(1+Q^{*2}_{L_{eq}}\right)}$$

All other, higher-order harmonics in the output power $P_R$ are dissipated in the coil resistor $R_{S_1}$ and do not reach the secondary[7].

[7] This is a conservative approximation as the derivation of $P_R$ assumed a constant tank resistor over the entire frequency band. This is not true because there is very little power transfer away from the secondary resonance (so $R_{eq} = 0$) and because the coil losses vary with the frequency. A frequency-dependant tank resistor would however render a mathematical derivation impossible.

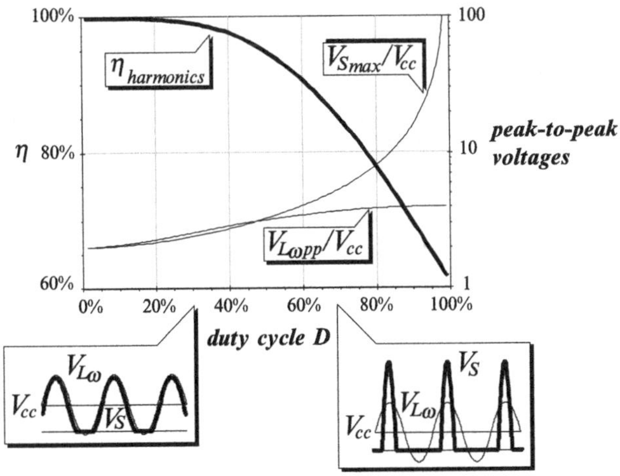


Figure 4.14. *The ratio $\eta_{harmonics}$ between $P_{link\ in}$ and $P_R$ of a saturating-class-C driver. This ratio decreases with the duty cycle since the amplitude $\hat{V}_{L_\omega}$ of the first harmonic of the output voltage is not following the peak value of the switch voltage. Both voltages are normalised to the supply voltage and plotted on the diagram.*

The ratio between $P_{link\ in}$ and $P_R$ can be seen as yet another efficiency. It expresses how much of the driver output is fed into the link and how much is lost in unusable harmonics (Figure 4.14).

$$\eta_{harmonics} \equiv \frac{P_{link\ in}}{P_R} = 12\frac{Q_{L_{eq}}^{*2} W^4 \left[\pi D \cos(\pi D) - \sin(\pi D)\right]^2}{\left(1+Q_{L_{eq}}^{*2}\right)\pi^2 \left(W^2-1\right)^2 \left[\pi^2 D^2 (3-D) + \frac{3}{W^2}\right]}$$

$$= 12\frac{W^4 \left[\pi D \cos(\pi D) - \sin(\pi D)\right]^2}{\pi^2 \left(W^2-1\right)^2 \left[\pi^2 D^2 (3-D) + \frac{3}{W^2}\right]} \quad (\text{high } Q_{L_{eq}}^*)$$

The overall driver efficiency is the product

$$\eta_{driver} = \eta_{sat.\ class\ C} \cdot \eta_{harmonics}$$

All circuit parameters are now expressed in terms of the primary coil inductance for the use in the design procedure of chapter 5 (cf. the design flow in the frame overleaf).

Augmenting the primary coil inductance is again beneficial for the driver efficiency as $R_{eq}$ and $R_{S_1}$ are still proportional to $L_{S_1}$, so the $p_{R_{ON}}$ term is reduced. The larger $V_{cc}$ needed to increase $V_{prim}$, yields a lower $p_{V_{S_{sat}}}$.

The output characteristic of a saturating-class-C amplifier is not so nicely defined as for the class-C and class-D drivers. It has a current-type character during switch-off and a voltage-type character during switch-on. An inductive link powered by a saturating-class-C driver corresponds to the non-resonant topology depicted at the top of Figure 3.1, because the primary tank is not actually resonating.

**THE SATURATING-CLASS-C DESIGN FLOW ($Q^*_{L_{eq}} > 15$ assumed)**

The link output power and voltage are given.

The secondary components, the transfer frequency $\omega$ and the required link output $V_{sec}$, result from the secondary optimisation.

Choose $L_{S_1}$ and the duty cycle $D$.

Compute $R_{eq}$ and the link gain $A$ with the equations of chapter 3 ($L_{eq} = L_{S_1}$ at secondary phase resonance).

$$\hat{V}_{prim} = \hat{V}_{sec} / A$$

$$W = \frac{1}{1-D}\left[\frac{1}{2} + \frac{1}{\pi}\arctan\left(\frac{1}{\pi D W}\right)\right]$$

$$C_1 = \frac{1}{\omega^2 W^2 L_{eq}}$$

$$V_{cc} = \frac{\hat{V}_{prim}\pi\left(W^2 - 1\right)}{2W^2\left|\pi D\cos(\pi D) - \sin(\pi D)\right|}$$

$$\eta_{driver} = \eta_{sat.\,class\,C} \cdot \eta_{harmonics}$$

$$\eta_{sat.\,class\,C} = \frac{p_R}{1 + p_{R_{ON}} + p_{V_{S_{sat}}} + p_{t_F} + p_{t_R}}$$

where

$$\begin{cases} p_R \equiv \dfrac{P_R}{P_{V_{cc}}} = 1 \\ p_{R_{ON}} \equiv \dfrac{P_{R_{ON}}}{P_{V_{cc}}} = A\dfrac{2\pi D R_{ON}}{\left(R_{eq} + R_{S_1}\right)} \\ p_{V_{S_{sat}}} \equiv \dfrac{P_{V_{S_{sat}}}}{P_{V_{cc}}} = 3A\,Q^*_{L_{eq}}\dfrac{V_{S_{sat}}}{V_{cc}} \quad \text{and } A \equiv \dfrac{\pi D^2 W^2}{3 + (3-D)\pi^2 D^2 W^2} \\ p_{t_F} \equiv \dfrac{P_{t_F}}{P_{V_{cc}}} = \dfrac{A}{8}\omega^2 W^2 t_F^2 Q^*_{L_{eq}} \\ p_{t_R} \equiv \dfrac{P_{t_R}}{P_{V_{cc}}} = \dfrac{A}{8}\omega^2 W^2 t_R^2 Q^*_{L_{eq}} \end{cases}$$

$$\eta_{harmonics} = 12\frac{W^4\left[\pi D\cos(\pi D) - \sin(\pi D)\right]^2}{\pi^2\left(W^2-1\right)^2\left[\pi^2 D^2(3-D) + \dfrac{3}{W^2}\right]}$$

$$I^*_{cc} = \frac{I_{cc}}{\eta_{sat.\,class\,C}} \quad \text{where} \quad I_{cc} = \frac{P_{link\,in}}{V_{cc}\eta_{harmonics}}$$

$$i_{S_{max}} = I_{OFF} = \frac{V_{cc}\pi D}{\omega L_{eq}}$$

$$v_{S_{max}} = V_{cc}\left(1 + \frac{1}{\sin\theta}\right) = V_{cc}\left(1 + \sqrt{1 + \pi^2 D^2 W^2}\right)$$

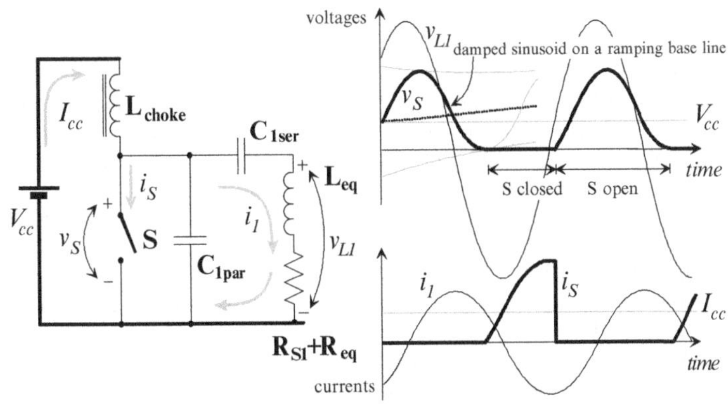


*The class-E primary coil driver.* *Figure 4.15.*

However, that doesn't render driver operation independent of coil coupling. As the equivalent coil inductance $L_{eq}$ and resistance $R_{eq}$ vary with the coupling, the amplifier tuning changes accordingly. It was shown in chapter 3 that secondary phase resonance offers a solution here. In this mode, $L_{eq} = L_{S_1}$ regardless coupling and load variations. Nonetheless, it is impossible to cancel their influence on $R_{eq}$.

## 4.7 CLASS E

Class-E drivers are fitted with a double-tuned output circuit (Figure 4.15). As it was already stated before, class-E circuits combine in this way, the advantages of the class D's with their series-tuned outputs and those of the saturating class C's with the parallel-tuned amplifier tanks (TROYK and SCHWAN, 1992a, 1992b). Inductive powering asks for high primary coil currents to generate the required magnetic fields, especially for weakly coupled links. The large primary coil leakage inductance pushes the driver output voltage to such levels that peak amplitudes of hundreds of volts are not uncommon. Class D solves this problem by series-tuning the primary coil to cancel its leakage inductance. However, the driver switches are still subjected to the intense coil currents. This means, on the one hand, that bipolar switches need large base drive currents. The small current gain, typical for high power transistors, makes things even worse. On the other hand, MOSFET switches need large die sizes to sustain the high current levels and have consequently large gate capacitors to be driven.

Parallel-tuning the primary coil doesn't solve the problem either. The switch currents are reduced as the coil current is now looping in the amplifier tank, but the switches are still exposed to the large coil voltage. High voltage transistors have again large gate capacitors to be driven.

The series-parallel tuned output network of the class E solves both deficiencies. The series capacitor acts as an a.c. coupling capacitor that separates the switch from the coil. The parallel capacitor bypasses the switch and takes up the larger part of the coil current.

The class-E output circuit is a multi-frequency network where the operation frequency is located right between the series and parallel resonances. Circuit operation is determined by the switch when on and by the transient response of the load network when the switch is off, just like for saturating class C. This transient is a damped sinusoidal current looping in the $C_{1ser}$-$C_{1par}$-$L_{eq}$ tank. The switch voltage is thus also sinusoidal. But, the d.c. current coming from the supply and passing through the choke coil $L_{choke}$, is simultaneously charging the $C_{1par}$ capacitor. The resulting switch voltage is a combination of both effects. It is a damped sinusoid with its baseline ramping up in time. A proper combination of the sine period, tank damping, $C_{1par}$ capacitance and choke-coil current, can be matched such that the subsequent voltage minimum occurs exactly at a zero switch voltage. Closing the switch at this moment, results in class-E operation: the switch voltage and its slope are both zero at turn-on[8]. A consequence of the zero turn-on slope is that the switch current also starts at zero. The result is that even considerable current rise times hardly affect the amplifier operation and do not cause additional switch dissipation.

Note that the influence of the tank damping on the switch timing disappears with low tank losses. The amplitude of the transient swing does not change much for quality factors above 10 or 20. The shaping of the switch voltage to reach its minimum at zero, is then only controlled by the integration of the supply current on the capacitor $C_{1par}$.

It should be emphasised that class-E drivers are running at neither of the resonance frequencies of their output network (i.e., the series-tuned branch is slightly inductive). A class-E driven link therefore corresponds to the non-resonant topology shown at the top of Figure 3.1. The situation is somewhat similar to the saturating-class-C drivers, as the coil coupling again influences $R_{eq}$ and thus the driver operation. The influence on $L_{eq}$ is also in this case, cancelled by secondary phase resonance.

The inspiring results obtained with class-E amplifiers were published in some excellent papers. Mainly five groups were involved in this fundamental research:

- Nathan O. and Alan D. Sokal were the original inventors of the class E circuit (SOKAL and SOKAL, 1975a, 1975b; SOKAL N.O., 1976, 1977). Their circuit analysis was based on a first-order circuit simplification, an infinite choke coil, and a 50 % switch duty cycle. They provided empirical design formulae that included the influence of low tank Q's. Yet, it was the author's experience that their method did not work for the high tank Q's that typically occur with inductive links.
- The first analytic analysis was carried out by Frederick H. RAAB (1977, 1978). He also started from a first-order simplified circuit with infinite choke coil, and used Fourier techniques to find the design equations

[8] Remark: the node between $C_{1ser}$ and $L_{eq}$ is a high-voltage, high-impedance node. If a cable connection is desired between the link coil and the driver, it should not be placed between $C_{1ser}$ and $L_{eq}$, but rather between $C_{1par}$ and $C_{1ser}$ (FORSTER, 1986). In this way, the cable carries a lower voltage and its capacitance is included in $C_{1par}$ so it does not disturb the driver operation.

assuming a high tank Q. He further derived first-order approximations of the transistor losses (RAAB and SOKAL, 1978)[9].

- Robert E. Zulinski *et al.* extended the work of F.H. Raab to class-E frequency multipliers using the very same assumptions and techniques (ZULINSKI and STEADMAN, 1986). They then showed that class-E operation can be made independent of the load resistor, by replacing the original infinite choke coil by a smaller coil (ZULINSKI, 1986; ZULINSKI and STEADMAN, 1986, 1987; ZULINSKI and GRADY, 1990). He also proved that this form of self-regulation couldn't be achieved with an infinite choke coil (ZULINSKI *et al.*, 1990). Self-regulation of the output current instead of the output voltage was also found impossible (GANDHI *et al.*, 1992). Zulinski's group then further analysed the first-order simplified circuit with finite d.c. feed inductance but arbitrary tank Q, using Laplace transforms (SMITH and ZULINSKI, 1990).
- Marian Kazimierczuk proposed in 1981 the dual class-E circuit topology where an inductor shunted the switch instead of a capacitor. He assumed a high tank Q, an ideal switch and a duty cycle of 50 %. He then improved the high-Q model of the original class-E topology by modelling the influence of the transistor-current fall time (KAZIMIERCZUK, 1983a). His research then moved to the low-Q analysis of class-E drivers with infinite choke inductance (KAZIMIERCZUK, 1986b; KAZIMIERCZUK and PUCZO, 1987, 1989a, 1989b) for the use in d.c./d.c. converters (KAZIMIERCZUK and JÓZWIK, 1989a, 1989b; KAZIMIERCZUK and BUI, 1989a, 1989b).
- Charalampos P. Avratoglou *et al.* applied numerical techniques to solve the circuit net equations in the "on" and "off" states (AVRATOGLOU and VOULGARIS, 1987). This allowed them to find exact solutions for arbitrary tank Q's. They found that the high-Q, approximate methods fail for tank Q's below 15. Sokal's empirical formulae also proved to approximate the exact solution well at low Q. They also presented a modified class-E topology with finite d.c. feed and without parallel capacitor (AVRATOGLOU and VOULGARIS, 1988), that better fitted thyristor switches (VOULGARIS and AVRATOGLOU, 1987). Later on, they extended their numerical method to the general class-E circuit with a finite d.c. feed inductance, arbitrary tank Q and non-zero switch-on resistance (AVRATOGLOU *et al.*, 1989). This analysis did, however, not indicate how the steady-state boundary conditions were found. This gap was then filled by Zulinski's group and resulted in a MatLab programme for the calculations (MANDOJANA *et al.*, 1990). Their analysis was based on a state-space circuit description (LIOU, 1972; SOKAL 1977).

9 The analysis of class-E drivers for inductive powering published by Philip R. TROYK and Martin A.K. SCHWAN (1992a, 1992b), also started from a high-Q approximation but incorrectly ignored the integrating effect of the d.c. supply current on $C_{1par}$. This error was not observed because they only calculated the driver components and did not investigate the driver consumption and its output power. As the current integration and the resulting ramping baseline have no substantial effect on the transient frequency, their approximations for the driver components matched well with their measurements.

The approximate approach of Frederick H. Raab, is further used in this study because it is useful at any switch duty cycle and because it provides analytic expressions for the components and transistor losses. Such analytic expressions are preferred as they easier fit in an optimisation procedure than transcendent or procedural relations between the design parameters. Raab's method also shows a nice balance between accuracy and complexity. When designing weakly coupled inductive links, the high tank quality factor automatically results from the need of high-Q coils for good link efficiency and from the poor coupling.

Raab's expressions for the circuit parameters also describe the operation mode with a non-zero switch-voltage slope at turn-on. This allows checking the class-E circuit running in a saturating-class-C mode. This may occur when the tank Q drops because the primary coil is held close to a metal object and eddy currents are induced. There is even a distinctive reason to deliberately opt for a saturating-class-C mode, especially when the driver is part of an auto-tuned induction system. A class-E driver needs tuning of two parameters when for instance its frequency is changed: one to keep the switch voltage zero at turn-on and one for a zero slope at turn-on (e.g., both tank capacitors, or one of them in combination with the duty cycle). Saturating-class-C operation involves the tuning of only one parameter for a zero switch voltage at turn-on. The efficiency of the saturating-class-C mode is lower compared to the class-E mode, since the turn-on current is not zero and the transistor rise time is limited. The saturating-class-C mode is also more sensitive to errors on the switch-on timing. However, the strongest benefit of a class-E circuit topology with respect to inductive powering, that is that the switch transistor is separated from the primary coil, remains.

Raab's analysis starts with the assumption of a lossfree switch ($R_{ON} = V_{Ssat} = t_F = t_R = 0$) and a high tank Q. The series capacitor $C_{1ser}$ and the coil $L_{eq}$ are not exactly at resonance. Their ***residual reactance*** is marked as ***x***:

$$x \equiv \omega L_{eq} - \frac{1}{\omega C_{1ser}}$$

Because of the high-Q assumption, $x$ is low at the operation frequency $\omega$ and high at its harmonics. The coil current is therefore sinusoidal:

$i_1(\theta) = gI_{cc}\sin(\theta + \phi)$, where $\theta = \omega t$ is the angular time and

$y = \pi(1 - D)$ the half off-time expressed in radians.

The amplitude $\hat{V}_{L_1}$ of the primary coil voltage is found as (Figure 4.15):

$$\hat{V}_{L_1} = \hat{I}_1\sqrt{R_{S_1}^{*2} + \omega^2 L_{eq}^2} = gI_{cc}R_{S_1}^*\sqrt{1 + Q_{L_{eq}}^{*2}} \text{ with } R_{S_1}^* = R_{S_1} + R_{eq}$$

The switch is open from $\theta_{OFF} = \dfrac{\pi}{2} - y$ to $\theta_{ON} = \dfrac{\pi}{2} + y$.

Since the choke coil forces a d.c. input current and the branch $C_{1ser}$-$L_{eq}$-$R$ forces a sinusoidal output current, the difference between those currents must flow into the switch-capacitor combination. When the switch is open, the difference flows into the capacitor $C_{1par}$. When the switch is closed, the difference flows through the switch. If a non-zero capacitor voltage is present at the moment of turn-on, the switch discharges that voltage to zero, thereby dissipating the stored energy $C_{1par}V^2/2$; a situation that must be avoided at any time.

The switch voltage and current are described by:

$$\begin{cases} v_{S_{OFF}}(\theta) = \dfrac{I_{cc}}{\omega C_{1par}}\left[y - \dfrac{\pi}{2} + \theta + g\left[\sin(\phi - y) + \cos(\phi + \theta)\right]\right] \\ i_{S_{OFF}}(\theta) = 0 \end{cases}$$

$$\begin{cases} v_{S_{ON}}(\theta) = 0 \\ i_{S_{ON}}(\theta) = I_{cc}\left[1 - g\sin(\phi + \theta)\right] \end{cases}$$

Lossfree operation requires that the switch voltage is zero at turn-on. This allows finding conditions for the unknown variables $\phi$ and $g$:

$$\phi = \arctan\left[\frac{\dfrac{\sin y}{y} - \cos y}{\dfrac{\zeta y}{\pi}\cos y - \left(1 + \dfrac{\zeta}{\pi}\right)\sin y}\right] \quad \text{and} \quad g = \frac{y}{\cos\phi \sin y}$$

*Equation 4.6.*

The parameter $\zeta$ ***is the normalised switch-voltage slope at turn-on***:

$$\zeta \equiv \frac{1}{V_{cc}} \frac{dv_S(\theta)}{d\theta}\bigg|_{\theta = \frac{\pi}{2} + y}$$

$\zeta$ is zero for class-E and negative for saturating-class-C operation. Note that the calculation of $\phi$ requires a four quadrant inverse tangent. Errors resulting in values that are shifted by an amount of $\pi$ are easily sorted out, as they result in a negative value for $g$. The circuit components are now determined as all signals are known (apart from the supply current $I_{cc}$, which follows from the desired output power)[10].

$$B \equiv \omega C_{1par} = \frac{2y^2 + 2yg\sin(\phi - y) - 2g\sin\phi\sin y}{\pi g^2 R^*_{S_1}}$$

$$\frac{x}{R^*_{S_1}} \equiv \tan\psi = \frac{q_1\sin\phi + q_2\cos\phi + q_3\cos 2\phi + gy}{q_2\sin\phi + q_3\sin 2\phi - q_1\cos\phi}$$

$$\text{where} \begin{cases} q_1 \equiv -2g\sin(\phi - y)\sin y - 2y\sin y \\ q_2 \equiv 2y\cos y - 2\sin y \\ q_3 \equiv -\dfrac{g}{2}\sin 2y \end{cases}$$

The harmonic content of the primary coil current reduces with the tank quality factor. For values above 3 and a 50 % duty cycle, the harmonic current components are approximated according to Table 4.2. (SOKAL and RAAB, 1977).

[10] The expression for $B$ corrects a minor error in the original paper (RAAB, 1978) where the resistive load was incorrectly written in the numerator instead of the denominator. Zulinski also corrected this error in his extension on Raab's analysis (ZULINSKI and STEADMAN, 1987).

*TABLE 4.2. THE HARMONIC CONTENT OF THE PRIMARY COIL CURRENT IN A CLASS E DRIVEN LINK, FOR A 50 % DUTY CYCLE AND A TANK QUALITY FACTOR $Q^*_{L_{eq}}$ HIGHER THAN 3. THE HARMONIC TERMS $I_n$ AND $P_n$ ARE, NORMALISED TO THE FUNDAMENTAL COMPONENTS $I_1$ AND $P_1$.*

| $n$ | 1 | 2 | 3 | 4 | 5 | 6 |
|---|---|---|---|---|---|---|
| $I_n/I_1$ | 1 | $\approx \frac{0.51}{Q^*_{L_{eq}}}$ | $\approx \frac{0.080}{Q^*_{L_{eq}}}$ | $\approx \frac{0.037}{Q^*_{L_{eq}}}$ | $\approx \frac{0.016}{Q^*_{L_{eq}}}$ | $\approx \frac{0.010}{Q^*_{L_{eq}}}$ |
| $P_n/P_1$ | 1 | $\approx \frac{0.26}{Q^{*2}_{L_{eq}}}$ | $\approx \frac{0.0064}{Q^{*2}_{L_{eq}}}$ | $\approx \frac{0.0014}{Q^{*2}_{L_{eq}}}$ | $\approx \frac{0.00026}{Q^{*2}_{L_{eq}}}$ | $\approx \frac{0.0001}{Q^{*2}_{L_{eq}}}$ |

The voltage amplitude of the harmonics increases with the duty cycle, but never exceeds 1% for quality factors above 10 (cf. Figure 11 in KAZIMIERCZUK and PUCZKO, 1987). The power content of the harmonics is thus negligible at practical coil quality factors and at any duty cycle. Hence, $P_{class\,E} \approx P_{link\,in}$ and $\eta_{harmonics} \approx 1$ and

$$P_{link\,in} = \frac{I_{cc}^2 R_{cc}}{2} = \frac{V_{cc}^2}{2R_{cc}} \text{ where } R_{cc} \equiv \frac{V_{cc}}{I_{cc}} = \frac{g^2 R^*_{S_1}}{2}$$

***$R_{cc}$ is the d.c. resistance that the loaded class-E driver represents to the power supply.***

The driver efficiency is again found as (RAAB and SOKAL, 1978)[11]:

*Equation 4.7.*

$$\eta_{class\,E} = \frac{p_R}{1 + p_{R_{ON}} + p_{V_{S_{sat}}} + p_{t_F} + p_{t_R} + p_{L_S}}$$

$$\begin{cases} p_R \equiv \dfrac{P_R}{P_{V_{cc}}} = 1 \\[2ex] p_{R_{ON}} \equiv \dfrac{P_{R_{ON}}}{P_{V_{cc}}} \\[2ex] \quad = \dfrac{R_{ON}}{2\pi R_{cc}} \left[ (g^2+2)(\pi - y) + 4g \sin y \cos\phi - \dfrac{g^2}{2} \sin(2y)\cos(2\phi) \right] \\[2ex] p_{V_{S_{sat}}} \equiv \dfrac{P_{V_{S_{sat}}}}{P_{V_{cc}}} = \dfrac{V_{S_{sat}}}{V_{cc}} \\[2ex] p_{t_F} \equiv \dfrac{P_{t_F}}{P_{V_{cc}}} = \dfrac{I_{OFF}^2 \omega^2 t_F^2}{48\pi B} \\[2ex] p_{t_R} \equiv \dfrac{P_{t_R}}{P_{V_{cc}}} = \dfrac{I_{ON}^2 \omega^2 t_R^2}{48\pi B} \\[2ex] p_{L_S} \equiv \dfrac{P_{L_S}}{P_{V_{cc}}} = \dfrac{L_S I_{OFF}^2}{2} f \end{cases}$$

The dissipation $P_{t_F}$ caused by the non-zero fall time, is calculated by assuming a constant output current in the series branch and a linearly falling switch current. The voltage on the capacitor $C_{Ipar}$ is thus parabolic. The corresponding dissipation is then found as the average of the product

[11] Note that expression (5) for $P_{R_{ON}}$ in (RAAB and SOKAL, 1978) contained a minor error that is corrected here.

of the parabolic switch voltage and the linear switch current. This approximation is valid for fall times that are short compared to the switching period. A more accurate but more complex analysis, valid for larger fall times, is found in (KAZIMIERCZUK, 1983a). Class-E circuits running in saturating-class-C mode, show an additional loss term $P_{t_R}$ caused by the limited rise time. It is computed in the very same way.

The last term $P_{L_S}$ represents the dissipation at turn-off, of the energy stored in the parasitic inductance $L_S$ between the switch and the capacitor $C_{1par}$. This energy is dissipated in the switch through simultaneous non-zero voltage and non-zero switch current. This loss term is often very small but since it was mentioned in the original paper, it is kept also here. Similar terms can also be included in the previous amplifier classes.

The class-E driver is fed by a current $I_{cc}$ flowing through the choke coil. This current sets the circuit signals almost independent of the switch losses. The supply voltage is found as the average voltage across the switch. Since the value $V_{cc}$ for the first-order simplified circuit does not include the switch parasitics, the average for a real switch is slightly larger. This "real" supply voltage $V_{cc}^*$ is found as:

$$V_{cc}^* = V_{cc} / \eta_{class\ E}$$

A suitable value for the choke inductance is found from the fact that the supply load $R_{cc}$ must be much smaller than the supply impedance. Hence,

$$\omega L_{choke} >> R_{cc}$$

The circuit equations are again written in terms of the primary coil inductance for the optimisation procedure of chapter 5 (see overleaf). Increasing the primary coil inductance again improves the efficiency.

The output power capability $p$ is again found as the ratio between $P_R$ and the product of the switch peak voltage with the maximal switch current (Figure 4.16, top left). It is function of the duty cycle $D$ and the normalised slope $\zeta$, but it is independent of $Q_{L_{eq}}^*$. The maximal output power capability is found for class-E mode and a 48.9 % duty cycle. The amount of harmonics in the coil current is also minimal at this point. This is why class-E drivers are mostly designed for a 50 % duty cycle. However, the situation is different with inductive links. Figure 4.16 depicts normalised curves for the supply voltage and the peak switch voltage and current. The voltage curves are normalised to $\hat{V}_{prim}$, the current curves to $\hat{V}_{prim}/(\omega L_{eq})$. This shows that the high coil Q typical for inductive links, moves the supply voltage to such low values (below 1 V) that it becomes impractical and leads to poor driver efficiency when using switches with a non-zero saturation voltage. A low supply voltage also means that the supply and switch currents are high to deliver the specified power. This causes a high dissipation in the switch-on resistance and further lowers the efficiency.

> Class-E drivers for high-Q inductive links run best at short duty cycles for a higher supply voltage and lower switch current.

**THE CLASS-E DESIGN FLOW ($Q^*_{L_{eq}}$ > *15* assumed)**

The link output power and voltage are given.

The secondary components, the transfer frequency $\omega$ and the required link output $V_{sec}$, result from the secondary optimisation.

Choose the duty cycle $D$, the normalised turn-on slope $\zeta$ and $L_{S_1}$.

Compute $R_{eq}$ and the link gain $A$ with the link equations of chapter 3 ($L_{eq} = L_{S_1}$ at secondary phase resonance).

$$\hat{V}_{prim} = \hat{V}_{sec} / A$$

$$y = \pi(1\text{-}D)$$

$$R^*_{S_1} = R_{S_1} + R_{eq}$$

$$\phi = \arctan\left[\frac{\dfrac{\sin y}{y} - \cos y}{\dfrac{\zeta y}{\pi}\cos y - \left(1 + \dfrac{\zeta}{\pi}\right)\sin y}\right]$$

$$g = \frac{y}{\cos\phi \sin y}$$

$$\tan\psi = \frac{q_1 \sin\phi + q_2 \cos\phi + q_3 \cos 2\phi + gy}{q_2 \sin\phi + q_3 \sin 2\phi - q_1 \cos\phi}$$

$$\text{where} \begin{cases} q_1 \equiv -2g\sin(\phi - y)\sin y - 2y\sin y \\ q_2 \equiv 2y\cos y - 2\sin y \\ q_3 \equiv -\dfrac{g}{2}\sin 2y \end{cases}$$

$$C_{1par} = \frac{2y^2 + 2yg\sin(\phi - y) - 2g\sin\phi\sin y}{\omega\pi g^2 R^*_{S_1}}$$

$$C_{1ser} = \frac{1}{\omega\left(\omega L_{eq} - R^*_{S_1}\tan\psi\right)}$$

$$I_{cc} = \frac{\hat{V}_{prim}}{gR^*_{S_1}\sqrt{1 + Q^{*2}_{L_{eq}}}} \approx \frac{\hat{V}_{prim}}{gR^*_{S_1}Q^*_{L_{eq}}} = \frac{\hat{V}_{prim}}{g\omega L_{eq}}$$

$$V_{cc} = R_{cc}I_{cc} \quad \text{where} \quad R_{cc} = \frac{g^2 R^*_{S_1}}{2}$$

$$\approx \frac{g\hat{V}_{prim}}{Q^*_{L_{eq}}}$$

an “infinite choke” $\Leftrightarrow$ $\omega L_{choke} >> R_{cc}$

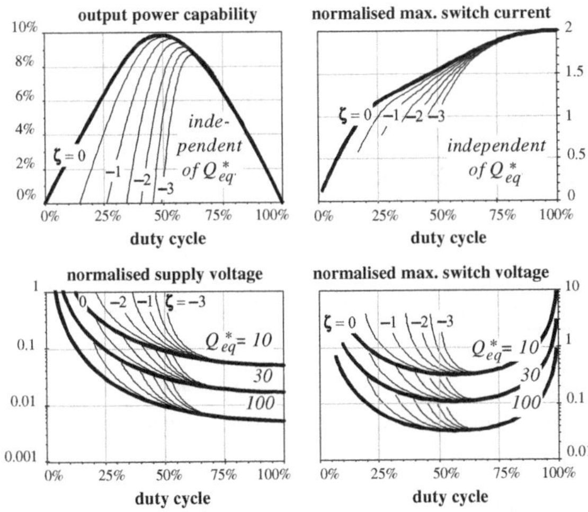


*Top left: the output power capability of the class-E driver in function of the duty cycle and the normalised slope at turn-on. The three other diagrams show the maximal switch current, normalised to $\hat{V}_{prim}/(\omega L_{eq})$, the maximal switch voltage and the supply voltage (both normalised to $\hat{V}_{prim}$). The bold lines are for class-E operation ($\zeta = 0$).* *Figure 4.16.*

$$\eta_{driver} \approx \eta_{class\,E} = \frac{p_R}{1 + p_{R_{ON}} + p_{V_{S_{sat}}} + p_{t_F} + p_{t_R} + p_{L_S}}$$

$$\begin{cases} p_R \equiv \dfrac{P_R}{P_{V_{cc}}} = 1 \\ p_{R_{ON}} \equiv \dfrac{P_{R_{ON}}}{P_{V_{cc}}} = \dfrac{R_{ON}}{\pi R^*_{S_1}}\left[\left(1+\dfrac{2}{g^2}\right)(\pi - y) + \dfrac{4}{g}\sin y \cos\phi - \dfrac{\sin(2y)\cos(2\phi)}{2}\right] \\ p_{V_{S_{sat}}} \equiv \dfrac{P_{V_{S_{sat}}}}{P_{V_{cc}}} = \dfrac{V_{S_{sat}}}{V_{cc}} \\ p_{t_F} \equiv \dfrac{P_{t_F}}{P_{V_{cc}}} = \dfrac{I^2_{OFF} t^2_F}{24 C_{1par}} f \quad \text{where} \quad I_{OFF} = I_{cc}\left[1 - g\cos(\phi - y)\right] \\ p_{t_R} \equiv \dfrac{P_{t_R}}{P_{V_{cc}}} = \dfrac{I^2_{ON} t^2_R}{24 C_{1par}} f \quad \text{where} \quad I_{ON} = I_{cc}\left[1 - g\cos(\phi + y)\right] \\ p_{L_S} \equiv \dfrac{P_{L_S}}{P_{V_{cc}}} = \dfrac{L_S I^2_{OFF}}{2} f \end{cases}$$

$$V^*_{cc} = V_{cc} / \eta_{class\,E}$$

$$\begin{cases} v_{S_{max}} = v_{S_{OFF}} @ \theta_{max} \quad \text{where} \quad \theta_{max} = \arcsin\left(\dfrac{1}{g}\right) - \phi \\ \text{if} \quad y - \pi < \phi \quad \text{then} \quad i_{S_{max}} = I_{cc}(1+g) \quad \text{else} \quad i_{S_{max}} = I_{OFF} \end{cases}$$

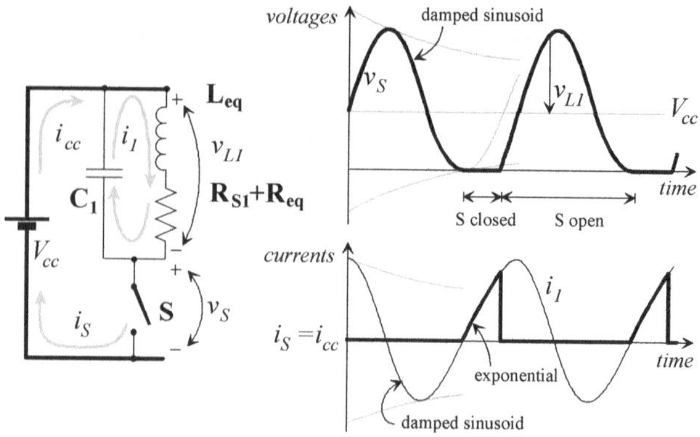


*Figure 4.17. A saturating-class-C driver where the switch duty cycle is adjusted to run in class E.*

## 4.8 CLASS E WITH 1 COIL AND 1 CAPACITOR

The saturating-class-C topology presented in section 4.6 can be tuned to operate in a class-E mode (i.e., zero voltage and zero slope at turn-on) (SOKAL, 1981; KAZIMIERCZUK, 1983b). Gradually reducing the duty cycle of a given saturating-class-C-tuned driver decreases the amplitude of its switch-voltage transient. At a certain point, switch transient remains positive and its first minimum just reaches zero. The driver is now running in class-E (Figure 4.17).

Note that there is tank damping required to arrive at the class-E mode. At the moment of turn-off, the tank capacitor is charged to $V_{cc}$ and the tank coil conducts a current $I_{OFF}$. The amplitude of the subsequent voltage transient is thus always larger than $V_{cc}$. There is thus tank dissipation needed to reduce the amplitude to exactly $V_{cc}$ when the transient swings down to its minimum. It is obvious that this class-E mode is only described by low-Q expressions that include the tank damping. These are found in appendix A3.

There is a fundamental difference in the way the classic class E and the class E with one coil and one capacitor acquire a zero slope at turn-on. For the classic class E, the zero turn-on slope mainly results from a properly timed frequency and duty cycle in combination with a switch-voltage baseline that ramps up in time due to the current $I_{cc}$ flowing into the capacitor $C_{1par}$. The $I_{cc}$ and $C_{1par}$ are independent of the tank damping as long as it is small enough.

The alternative class E controls its turn-on slope, by a delicate combination of tank damping, frequency and duty cycle. This renders them almost unsuitable for inductive powering owing to the fact that variations in the coil coupling or the secondary load directly affect the tank damping and thus the turn-on slope accordingly. Including the class-E amplifier with one coil and capacitor in this overview of coil drivers is thus a mere academic exercise.

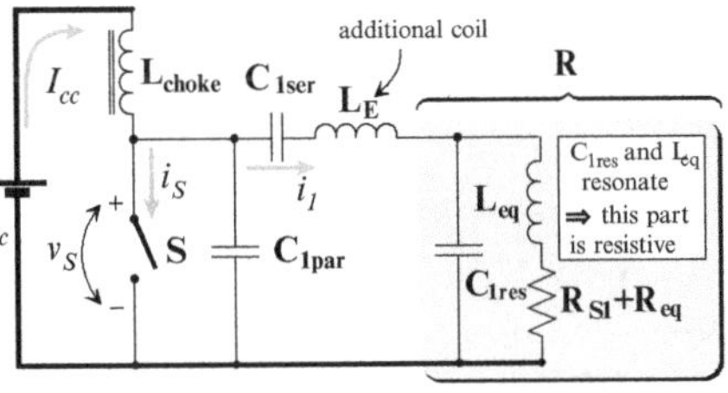


*The class-E driver with a parallel-resonant primary coil.* *Figure 4.18.*

## 4.9 DRIVING WEAKLY COUPLED LINKS

There is an important difference in driving a well-coupled link and a weakly coupled link. The reason is that weak coupling leaves the driver sort of freewheeling, whereas the driver of a well-coupled link is appropriately loaded, even for a high-Q primary, since the secondary power consumption adds a considerable resistance to the primary coil.

It was already mentioned before, that increasing the primary coil inductance improves the driver efficiency. However, the class-C amplifiers are rapidly outplayed here since the coil-to-coil voltage transfer is falling with increasing $L_{S1}$ and the switches are directly exposed to the coil voltages. Also, as the peak-to-peak coil voltage equals about twice the supply amplitude, unrealistic supply voltages are needed. A ratio between the coil and the supply voltage of 10 is feasible with class-E drivers. They are therefore better suited to drive weakly coupled links.

There is a limit on enlarging primary coils since their resonance frequency must remain above the driver frequency (this is also true for the choke coil). Besides, a voltage ratio of 10 may be not enough for extremely low coupled links ($k < 1\%$). Moreover, systems with elevated voltages at r.f. frequencies are hard to comply to EMC regulations (section 1.2.2.1.D). An ideal solution would be to raise the resistive driver load without increasing the coil inductance (to keep the circuit voltages low) or reducing its quality factor (to maintain the link efficiency). This is done by applying a capacitor in parallel to the primary coil and tuning it into resonance (Figure 4.18). This parallel tank behaves like a resistor because it is resonating, but it exhibits a higher resistance than the coil itself (Figure 4.19). In fact, the link coil and the parallel capacitor form an impedance transformer. The relation between the apparent resistor $R$ and the tank resistor $R_{S1} + R_{eq}$ is derived from Figure 3 in GUTMANN (1980)[12]:

$$R = \frac{L_{eq}}{C_{1res}\left(R_{S_1} + R_{eq}\right)} \Leftrightarrow R = \frac{Q^*_{L_{eq}}}{\omega\, C_{1res}}$$

As the parallel resonance dissolves the inductance in the class-E series branch, an extra coil $L_E$ is to be added.

---

[12] Z. HAMICI *et al.* (1996) applied series resonance to cancel out the primary coil inductance. This method doesn't provide and impedance transformation and is therefore less suitable for driver optimisation.

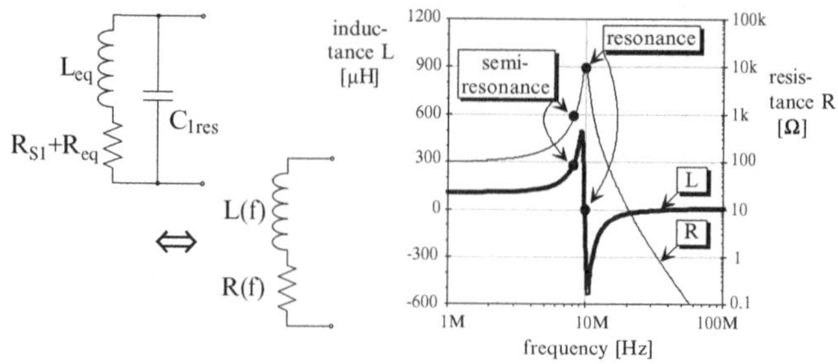


*Figure 4.19.* *Primary coil resonance or semi-resonance, improves class-E efficiency of weakly coupled links. The example shown here is for $L_{eq}$ = 100 μH and $R_{S1} + R_{eq}$ = 100 Ω.*

This has the benefit that the class-E series coil and the inductive-link primary can be optimised individually. The drawback is that another tank is to be tuned to the transfer frequency.

It is not necessary to tune the parallel capacitor and the link primary into full resonance, to obtain a resistance increase. If the capacitor $C_{1res}$ and the primary resonate at a frequency $\omega_0$, slightly above the operation frequency $\omega$, their apparent resistance and inductance are found as (cf. expressions A2.13 and A2.14 in appendix 2):

$$R \approx \frac{R_{S_1} + R_{eq}}{\left[1 - \left(\frac{\omega}{\omega_0}\right)^2\right]^2} \quad \text{and} \quad L \approx \frac{L_{eq}}{1 - \left(\frac{\omega}{\omega_0}\right)^2}$$

This operation mode is named ***semi-resonance*** in this text, because on one hand, it is based on tank-resonance, but on the other, the tank is running somewhat below resonance. The technique has the benefit that there is no additional coil in the series branch. The previous drawback of an extra tank that has to be tuned to the transfer frequency remains.

**THE DESIGN OF A CLASS-E WITH RESONANT PRIMARY ($Q_{L_E} > 15$)**

The link output power and voltage are given.

The secondary components, the transfer frequency $\omega$ and the required link output $V_{sec}$, result from the secondary optimisation.

Choose the duty cycle $D$, the normalised turn-on slope $\zeta$ and $L_{S_1}$.

Compute $R_{eq}$ and the link gain $A$ with the equations of chapter 3 ($L_{eq} = L_{S_1}$ at secondary phase resonance).

$$C_{1res} = \frac{L_{eq}}{\omega^2 L_{eq}^2 + \left(R_{S_1} + R_{eq}\right)^2} \quad \text{and} \quad R = \frac{\omega^2 L_{eq}^2 + \left(R_{S_1} + R_{eq}\right)^2}{R_{S_1} + R_{eq}}$$

Choose the additional coil $L_E$ and calculate the class E according to the flow on page 134, but replace $L_{eq}$ by $L_E$ and $R_{S_1}^*$ by $R$.

**DESIGN OF A CLASS-E WITH SEMI-RESONANT PRIMARY ($Q_L > 15$)**

The link output power and voltage are given.

The secondary components, the transfer frequency $\omega$ and the required link output $V_{sec}$, result from the secondary optimisation.

Choose the duty cycle $D$, the normalised turn-on slope $\zeta$ and $L_{S_1}$.

Compute $R_{eq}$ and the link gain $A$ with the equations of chapter 3 ($L_{eq} = L_{S_1}$ at secondary phase resonance).

Choose a resonance frequency $\omega_0$

$$C_{1res} = \frac{L_{eq}}{\omega_0^2 L_{eq}^2 + \left(R_{S_1} + R_{eq}\right)^2}$$

$$R \approx \frac{R_{S_1}^*}{\left[1 - \left(\frac{\omega}{\omega_0}\right)^2\right]^2} \quad \text{and} \quad L \approx \frac{L_{eq}}{1 - \left(\frac{\omega}{\omega_0}\right)^2}$$

Calculate the class-E driver according to the procedure on page 134, but replace $L_{eq}$ by $L$ and $R_{S_1}^*$ by $R$.

Table 4.3. summarizes the example of the same inductive link driven by the three class-E drivers described in this chapter. The link transfers 20 mW to an a.c. equivalent load of 1 kΩ over 1 % coupled coils. The link components are designed according to chapter 3 for maximum link efficiency. The primary coil is chosen for the best driver performance within the maximal switch ratings of 1 A and 50 V peak.

The classic class-E circuit requires a large 250 μH primary, to obtain optimal efficiency. It is difficult to achieve such a value and still keep the parasitic resonances above 1 MHz. The topology with the resonant output tank offers a comparable efficiency for smaller coil inductances and lower circuit voltages (values indicated in bold). However, the choke coil now tends to be too large because of the larger resistive load. With a properly chosen $\omega_0$, the semi-resonant circuit sits nicely in between.

# 4.10 CONCLUSIONS

Switch-mode amplifiers offer theoretically loss-free operation thanks to tuned LC tanks, low-resistance switches and meticulous switch timing. Two driver classes, namely the saturating class C and the class E, were found appropriate for powering inductive links. The saturating class C is suited to drive moderately coupled links that do not require high primary coil voltages. It is a simple circuit with only one LC tank to tune. It is applicable up to several MHz as the parasitic switch and wiring capacitors are included in the tank capacitor.

Class-E is the only topology that can cope with the stringent demands of weakly coupled links. It produces the smoothest signals, is insensitive to slow switch-turn-on and brings as such several tens of MHz into reach.

*TABLE 4.3. A COMPARISON OF THE THREE CLASS-E TOPOLOGIES DRIVING THE SAME LINK (Switch specifications: $V_S < 50\ V_{pk}$, $I_S < 1\ A_{pk}$, $t_F = t_R = 10\ ns$, $R_{ON} = 1\ \Omega$, $V_{Ssat} = 0.5\ V$, $L_S = 25\ nH$).*

| | classic | resonant prim. | semi-res. prim. | units |
|---|---|---|---|---|
| $R_{load2}$ | 1 | 1 | 1 | kΩ |
| $P_{link\ out}$ | 20 | 20 | 20 | mW |
| $k$ | 1 % | 1 % | 1 % | - |
| $f$ | 1 | 1 | 1 | MHz |
| $C_2$ | 7.85 | 7.85 | 7.85 | nF |
| $L_{S2}$ | 2.84 | 2.84 | 2.84 | μH |
| $Q_{L_{S_2}}$ | 60 | 60 | 60 | |
| $V_{sec}$ | 4.47 | 4.47 | 4.47 | $V_{RMS}$ |
| $\eta_{link}$ | 9.77 % | 9.77 % | 9.77 % | - |
| $P_{link\ in}$ | 205 | 205 | 205 | mW |
| $L_{S1}$ | **250** | **1.3** | **25** | μH |
| $Q_{L_{S_1}}$ | 80 | 80 | 80 | - |
| $V_{prim}/V_{sec}$ | 0.0307 | 0.426 | 0.0972 | - |
| $V_{prim}$ | **145** | **10.5** | 46 | $V_{RMS}$ |
| $D$ | 16 % | 50 % | 20 % | - |
| $\zeta$ | 0 | 0 | 0 | - |
| $C_{1par}$ | 816 | 54.4 | 490 | pF |
| $C_{1ser}$ | 114 | 265 | 339 | pF |
| $\omega_0$ | **>> 1** | 1 | 1.125 | MHz |
| $C_{1res}$ | **<< 0.1** | 19.5 | 0.800 | nF |
| $L_{choke}$ (@ 1 MHz) | 0.38 | **8.6** | **0.86** | mH |
| $V^*_{cc}$ | 21.3 | 14.3 | 21.1 | V |
| $I_{cc}$ | 9.9 | 14.8 | 10.0 | mA |
| $V_{Smax}$ | 45 | 49.2 | 46.7 | V |
| $I_{Smax}$ | 115 | 42.5 | 89 | mA |
| $P_{switch}$ | 6.3 | 8.1 | 5.9 | mW |
| $\eta_{driver}$ | 97 % | 96.2 % | 97.2 % | - |
| $\eta$ | 9.48 % | 9.40 % | 9.50 % | - |

Class-D stages tend not to perform well above about 1 MHz. Their two-pole structure makes switch timing difficult and dissipation hard to control. But, they are the only choice for driving the out-of-resonance loads of e.g. stagger-tuned links. Yet, such links were not studied here since they only work for coupling factors above 5.10 %. This review also mentioned the non-saturating class C for historic reasons. It has been frequently applied for inductive links, at a time when the other classes were less well understood.

Driving weakly coupled links is difficult because of the high Q's involved. High quality factors compromise class-E efficiency since they lead to low circuit voltages and high currents. Class-E amplifiers operate better at a shorter than 50 % duty cycle in these cases. Reducing the duty cycle cuts down $V_{Ssat}$ and $R_{ON}$ losses as it increases the supply voltage and reduces the switch current. Larger primary coil inductances are beneficial for the efficiency for the same reasons, and this counts for all driver classes. Finally, a new class-E circuit with a parallel-tuned primary is proposed. It has a lower load Q, so the driver performance is better, for the same coil Q so the link efficiency is not compromised.

## 4.11 REFERENCES


**Avratoglou Ch.P.** and N.C Voulgaris, "A new method for the analysis and design of the class E power amplifier taking into account the $Q_L$ factor," *IEEE Trans. Circuits Syst.*, vol. CAS-34, pp. 687–691, 1987.

**Avratoglou Ch.P.** and N.C Voulgaris, "A class E tuned amplifier configuration with finite DC-feed inductance and no capacitance in parallel with switch," *IEEE Trans. Circuits Syst.*, vol. CAS-35, pp. 416–422, 1988.

**Avratoglou Ch.P.**, N.C Voulgaris and I.I. Fotini, "Analysis and design of a generalized class E tuned power amplifier," *IEEE Trans. Circuits Syst.*, vol. CAS-36, pp. 1068–1079, 1989.

**Forster I.C.**, "Preliminary development of a radiotelemetry system for biological applications," *Med. & Biol. Eng. & Comput.*, vol. 24, pp. 281–291, 1986.

**Gandhi S.**, Zulinski R.E. and J.C. Mandojana, "On the feasibility of load-independent output current in class E amplifiers," *IEEE Trans. Circuits Syst.*, vol. 39, pp. 564–567, 1992.

**Gutmann R.J.**, "Application of RF circuit design principles to distributed power converters," *IEEE Trans. Ind. Electron. Contr. Instrum.*, vol. IECI-27, pp. 156–164, 1980.

**Hamici Z.**, R. Itti and J. Champier, "A high-efficiency power and data transmission system for biomedical implanted electronic devices," *Meas. Sci. Technol*, vol. 7, pp. 192–201, 1996.

**Hansen B.**, K. Aabo and J. Bojsen, "An implantable, externally powered radiotelemetric system for long-term ECG and heart rate monitoring," *Biotelemetry Patient Monitg.*, vol. 9, pp. 227–237, 1982.

**Herman K.J.**, and R.E., Zulinski, "The infeasibility of a zero-current switching class E amplifier," *IEEE Trans. Circuits Syst.*, vol. 37, pp. 152–154, 1990.

**Kazimierczuk M.K.**, "Class E tuned power amplifier with shunt inductor," *IEEE J. Solid-State Circuits*, vol. SC-16, pp. 2–7, 1981.

**Kazimierczuk M.K.**, "Effects of the collector current fall time on the class E tuned power amplifier," *IEEE J. Solid-State Circuits*, vol. SC-18, pp. 181–193, 1983a.

**Kazimierczuk M.K.**, "Exact analysis of class E tuned power amplifier with only one inductor and one capacitor in load network," *IEEE J. Solid-State Circuits*, vol. SC-18, pp. 214–221, 1983b.

**Kazimierczuk M.K.**, "Collector amplitude modulation of the class E tuned power amplifier," *IEEE Trans. Circuits Syst.*, vol. CAS-31, pp. 543–549, 1984.

**Kazimierczuk M.K.**, "Generalization of conditions for 100-percent efficiency and nonzero output power in power amplifiers and frequency multipliers," *IEEE Trans. Circuits Syst.*, vol. CAS-33, pp. 805–807, 1986a.

**Kazimierczuk M.K.**, "Class E tuned power amplifier with nonsinusoidal output voltage," *IEEE J. Solid-State Circuits*, vol. SC-21, pp. 575–581, 1986b.

**Kazimierczuk M.K.** and K.Puczko, "Exact analysis of class E tuned power amplifier at any Q and switch duty cycle," *IEEE Trans. Circuits Syst.*, vol. CAS-34, pp. 149–159, 1987.

**Kazimierczuk M.K.** and K.Puczko, "Power-output capability of class E amplifier at any loaded Q and switch duty cycle," *IEEE Trans. Circuits Syst.*, vol. CAS–36, pp. 1142–1143, 1989a.

**Kazimierczuk M.K.** and K.Puczko, "Class E tuned power amplifier wityh antiparallel diode or series diode at switch, with any loaded Q and switch duty cycle," *IEEE Trans. Circuits Syst.*, vol. CAS-36, pp. 1201–1209, 1989b.

**Kazimierczuk M.K.** and J. Józwik, "Optimal topologies of resonant DC/DC converters," *IEEE Trans. Aerosp. Electron. Syst.*, vol. AES-25, pp. 363–372, 1989a.

**Kazimierczuk M.K.** and J. Józwik, "Class $E^2$ narrow-band resonant DC/DC converters," *IEEE Trans. Instr. Meas.*, vol. 38, pp. 1064–1068, 1989b.

**Kazimierczuk M.K.** and X.T. Bui, "Class E DC/DC converters with an inductive impedance inverter," *IEEE Trans. Power Electron.*, vol. PE-4, pp. 124–135, 1989a.

**Kazimierczuk M.K.** and X.T. Bui, "Class E DC/DC converters with an capacitive impedance inverter," *IEEE Trans. Ind. Electron.*, vol. IE-36, pp. 425–433, 1989b.

**Liou M.-L.**, "Exact analysis of linear circuits containing periodically operated switches with applications," *IEEE Trans. Circuit Theory*, vol. CT-19, pp. 146–154, 1972.

**Mandojana J.C.**, K.J. Herman and R.E. Zulinski, "A discrete/continuous time-domain analysis of a generalized class E amplifier," *IEEE Trans. Circuits Syst.*, vol. CAS-37, pp. 1057–1060, 1990.

**Molnár B.**, "Basic limitations on waveforms achievable in single-ended switching-mode tuned (class E) power amplifiers," *IEEE J. Solid-State Circuits*, vol. SC-19, pp. 144–146, 1984.

*__Raab F.H.__, "Idealized operation of the class E tuned power amplifier," *IEEE Trans. Circuits Syst.*, vol. CAS-24, pp. 725-735, 1977.

*__Raab F.H.__, "Effects of circuit variations on the class E tuned power amplifier," *IEEE J. Solid-State Circuits*, vol. SC-13, pp. 239–247, 1978.

*__Raab F.H.__ and N.O. Sokal, "Transistor power losses in the class E tuned power amplifier," *IEEE J. Solid-State Circuits*, vol. SC-13, pp. 912–914, 1978.

**Redl R.**, B. Molnár and N.O. Sokal, "Class E resonant regulated DC/DC power converters: analysis of operations, and experimental results at 1.5MHz," *IEEE Trans. Power Electron.*, vol. PE-1, pp. 111–120,1986a.

**Redl R.**, B. Molnár and N.O. Sokal, "Small-signal dynamic analysis of regulated Class E DC/DC converters," *IEEE Trans. Power Electron.*, vol. PE-1, pp. 121–128,1986b.

*__Schwan M.,__ P. Troyk and G. Loeb, "Suspended carrier modulation for transcutaneous telemetry links," in: *Biotelemetry XIII (Proc. 13th Int. Symposium on Biotelemetry)*, Eds. C. Cristalli, C.J. Amlaner Jr. and M.R. Neuman, Williamsburg, VI, pp. 27–32, 1996.

**Smith G.H.** and Zulinski R.E., "An exact analysis of class E amplifiers with finite DC-feed inductance at any output Q," *IEEE Trans. Circuits Syst.*, vol. 37, pp. 530–534, 1990.

**Sokal A.D.**, "A remark on 'Exact analysis of linear circuits containing periodically operated switches with applications'," *IEEE Trans. Circuits Syst.*, vol. CAS-24, pp. 575–577, 1977.

**Sokal N.O.**, "RF power transistor storage time: theory and measurements," *IEEE J. Solid-State Circuits*, vol. SC-11, pp. 344–346, 1976.

*__Sokal N.O.__, "Class E can boost the efficiency," *Electronic Design*, vol. 20, pp. 96–102, Sept. 1977.

**Sokal N.O.**, "Class E high-efficiency switching-mode tuned power amplifier with only one inductor and one capacitor in load network - approximate analysis," *IEEE J. Solid-State Circuits*, vol. SC-16, pp. 380–384, 1981.

**Sokal N.O.** and F.H. Raab, "Harmonic output of Class-E power amplifiers and load coupling network design," *IEEE J. Solid-State Circuits*, vol. SC-12, pp. 86–88, 1977.

*__Sokal N.O.__ and A.D. Sokal, "Class E - a new class of high-efficiency tuned single-ended switching power amplifiers," *IEEE J. Solid-State Circuits*, vol. SC-10, pp. 168–176, 1975a.

**Sokal N.O.** and A.D. Sokal, "High efficiency switching power amplifier," *U.S. Patent 3 919 656*, Nov. 11, 1975b.

**Sowlati T.,** C.A.T. Salama, J. Sitch, G. Rabjohn and D. Smith, "Low voltage, high efficiency GaAs class E power amplifiers for wireless transmitters," *IEEE J. Solid-State Circuits.*, vol. SC-30, pp. 1074–1080, 1995.

***Troyk P.R.** and M.A.K. Schwan, "Class E driver for transcutaneous power and data link for implanted electronic devices," *Med. & Biol. Eng. & Comput.*, vol. 30, pp. 69–75, 1992a.

***Troyk P.R.** and M.A.K. Schwan, "Closed-loop class E transcutaneous power and data link for MicroImplants," *IEEE Trans. Biomed. Eng.*, vol. BME-39, pp. 589–599, 1992b.

**Voulgaris N.C.** and Ch.P. Avratoglou, "The use of a thyristor as a switching device in a class E tuned power amplifier," *IEEE Trans. Circuits Syst.*, vol. CAS-34, pp. 1248–1250, 1987.

**Zulinski R.E.**, "A high-efficiency self-regulated class E power inverter/converter," *IEEE Trans. Ind. Electron.*, vol. IE-33, pp. 340–342, 1986.

**Zulinski R.E.** and J.W. Steadman, "Performance evaluation of Class E frequency multipliers," *IEEE Trans. Circuits Syst.*, vol. 33, pp. 343–346, 1986.

**Zulinski R.E.** and J.W. Steadman, "Class E power amplifiers and frequency multipliers with finite DC-feed inductance," *IEEE Trans. Circuits Syst.*, vol. 34, pp. 1074–1087, 1987.

**Zulinski R.E.,** J.C. Mandojana and K.J. Herman, "The infeasibility of self-regulating constant-current-fed class E power inverters," *IEEE Trans. Circuits Syst.*, vol. 37, pp. 558–561, 1990.

**Zulinski R.E.** and K.J. Grady, "Load-independent class E power inverters: part I - theoretical development," *IEEE Trans. Circuits Syst.*, vol. 37, pp. 1010–1018, 1990.

# 5 Optimisation of the driven inductive link

Inductive link design is characterised by about a dozen of interlinked variables and component values: coupling, material characteristics, inductances, quality factors, resonance capacitors, and so on. The foregoing analysis revealed that poor coupling even lengthens this list, as the driver design cannot be separated from the coil design in that case. The purpose of this chapter is to find a way through this maze of interrelated parameters.

A universal procedure is provided to optimise a driven inductive link. The technique allows gradually determining each of the individual link parameters. Emphasis is put on the minimal need for experimental data and a minimal number of design iterations. The method is general in a sense that it fits to a broad range of applications with either well or badly coupled coils, and low or high power transfer.



K. Van Schuylenbergh and R. Puers, *Inductive Powering: Basic Theory and Application to Biomedical Systems*, Analog Circuits and Signal Processing, 145–168. © Springer Science+Business Media B.V. 2009

# 5.1 OPTIMISATION OF THE DRIVEN LINK

The goal of the optimisation is a maximal overall efficiency $\eta = \eta_{link}*\eta_{driver}$ for the entire driven inductive link, possibly in combination with critical coupling. Maximising the link efficiency $\eta_{link}$ only is useless if it cannot be combined with a maximal driver efficiency $\eta_{driver}$. The problem is that it is impossible to find a closed analytical expression for the system optimum. This would require an analytical expression for the first derivative of the system efficiency $\eta$ (including the driver efficiency).

This is the reason why the foregoing link equations were shaped such the parameters that set the link efficiency have only a second order impact on the driver performance. Also, the other way around, the primary inductance $L_{S1}$ which has no influence on the link efficiency, is identified as a key parameter to control the driver performance. This approach makes it more likely for the link optimum to coincide with the global optimum. Moreover, the circuit topologies presented in the previous chapter are characterised by a nearly loss-free operation. Almost regardless of the link specifications, a driver efficiency above 90% is achievable with at least one of the presented topologies, just because $L_{S1}$ is available for the driver design. The differences in optimised driver efficiencies are thus not that large. They may shift from 90% for a complicated, to 97% for a simple design. Link efficiencies vary much more and are thus dominant.

The influence of the independent design parameters on the global efficiency is studied now more in detail. It is recommended to keep the design equations from the previous chapter at hand.

## 5.1.1 RELATION BETWEEN $k$ AND $\eta$

The coupling factor is the most important parameter for the link efficiency: the higher the coupling, the higher the link efficiency.

The coupling factor influences the driver efficiency via $R_{eq}$ and the link gains $A$ (for a saturating-class-C or class-E driver) or $A_{stp}$ (class-D).

The driver load $R_{S1} + R_{eq}$ increases with the coupling according to:

*Equation 5.1.*

$$R_{S_1} + R_{eq} = L_{S_1}\omega_{resP}\left(\frac{1}{Q_{L_{S_1}}} + k^2\frac{\alpha.Q_{L_{S_2}}}{\alpha + Q_{L_{S_2}}}\right) \quad \text{for parallel - reson. sec.}$$

or

$$R_{S_1} + R_{eq} = L_{S_1}\omega_{res}\left(\frac{1}{Q_{L_{S_1}}} + k^2\frac{Q_{L_{S_2}}}{Q_{L_{S_2}}\alpha + 1}\right) \quad \text{for series - resonant sec.}$$

A higher resistive driver load with respect to $R_{ON}$ improves the driver efficiency according to Equation 4.4. for saturating class C, Equation 4.1. for class D or Equation 4.7. for class E.

The link gains $A$, $A_{stp}$ or $r_{ptp}$ also grow with k (Equation 3.8. Equation 3.10. and Equation 3.40.–Equation 3.42.). But as the link output $V_{sec}$ should be kept equal to the specified value, either $L_{S1}$ must be increased to reduce the gain, or $V_{cc}^*$ must be lowered. Augmenting $L_{S1}$ further improves the driver efficiency via $R_{eq}$. Reducing

> A maximal coupling factor $k$ is beneficial for the overall efficiency

the supply voltage lowers the driver efficiency and is only accepted if it largely exceeds $V_{Ssat}$.

### 5.1.2 RELATION BETWEEN $Q_{L_{S_1}}$ AND $\eta$

Large coil quality factors are always preferable for good link efficiency.

The quality factors influence the driver performance via $R_{S1}$ and the link gains. A larger primary coil quality factor reduces the driver load via $R_{S1}$. But, since there is no influence on $R_{eq}$, this becomes only important at low coupling factors where $R_{eq}$ is small compared to $R_{S1}$. The solution in such case is to use a class-E driver with a (semi-)resonant primary.

The link gains increase with the primary coil quality factor and saturate at large Q values (the constants $a$, $b$ and $c$ are independent of $Q_{Ls1}$):

$$A \propto \frac{Q_{L_{S_1}}}{\sqrt{Q^2_{L_{S_1}} + aQ_{L_{S_1}} + b}} \qquad \text{for a non-resonant primary}$$

or

$$A_{stp} \propto r_{ptp} \propto \frac{Q_{L_{S_1}}}{Q_{L_{S_1}} + c} \qquad \text{for a resonant primary}$$

Larger link gains are advantageous for the driver efficiency for the same reasons as before, provided they are compensated by larger primary coil inductances.

A higher quality factor also cuts down the harmonics in the coil current. However, the effect on $\eta_{driver}$ is insignificant for Q factors above 10.

Note that the primary coil quality factor has an additional positive effect on the saturating-class-C efficiency via the losses associated with the switch fall and rise times $t_R$ and $t_F$ (Equation 4.4.).

The conclusion is that a high primary coil quality factor is good for the link efficiency and the link gain. It has a negative impact on the driver load at low coupling that may be compensated by selecting the proper driver circuit.

> Maximal coil quality factors are favourable for the overall efficiency

### 5.1.3 RELATION BETWEEN $Q_{L_{S_2}}$ AND $\eta$

A larger secondary coil quality factor increases the driver load (Equation 5.1.) and the link gains $A$, $A_{stp}$ and $r_{ptp}$ (the constants $a$, $b$ and $c$ are independent of the secondary coil quality factor):

$$A \propto \frac{Q_{L_{S_2}}}{\sqrt{Q^2_{L_{S_2}} + dQ_{L_{S_2}} + e}} \qquad \text{for a non-resonant primary}$$

or

$$A_{stp} \propto r_{ptp} \propto \frac{Q_{L_{S_2}}}{Q_{L_{S_2}} + f} \qquad \text{for a resonant primary}$$

Both effects improve the driver efficiency as explained above.

*TABLE 5.1. THE RELATION BETWEEN $\alpha$ AND THE LINK GAINS OR $R_{eq}$.*

| | Parallel-resonant secondary | Series-resonant secondary |
|---|---|---|
| | $R_{eq} \propto \dfrac{\alpha}{\alpha + Q_{L_{S_2}}}$ | $R_{eq} \propto \dfrac{1}{\alpha\, Q_{L_{S_2}} + 1}$ |
| **Non-resonant primary** | $A \propto \sqrt{\dfrac{\alpha^2+1}{\alpha^2+g\alpha+h}}$ | $A \propto \dfrac{\alpha}{\sqrt{\alpha^2+l\alpha+m}}$ |
| **Resonant primary** | $A_{stp} \propto r_{ptp} \propto \dfrac{\sqrt{\alpha^2+1}}{i\alpha+j}$ | $A_{stp} \propto r_{ptp} \propto \dfrac{\alpha}{n\alpha+p}$ |

### 5.1.4 RELATION BETWEEN $\omega$ AND $\eta$

The driver efficiency plays a minor role in determining the optimal transfer frequency. The switching frequencies applied for inductive links –from some kilohertz to some 10 MHz at most – are not so high that they are problematic for the class-E or saturating-class-C topology.

The driver losses from the parasitic switch capacitors ($\propto f\,C$), the limited switching speeds ($\propto f^2 t_F$ or $\propto f^2 t_R$) and the gate drive ($\propto f\,C_G$ for MOSFET switches) all increase with the frequency. Small (low $C_G$) and fast (low $t_R$ and $t_F$) MOSFET's are therefore favoured at higher frequencies instead of high-power devices (low $R_{ON}$, but large $C_G$, $t_F$ and $t_R$).

### 5.1.5 RELATION BETWEEN $D$ AND $\eta$

The duty cycle has no influence on the link performance. It is chosen for optimal driver efficiency.

### 5.1.6 RELATION BETWEEN THE COIL INDUCTANCES AND $\eta$

The primary coil inductance has no influence on the link efficiency. Larger values are advantageous for the driver efficiency.

The secondary coil inductance is no independent design parameter. It is set by the load $R_{load2}$, the optimal $\alpha$ and the transfer frequency.

### 5.1.7 RELATION BETWEEN $\alpha$ AND $\eta$

The link performance was optimised by selecting the $\alpha_{opt}$ for maximal link efficiency. There must still be verified if this operation mode can be combined with a high driver efficiency to yield an optimal system.

The parameter $\alpha$ appears in the driver equations in the expressions for $R_{eq}$ and the link gain that sets the value for $V_{prim}$. $R_{eq}$ goes up with $\alpha$ in case of a parallel-resonant secondary, and goes down in case of series resonance (Table 5.1.). The link gain increases with $\alpha$ in all cases. As a result, the driver efficiency increases with $\alpha$ for a link with a parallel-resonant secondary, and decreases for a series- resonant secondary. However, the driver efficiency varies only slowly with $\alpha$, whereas the link efficiency has a rather peaky optimum at $\alpha_{opt}$. The system optimum is thus dominated by the peak in the link efficiency.

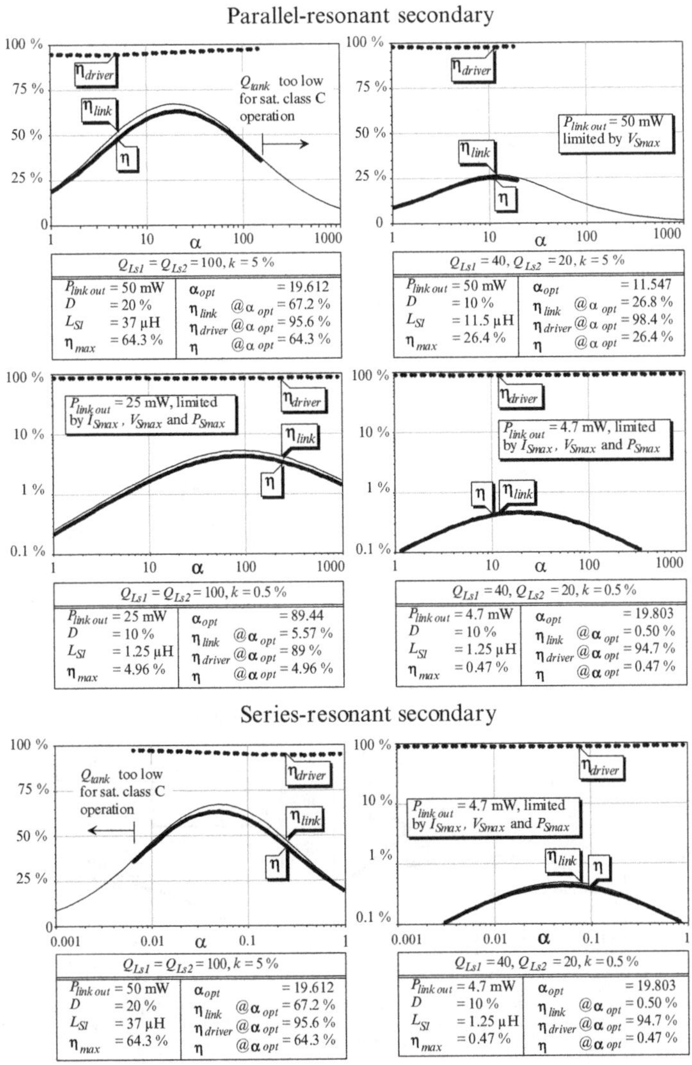


*Figure 5.1.* *System efficiency as a function of $\alpha$ for a saturating-class-C-driven link. Four extremes are studied: a high (5%) and low (0.5%) coil coupling in combination with a high ($Q_{Ls1} = Q_{Ls2} = 100$) or low ($Q_{Ls1} = 40$, $Q_{Ls2} = 20$) coil quality factor. The transfer frequency is 1 MHz. The transferred power is 50 mW where possible. Otherwise, the maximum power obtainable within the coil and switch limitations, is taken ($L_{S1} < 50\ \mu H$, $V_S < 50\ V$, $I_S < 1\ A$, $P_S < 50\ mW$). The primary coil inductance is set to that value that produces the maximum system efficiency at $\alpha_{opt}$. The driver capacitor is adjusted for each $\alpha$.*

*The two bottom diagrams show the high-Q, high-k and the low-Q, low-k cases for a series-resonant secondary. These diagrams correspond to the parallel-resonant case where $\alpha$ is replaced by $1/\alpha$.*

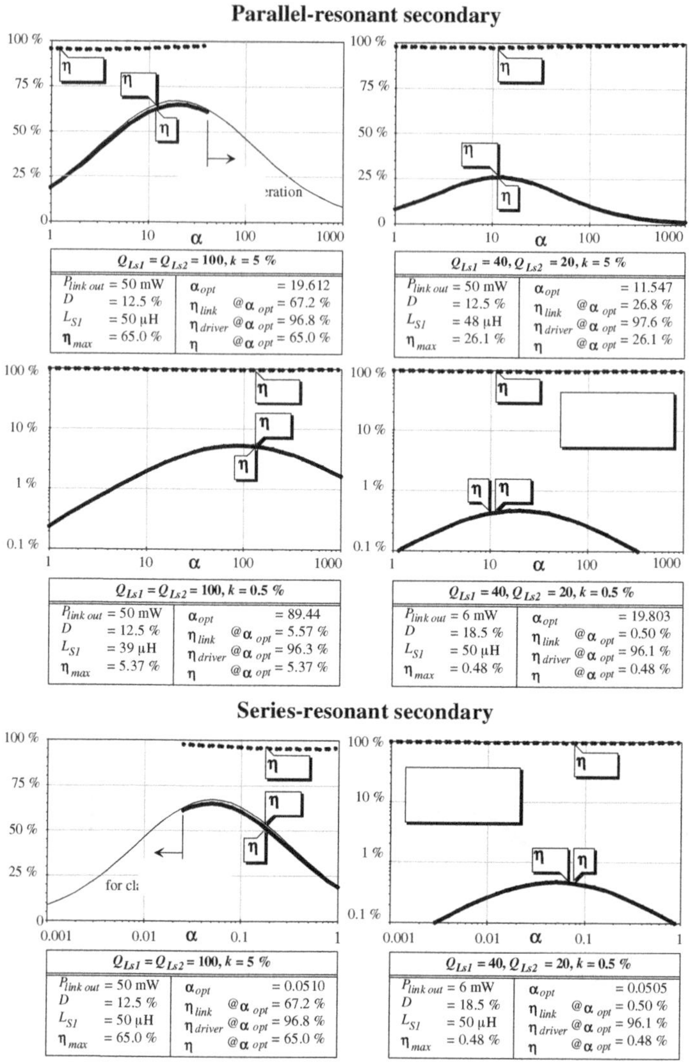


*Figure 5.2. System efficiency as a function of $\alpha$ for a link fed by a class-E driver. The four link extremes and the coil and switch limits are identical to Figure 5.1. The bottom diagrams show again the high-Q, high-k and the low-Q, low-k cases for a series-resonant secondary. These diagrams correspond once more to the parallel-resonant case where $\alpha$ is replaced by $1/\alpha$.*

This is demonstrated in Figure 5.1 and Figure 5.2 for different combinations of circuit topologies, coupling and quality factors. These diagrams depict the link efficiency, the driver efficiency and their product, the system efficiency, as a function of the parameter $\alpha$. All curves are computed with the link formulae of Chapter 3 and the first-order simplified driver equations of Chapter 4. There are four cases studied: a high (5%) coil coupling in combination with high ($Q_{Ls1} = Q_{Ls2} = 100$) or with low

($Q_{Ls1}$ = 40, $Q_{Ls2}$ = 20) coil quality factors, and low (0.5%) coil coupling in combination with the same high or low quality factors. For each case, the primary inductance is set to that value that produces the maximum system efficiency at $\alpha_{opt}$. The driver capacitor is adjusted for each $\alpha$. It is tried to achieve a net link output of 50 mW with coils of maximally 50 μH and switch voltages and currents below 50 V and 1 A, respectively. The maximal switch dissipation is set to 50 mW.

The diagrams indicate that a high and fairly constant driver efficiency is obtained over the entire $\alpha$ range. The $\alpha_{opt}$ for maximal link efficiency is thus very near the system optimum.

Operation at $\alpha_{opt}$ is OK for maximal overall efficiency

Note also that the saturating-class-C driven links have a lower maximal output than the class-E links, because the driver switch is straightway connected to the primary coil.

The foregoing explanation on the optimisation of $\alpha$ does not hold for critically coupled systems. Critical coupling occurs only at an $\alpha$ value equal to $\alpha_{crit}$. This means that $\alpha$ is not available anymore for the system optimisation. Maximising the efficiency is then accomplished by optimising the remaining design parameters, $k$, $L_{S_1}$, $Q_{L_{S_1}}$, $Q_{L_{S_2}}$, $\omega$ and $D$. The critically coupled, driven inductive link is discussed below.

## 5.1.8 CRITICAL COUPLING OF THE DRIVEN INDUCTIVE LINK

The question is here if the gain maximum still occurs at $\alpha = \alpha_{crit}$ when the link is connected to its coil driver. Or otherwise stated, "Is the driver output independent of the coupling factor?" The answer is clearly positive for the saturating-class-C drivers since their output voltage $V_{prim}$ equals the supply voltage times a factor that is only related to the switch duty cycle (Equation 4.5. and Equation 4.3.). In fact, the switch voltage does not change with the coupling as long as the tank Q remains high. This makes the driver very suitable for links with a moderate – as the output voltage is rather limited – but highly variable coupling. Note also that critical coupling is only possible for coupling factors above 1% where the saturating-class-C drivers are often applicable.

The class-D-driver output is not perfectly but nearly independent of the coupling, as long as the resistive load stays well above the switch-on resistance (Equation 4.2.).

The class-E drivers are the most sensitive to coupling variations. Their output is approximated for high Q's by:

$$\hat{V}_{prim} \approx \frac{V_{cc} Q^*_{L_{eq}}}{g} = \frac{V_{cc}\omega L_{S_1}}{g\left(R_{S_1} + R_{eq}\right)}$$

where $g$ is related to the duty cycle $D$ and the slope $\zeta$ of the turn-on voltage (Equation 4.6.). The resistance $R_{eq}$ is very sensitive to coupling variations ($R_{eq} \propto k^2$). Moreover, operation at critical coupling makes that $R_{eq}$ cannot be neglected with respect to $R_{S1}$ (Equation 3.24. and Equation 3.28.).

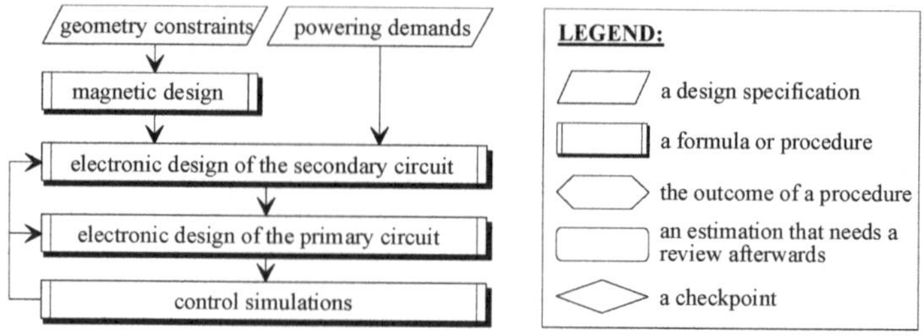


*Figure 5.3.* *Simplified flowchart of the design procedure and the list of flowchart symbols used throughout this text.*

The variations on $R_{S1} + R_{eq}$ not only change the driver output, they also take the driver out of its class-E operation mode. In fact, $C_{1ser}$ and especially $C_{1par}$ are related to the driver load and should thus be adjusted when the coupling changes.

Class-E drivers should thus only be utilized at high coupling if they feature a feedback mechanism that automatically adjusts their tuning to coupling variations. Straightforward class-E circuits are only applicable for driving efficiency-optimised links with a constant or a low coupling (where $R_{eq}$ is much smaller than $R_{S1}$, cf. Equation 3.19.).

## 5.2 THE OPTIMISATION STRATEGY

All parts of the large optimisation puzzle are now on the table; there is a complete list of link formulae, there is a design procedure for each of the driver types and there is an exhaustive catalogue of rectifiers and regulators. It now comes to piecing the puzzle together.

The optimisation procedure is structured in two parts; the magnetic and the electronic link design (Figure 5.3; VAN SCHUYLENBERGH and PUERS, 1993). The electronic design is further split in three sub-parts which brings the number of major design steps to four. A single-step optimisation of all design parameters at once is not only impossible, it is also not desirable. A one-by-one optimisation is easier to oversee and enables a more transparent design flow.

The procedure starts with the magnetic link design, because the magnetic coupling is only a function of coil placement, geometry and material properties. The magnetic design attempts to determine the dimensions, the placement and the shape of the coils, the cores and the packaging. The design target is to maximise coil coupling and minimise magnetic losses, within the given geometrical constraints. Better coupling and lower magnetic losses always improve the link efficiency regardless whether the link needs optimisation towards efficiency or towards misalignment tolerance (Figures 3.5 and 3.9).

The main outcome of the magnetic design is a value for the coupling factor $k$. This value is independent of the circuit topology or the powering demands, such as the required output power and voltage. The magnetic design step shall also indicate a useful transfer-frequency band because the losses in the surrounding media are field related and frequency dependent.

The second design step deals with the electronic link aspects. It is divided in three sub-blocks:

- Calculation of the transfer frequency and the secondary circuit components (coil, capacitors, rectifier and regulator)
- Determination of the primary circuit (including the driver), and
- Verification of the electronic design by Spice simulations

The electronic and the magnetic design are well separated: once the magnetic aspects are optimised, they need no further revision during the electronic optimisation.

This structured approach splits the design process as much as possible into sub-designs with minimal interaction. The rationale is twofold. On one hand, the component values for the driven link are tightly related to the actual coil data. On the other hand, coil prototyping and measuring their parameters is elaborate and time consuming. The procedure is therefore structured in such a way that each coil is individually fine-tuned. This is to avoid that for instance, the secondary coil is to be redesigned if one aspect of the primary coil is altered. Otherwise, one would be bouncing back and forth between the primary and the secondary coil and making each time new prototypes and measurements, in search of an optimal set.

The breaking up of the design process goes even beyond single components as the coils are further divided into their geometry and the number of turns. The coil dimensions are first optimised in the magnetic design phase for maximal coupling and minimal magnetic losses. There is no prototype data needed here. This step already reduces the number of coil candidates to one family with identical geometry. It also sets the coil inductance per winding, the useful frequency band and a range of typical quality factors (the coil dimensions and the applied materials tend to determine the maximal quality factor more than the actual number of turns).

The number of coil turns and the transfer frequency are then determined in the electronic design step, based on estimations for the coil quality factors. This gives a reliable clue about what is feasible for the given application in terms of efficiency, the optimal frequency and what kind of prototypes are needed. The number of required prototypes is limited this way, to something like five per coil. There is basically one iteration required to replace the estimated coil quality factors by actual measured data to further refine the calculations. This refinement changes the actual component values and not the concept of the link circuit, provided the initial estimations were somewhat realistic (e.g., 80–120 for centimetre-sized coils and not 500).

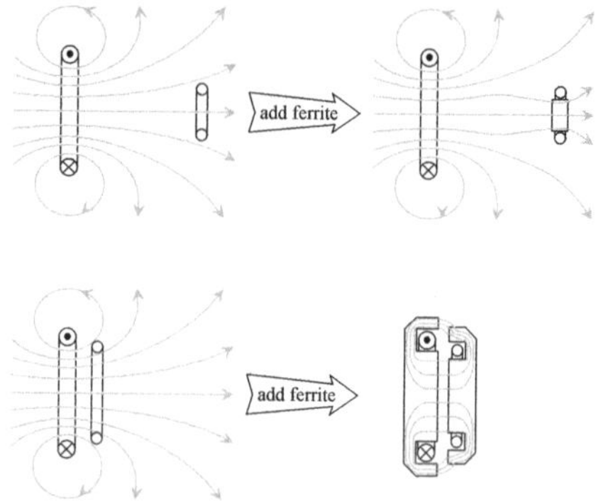


*Figure 5.4. Magnetic flux lines follow pathways of minimal magnetic resistance. Ferrite cores may thus conduct the flux lines through the centre of the secondary and improve the coil coupling (right). The top figure depicts a low-coupled situation. A well coupled coil set is shown at the bottom.*

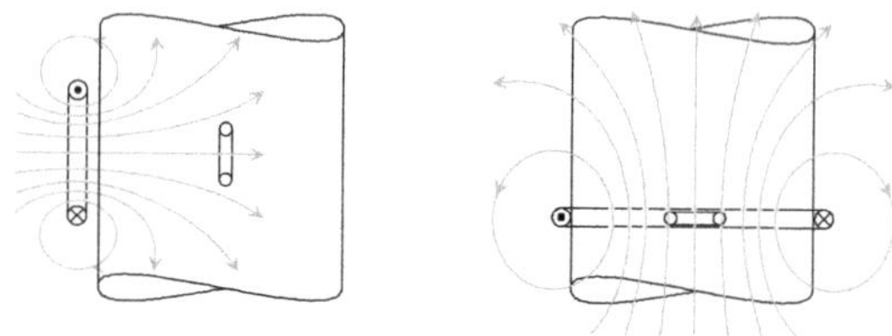

*Figure 5.5. A secondary embedded in a cylindrical object. The concentric set-up on the right has a better coupling than the parallel configuration on the left.*

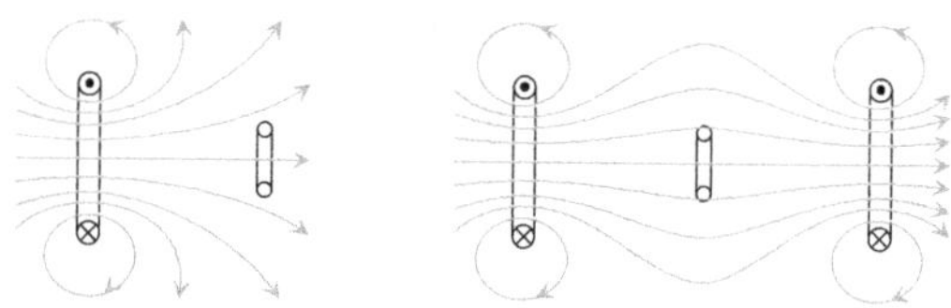

*Figure 5.6. The Helmholtz coil set-up (right) reduces the diversion of the flux lines at the level of the link secondary (FORSTER, 1986).*

### 5.2.1 MAGNETIC DESIGN

The magnetic link design strives to maximise the coil coupling and minimise the magnetic losses, within the specified geometrical constraints. This step determines the coil dimensions, the placement and the shape of the coils, the cores and the packaging. The design result is a value for the coupling factor $k$. This value is independent of the circuit topology or the powering demands.

There are two ways to control coil coupling: first, the sizing and positioning of the coils and second, the use of ferrite cores to guide the flux lines (Figure 5.4). Coil coupling is determined by the ratio of the two areas circumscribed by the coils, and above all, by the distance between both coil planes. A concentric coil positioning is therefore preferred over a parallel set-up (Figure 5.5). The Helmholtz configuration of Figure 5.6 is an alternative to improve the coupling when a concentric arrangement is not desired.

Ferrite coil cores are applied for two reasons. First, they can concentrate the magnetic flux lines where they are actually needed to increase the coupling (MATSUKI *et al.*, 1990). Second, they allow producing larger inductances in a smaller space. This is particularly useful for small devices or when the generation of eddy currents in a surrounding metal package forces the transfer frequency to very low values.

The latter aspect touches a second topic covered by the magnetic design. The fact that the magnetic losses in the surrounding media are field related and frequency dependent allows indicating a usable band for the transfer-frequency during the magnetic design. There is always a trade-off between the available space – a smaller device size pushes the frequency to higher values because of the matching of the secondary to the load – and the generation of eddy currents, the losses and the radiation that all increase with the frequency.

The skin and proximity effects are not considered in the magnetic design because their influence on the coil coupling is negligible. Their effect on the coil resistance is examined in the electronic design phase where they are included in the quality factor.

It is impossible to provide general parameterised models for the analytical optimisation of the link magnetics, apart from the simple cases summarised below. This is because the coil coupling and the magnetic losses are entirely linked to the three-dimensional distribution of the magnetic and the current fields. Only finite-element simulations of the actual set-up can cope with this complexity (MOHAMMED and BATINA, 1985, 1986; MOHAMMED, 1988).[1]

#### 5.2.1.1 Simple coils in a passive medium

***A. Self inductance***

The inductance of a single-turn circular loop is approximated for large loop radii $r$ compared to the wire radius $r_{wire}$, by (TERMAN, 1943):

$$L \approx \mu_0 r \left[ \ln\left(\frac{8r}{r_{wire}}\right) - 2 + \frac{\mu_r}{4} \right]$$ *Equation 5.2.*

The permeability of the medium surrounding the coil is $\mu_0$, $\mu_r$ the relative permeability of the wire material.

The inductance of coils constructed of multiple coaxial loops –like solenoids or pancake coils – is found as (ZIERHOFER and HOCHMAIR, 1996) :

$$L = \sum_{i=1}^{n} L_i + \sum_{i=1}^{n}\sum_{j=1}^{n} \delta_{ij} M_{ij}$$ *Equation 5.3.*

$M_{ij}$ is the mutual inductance between loop $i$ and loop $j$. The parameter $\delta_{ij}$ is one for $i \neq j$ and zero for $i = j$.

Remark: Coil inductances are proportional to the square of the number of coil turns.

[1] *MagNet 5.1* is a software package from the Infolytica Corporation, London and Montreal, that suits for this kind of finite element computations of electromagnetical problems.

Coils assembled of straight conductors are calculated in a similar way. The inductance of a round straight wire of length $l$ and radius $r_{wire}$ is:

Equation 5.4.
$$L \approx \frac{\mu_0 l}{2\pi}\left[\ln\left(\frac{8l}{r_{wire}}\right) - 1 + \frac{\mu_r}{4}\right]$$

The inductance of a complete coil is also calculated with Equation 5.3., where $M_{ij}$ equals the mutual inductance between wire $i$ and wire $j$. The parameter $\delta_{ij}$ is then 1 for $i \neq j$ provided that the currents in wires $i$ and $j$ flow in the same direction, and −1 if they flow in opposite directions. It is zero for $i = j$ (GROVER, 1946).

Some empirical formulae for other coil shapes are listed in Appendix A2.

### B. Mutual inductance

The mutual inductance between single-turn coils is calculated with Neumann's formula:

$$M = \frac{\mu_0}{4\pi}\oint\oint\frac{dl_1.dl_2}{r_{12}}$$

FLACK *et al.* (1971) have numerically solved this integral for various combinations of parallel but non-coaxial circular loops and summarised their results on normalised diagrams.

The mutual inductance between two coils of multiple loops is given by (ZIERHOFER and HOCHMAIR, 1996):

Equation 5.5.
$$M = \sum_{i=1}^{n_1}\sum_{j=1}^{n_2} M_{i,j}$$

> Mutual inductances are proportional to the number of primary coil turns times the number of secondary coil turns. The coupling factor $k$ is independent of the number of coil turns.

$M_{ij}$ is the mutual inductance between loop $i$ of coil 1 and loop $j$ of coil 2.

Solenoids, and disk or pancake coils are natural extensions of the ring shape. The mutual coupling between such coils is thus found as

$$M = \frac{1}{h_1 h_2}\int_{z_1=0}^{h_1}\int_{z_2=0}^{h_2} M(z_1,z_2)dz_1\,dz_2$$

$z_1$ and $z_2$ are the longitudinal (for a cylinder coil) or radial (for a pancake coil) variables (Figure 5.7). $h_1$ and $h_2$ are the cylinder length or pancake width. $M(z_1, z_2)$ is the mutual inductance of a ring at level $z_1$ on coil 1 and a ring at level $z_2$ on coil 2. HOCHMAIR (1984) numerically evaluated this integral for parallel cylindrical coils.

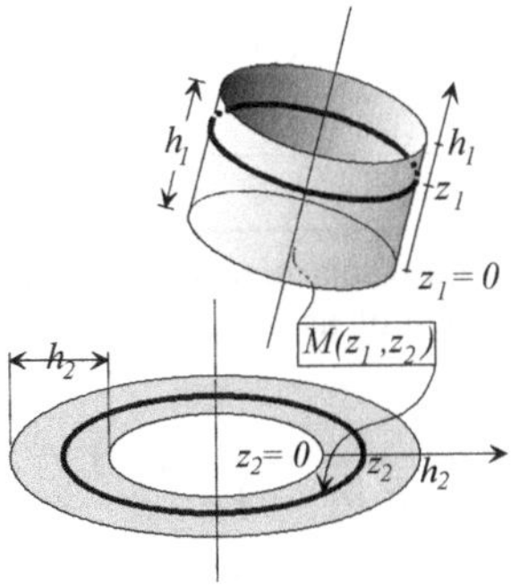


*Co-ordinate system for the calculation of the mutual inductance between two cylinder, disk or pancake coils.* *Figure 5.7.*

In the special case of coaxial loops, Neumann's formula is solved to (HARNWELL, 1949; SOMA *et al.*, 1987):

$$M = \mu_0 \sqrt{r_1 r_2}\, G(g)$$

$$\text{where } G(g) = \left(\frac{2}{g} - g\right).K(g) - \frac{2}{g}E(g) \text{ with } g = \sqrt{\frac{4r_1 r_2}{(r_1 + r_2)^2 + d^2}}$$

*Equation 5.6.*

$r_1$ and $r_2$ are the coil radii, $d$ is the mutual coil distance. $K(g)$ and $E(g)$ are the complete elliptical integrals of the first and second kind respectively (ABRAMOWITZ and STEGUN, 1965). The wire radii are assumed to be small compared to the coil radii.

TROYK and SCHWAN (1996) approximate the product $\mu_0.G(g)$ for well separated coils with significantly different diameters, by $2.467402498*g^3$ [nH/cm]. This estimation is valid within +0/–5% for $g$ values below 0.25. It is very useful for very small $g$ values where the exact expression for $G(g)$ is numerically unstable.

The factor $G$ is a monotonous rising function in $g$. The coil coupling is thus maximal for maximal $g$. Hence, the primary coil radius $r_1$ that maximises the coupling for a given coil spacing $d$ and secondary coil radius $r_2$ is (this expression was also found by KO *et al.*, 1977):

$$\frac{\partial g(r_1, r_2, d)}{\partial r_1} = 0 \Leftrightarrow r_1\Big|_{\text{max. coupling}} = \sqrt{r_2^2 + d^2}$$

*Equation 5.7.*

Coil misalignments change the coupling.[2] HOANG *et al.* (1993) compute the mutual inductance of two parallel rings, laterally displaced over a distance $\Delta$, as[3]:

$$M_{lat} = \mu_0 \frac{r_1 r_2^2}{\pi} \sqrt{2} \int_0^{2\pi} \frac{r_1 - \Delta\cos\phi}{\sqrt{Q^3}} G(r)\, d\phi$$

*Equation 5.8.*

(this formula continues overleaf)

2 Remark: the statistical analysis of coil misalignments presented in (JACOB, 1976) aims at data telemetry over very weakly coupled coils. It is not useful for the treatment of powering links.

3 Formula (2) in (ZIERHOFER and HOCHMAIR, 1996) is a more CPU hungry alternative for $M_{lat}$.

$$\text{where}\begin{cases} Q = 2r_2\sqrt{\Delta^2 - 2\Delta r_1\cos\phi + r_1^2} \\ r = \sqrt{\dfrac{2Q}{Q + \Delta^2 + r_1^2 + r_2^2 + d^2 - 2\Delta r_1\cos\phi}} \end{cases}$$

SOMA *et al.* (1987) approximate this formula for displacements $\Delta$ smaller than the coil spacing $d$ and $r_2/2$, by the average of its upper and lower bounds:

$$M_{lat} \approx \frac{M_{lat_{max}} + M_{lat_{min}}}{2} \quad \text{where}\begin{cases} M_{lat_{max}} = \mu_0 \dfrac{r_1 r_2}{\sqrt{r_1(r_2 - \Delta)}} G(g_{max}) \\ g_{max} = \sqrt{\dfrac{4r_1(r_2 + \Delta)}{(r_1 + r_2 + \Delta)^2 + d^2}} \\ M_{lat_{min}} = \mu_0 \dfrac{r_1 r_2}{\sqrt{r_1(r_2 + \Delta)}} G(g_{min}) \\ g_{min} = \sqrt{\dfrac{4r_1(r_2 - \Delta)}{(r_1 + r_2 - \Delta)^2 + d^2}} \end{cases}$$

We experienced that this estimation is only accurate for coil sets similar to the examples given in (SOMA *et al.*, 1987). These are coils of about equal size and spaced at about half a radius. The error rises rapidly to 5–10% for other combinations and even more if $\Delta \approx r_2$ at the same time. Some verification with the exact formula is thus strongly recommended.

SOMA *et al.* (1987) also predict the influence of a combination of lateral small angular misalignments ($\alpha < \pm 20°$):

$$M \approx \frac{M_{lat}}{\sqrt{\cos\alpha}}$$

This approximation models the inductance increase from tilting a coil. Tilting the secondary coil brings half of the coil closer to the perimeter of the primary, and since the magnetic field is larger at the coil perimeter, the coupling increase overcompensates some coupling loss due to the larger distance between the primary and the other half of the secondary. However, this formula largely overestimates the effect of coil tilting for coils with different sizes. Since the magnetic flux density is roughly constant over the section of smallest coil – supposing that the magnetic field is generated by the largest coil – the coupling is better estimated by

$$M \approx M_{lat}\cos\alpha$$

#### 5.2.1.2 Coils nearby electric materials

The eddy currents in nearby conductive parts have a negative influence on the link performance. They reduce the magnetic powering field and cause Ohmic dissipation in the media where they flow. Both effects are modelled as a reduction of the self and mutual inductances, and an increase of the coil resistances.[4]

[4] DONALDSON (1992) applies a three-coil model to estimate the influence of a metal weld. His formalism is however not compatible with the optimisation procedure presented in this study.

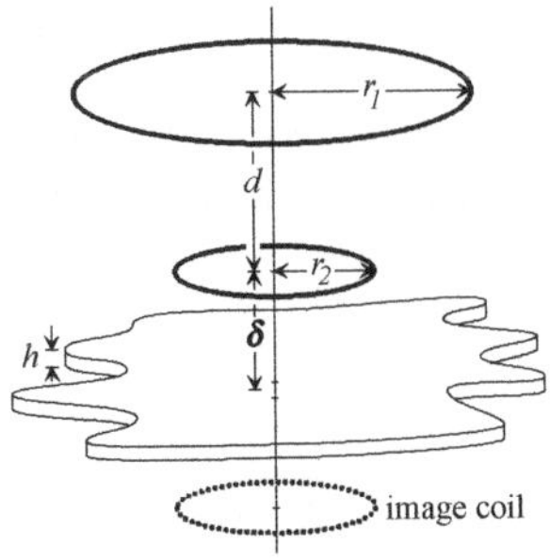


*Figure 5.8.* *Set-up for the calculation of the mutual inductance of two coaxial rings close to a parallel conductive sheet.*

The influence of the conductivity of body tissue on coil parameters has been studied in detail by KWOK and PEPPER (1991), PEPPER *et al.* (1991), SCHUDER *et al.* (1976), HART *et al.* (1988) and BRAMANTI (1982).

It is possible to calculate the eddy current effects for the simple set-up of a one-turn circular coil with radius $r$, held in parallel to a large conductive sheet at a distance $\delta$ (GESELOWITZ *et al.*, 1992). The presence of the sheet changes the inductance $L_S$ and the resistance $R_S$ by $\Delta L_S$ and $\Delta R_S$:

$$\Delta L_S \approx -r\mu_0 G\left(\frac{r}{\sqrt{r^2+\delta^2}}\right)$$

$$\Delta R_S \approx \frac{2\varsigma\delta}{\sqrt{r^2+\delta^2}}\left[-K\left(\frac{r}{\sqrt{r^2+\delta^2}}\right)+\frac{r^2+2\delta^2}{2\delta^2}E\left(\frac{r}{\sqrt{r^2+\delta^2}}\right)\right]$$

$\varsigma$ is the area resistivity of the sheet. It equals the volume resistivity $\rho$ divided by the sheet thickness or the skin depth depending of which of both is the smallest. The above approximations are valid for high enough frequencies where the induction of eddy currents dominates the sheet resistance: $\varsigma < 0.07\ \mu_0\omega\, r$. The inductance change $\Delta L_S$ is independent of the frequency under these conditions. The same holds for the resistance change $\Delta R_S$, apart from the indirect influence via the skin depth.

The mutual inductance also changes by the presence of the sheet:

$$\Delta M = -\mu_0\sqrt{r_1 r_2}\, G(g) \text{where } g = \sqrt{\frac{4r_1 r_2}{(r_1+r_2)^2+(d+2\delta)^2}}$$

Thus, the change in mutual inductance of a pair of ring-shaped coils near a conductive sheet may be evaluated by considering the mutual inductance between one coil and the image of the other coil (Figure 5.8). Note how this inductance disappears when the coil plane coincides with the sheet.

### 5.2.2 ELECTRONIC DESIGN

The electronic optimisation flow is described in Figure 5.9 and Figure 5.10. These diagrams are fully detailed and pretty self-explanatory. The design procedure starts with the computation of $R_{load2}$ from the powering demands and the rectifier circuit. The $R_{load2}$ formulas are found in Chapter 2. A series-resonant secondary is then selected for a small $R_{load2}$ to obtain a current-type output characteristic. The parallel-resonant set-up is for a high $R_{load2}$ and a voltage-type output.

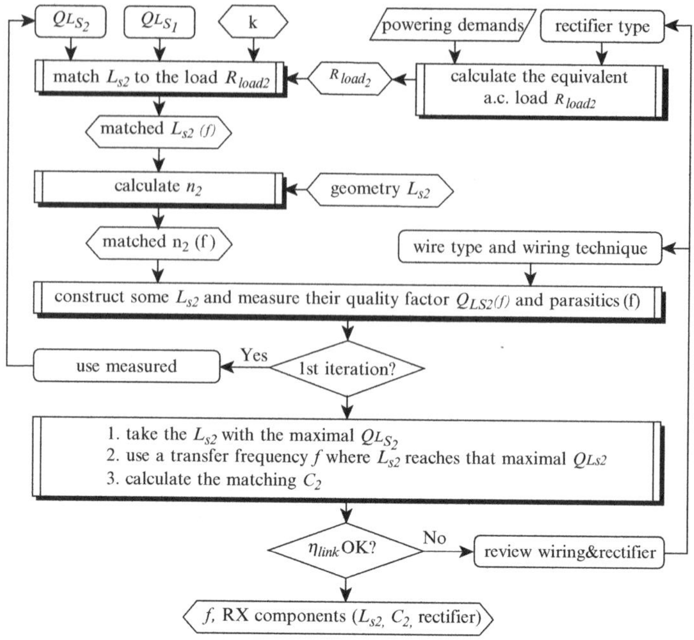


*Figure 5.9. Electronic design of the link secondary and calculation of the frequency.*

The $R_{load2}$ value is then combined with the coupling factor $k$ from the magnetic design and a realistic guess for the coil quality factors to calculate $\alpha_{opt}$ for maximal link efficiency or critical coupling. Typical quality factors are 50–100 for centimetre-size coreless solenoids, 100–200 for their cored counterparts and 10–20 for millimetre-size planar coils printed on a flat substrate. These numbers are for copper or silver conductors. Other metals lead to lower values. The $\alpha_{opt}$ figure allows expressing the secondary coil inductance that optimally matches the load, as a function of the frequency.

Since the coil geometry is already set in the magnetic design, a list of winding values with matching transfer frequencies can be produced. Two or three prototypes of the most realistic combinations are then assembled. The use of Litze wire may be considered for the 100 kHz to 3 MHz range to reduce wire losses. Bank winding helps reducing the interwinding capacitance (Appendix A3). Measurements of the quality factor and the parasitics as a function of the frequency are fed back into the link equations for each prototype. The prototype and transfer frequency that yields the highest $Q_{Ls2}$ are then selected. The secondary circuit may already need revision if the tentative value for the link efficiency is found unsatisfactory.

The circuit design of the link primary starts by selecting a driver topology. It will be a class-E type when the coupling is weak or constant, and a saturating class C or class D for a moderate and variable coupling. Class-E amplifiers may also be chosen in this case for their excellent efficiency, but an automatic regulating of the driver tuning is needed then.

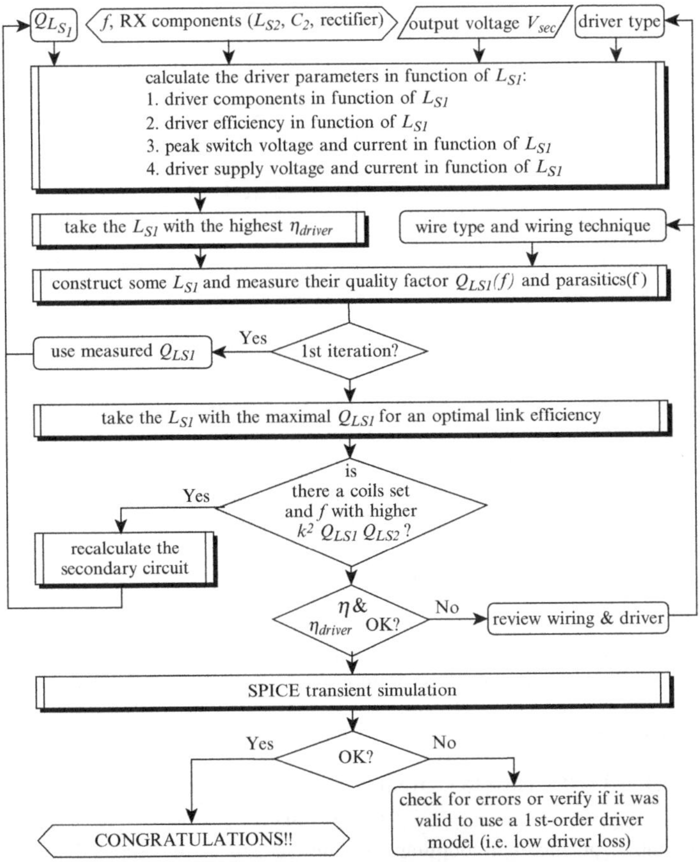


*Electronic design of the primary link circuit.* *Figure 5.10.*

The primary components, the driver efficiency, its supply specifications and the maximal switch signals are now expressed as a function of $L_S$ with the equations from the previous chapter. Note that the driver output $V_{prim}$ is also related to $L_{S1}$ via the link gain.

The primary inductance that yields the best driver efficiency within the switch and supply ratings, is now selected. The first-order formulas of Chapter 4 are reliable at driver efficiencies above 90–95%. Much lower efficiencies suggest that the component range or the driver type is incorrect. The simplified driver model is, however, still useful in this case to indicate how the driver components must be modified to improve the efficiency.

Some prototypes can now be assembled and measured. These coil data should then be imported in the design equations to obtain a better accuracy. The prototype that features the highest $Q_{Ls1}$ at the transfer frequency must be chosen. Check also if there is no other combination between the primary and secondary prototypes that yields a better $Q_{Ls1}Q_{Ls2}$ product. Eventually, the transfer frequency can e shifted somewhat if that further improves the $Q_{Ls}$ product.

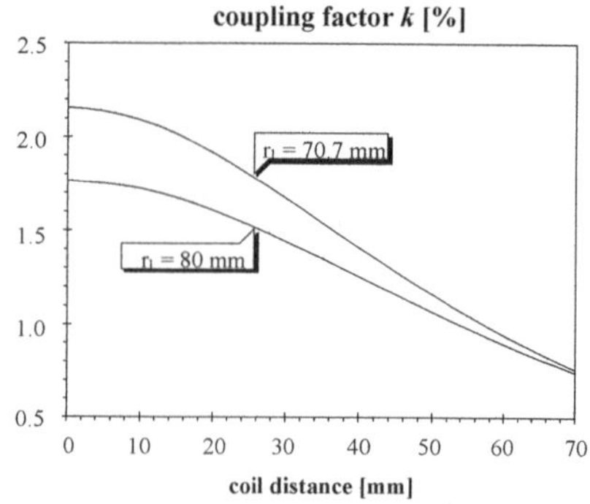


*Figure 5.11.* *Coupling factor as a function of the coil distance d for the primary coil radii 70.7 and 80 mm. The secondary coil measures 10 mm in radius.*

The primary and secondary components are then updated for the new frequency and quality factors. A second coil prototyping iteration is not needed since the component updates are usually minor at this stage.

A final Spice transient analysis completes the design to check for errors and validate the use of the first-order driver model.

## 5.3 DESIGN EXAMPLE

This example design powers a telemetry unit in a non-conductive housing that accepts a 6 mm solenoid with 20 mm diameter. This telemeter consumes a maximum of 4 mA from a 4 V-regulated supply. The maximum target distance is 70 mm. This example was originally designed to power a medical implant.

### 5.3.1 MAGNETIC LINK DESIGN

The magnetic design starts by fitting the telemeter an as large as possible solenoid for maximal coupling. A 6 mm secondary of 20 mm diameter has as an inductance of 27 nH per winding according to Equation A2.3.

Given the 10 mm secondary radius and the maximal coil distance of 70 mm, an optimal primary radius of 70.7 mm is found with Equation 5.7. We made our primary somewhat larger (80 mm) to desensitise the set-up somewhat for distance variations (Figure 5.11). Figure 5.12 shows the impact of lateral misalignment on the coil coupling for various coil distances. This diagram, calculated using the exact Equation 5.8, indicates that our coupling is at least 0.7% with this coil set. This 0.7% figure is now further used in the electrical optimisation. The primary inductance equals 433 nH per winding following Equation 5.2 assuming a wire diameter of 3 mm.

Electromagnetic losses in the surrounding media, radiation and EMC regulation determine the optimal transfer frequency. Body tissue losses set the upper frequency limit to about 10 MHz. The penetration depth of electromagnetic fields in living tissue is about a metre at 1 MHz and drops to 21.6 cm and 11.2 cm at 10 and 40 MHz respectively (Table 1.5). There is no metal or ferrite nearby that may further restrict the transfer frequency. Making the coils much smaller than the wavelength to avoid radiation is clearly a non-issue here since 10 MHz already corresponds to a wavelength of a 30 m.

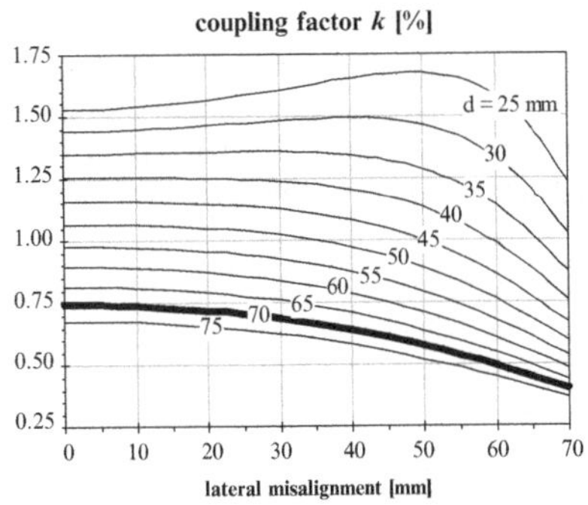


*Figure 5.12.* *Coupling factor as a function of the coil distance d and the lateral misalignment. The bold trace for d = 70 mm indicates the maximal coil distance with this application. The radius of the primary coil is 80 mm, that of the secondary is 10 mm.*

## 5.3.2 ELECTRONIC LINK DESIGN

The link is now further designed to transfer the target peak power with the highest possible system efficiency in worst-case conditions. The peak secondary power consumption of 4 mA at 4 $V_{d.c.}$ corresponds to an equivalent d.c. load $R_{d.c.}$ of 1,125 Ω, taking a 0.5 V regulator dropout into account (1,125 Ω = 4.5 V/4 mA). This translates to an equivalent a.c. load $R_{load2}$ of 625 Ω assuming a half-wave Schottky rectifier with a 0.5 V diode voltage drop $V_{diode}$ (Equation 2.5.Equation 2.5.). Such a high $R_{load2}$ works best with a parallel-tuned secondary. A series-tuned tank would require an unreasonably high inductance.

The coil calculation starts by assuming coil quality factors of e.g. 80. The optimal secondary inductance of 1.5 μH is then found using Equation 3.12. and Equation 3.13. We then assembled and characterised a couple of prototypes with 6–9 coil windings (i.e., 0.97–2.19 μH) to obtain actual quality factors. We measured a maximal quality factor of 72 around 1 MHz. Operation at a maximal 6.21% link efficiency is then found for $\alpha_{opt}$ = 63.58 (Equation 3.13.), $C_2$ = 16.19 nF (Equation 3.11.) and $L_{S_2}$ = 1.56 μH (Equation 3.12.).

A secondary inductance of 1.56 μH assumes 7.6 coil turns. Since uneven winding numbers are rather impractical, a coil of eight windings and thus 1.73 μH is used instead. The capacitor $C_2$ is then calculated from the requirement of phase resonance at the transfer frequency of 1 MHz: $C_2$ = 14.63 nF (Equation 3.1.), $\alpha_{opt}$ = 57.45 (Equation 3.3.).

Two types of coil drivers are now investigated: the saturating-class-C and the class-E driver. The driver design encompasses the transistor selection and the determination of the primary coil and tank capacitors.

### 5.3.2.1 The saturating-class-C driver

The driver transistor is selected based on five parameters: the on-resistance, the maximal voltage and current ratings, the transition speeds and the gate drive requirements. The switch may be a bipolar as well as a MOSFET transistor. MOSFET's may be a bit easier to drive from a low-power CMOS clock. The transfer frequency of 1 MHz and the small duty cycles encountered with low coupling, ask for transition times under 25 ns.

Power MOSFET's are not very suitable for the example application. Although they feature a very low on-resistance, their large gate capacitance requires a special buffer. Moreover, the average gate current can be as large as the actual driver consumption: operation at 1 MHz and a 12 V gate swing already takes 10 mA. They are also too slow with their transition times of 50 ns and more.

The small-signal BS170 is therefore proposed instead. This 800 mW n-channel MOSFET is fast ($t_F = t_R = 4$ ns), has a gate capacitance of only 25 pF and an acceptable 2.5 Ω on-resistance. It handles a maximum of 500 mA (1,200 mA pulsed) and 60 V.

The maximum transistor ratings allow determining the number of primary coil windings and the switch duty cycle. The coil inductance is taken as large as possible for optimal driver efficiency but not so large that the maximal switch voltage is exceeded. Two coil windings and an inductance of 1.73 μH (Equation 5.2) shows to be the limit. The measured quality factor of the best prototype is 115. The increase of the estimated value of 80 to the actual 115 moves $\alpha_{opt}$ to 59.97. The corresponding optimal secondary has 7.8 windings, which means that the eight turns are kept. The switch peak current with this set-up is 619 mA. Though this current is (borderline) acceptable for one BS170, two of them in parallel were used for additional safety.

The components and parameters of the complete driven inductive link are listed in Table 5.2. They are calculated with the high-Q expressions of Chapter 4 for a saturating-class-C driver. This system is then simulated in Spice[5] and prototyped. The results summarised in Table 5.2 show good agreement between the first-order simplified calculations used for the link optimisation, the Spice control simulations and the measured data. The lower efficiency and higher supply current in the measured data are due to the capacitor losses that are not included in the model.

This circuit formed the basis for the autotuned link driver described in Chapter 6.

#### 5.3.2.2 The class-E driver

The same strategy is followed to design a class-E version of the primary coil driver. Since the primary coil is not directly coupled to the driver switch but rather through the series capacitor $C_{1ser}$, the primary coil voltage can increased. The primary inductance increases accordingly to 54.96 μH or 12 windings (Equation 5.2 for a 5 mm wire diameter). The larger primary inductance lifts the quality factor up to 235, thereby improving the link efficiency. The optimal $\alpha_{opt}$ still corresponds to an 8-turn secondary. As the primary coil prototype resonates at 2.583 MHz, the driver formulae for a semi-resonant primary are used (cf. Section 4.9). The switch peak current now equals 223 mA and is thus easily handled by a single BS170. The switch dissipation is almost three times lower than for its saturating-class-C equivalent.

[5] PSpice transient simulation using a BS170 transistor model from the SIEMENS SIPMOS Spice library 1.0 and a BAT64 Schottky rectifier model from the SIEMENS Small Signal Semiconductors Spice library 3.2. The simulations were allowed to run long enough to reach the steady state solution and make sure that the start-up transients died out.

*TABLE 5.2.* *THE EXAMPLE LINK DRIVEN BY A SATURATING-CLASS-C AMPLIFIER. THE FIRST ORDER CALCULATIONS ARE FOR $V_{Ssat} = 0$ V, $R_{ON} = 1.25\ \Omega$, AND $t_F = t_R = 4$ ns, WHICH ARE THE TYPICAL VALUES FOR TWO BS170 MOSFET'S IN PARALLEL. THE SIMULATIONS ARE FOR k = 0.7% AND THE MEASURED QUALITY FACTORS 72 AND 115.*

<table>
<tr><th></th><th>1st order calculations</th><th>Spice simul. with 2 BS170s</th><th>Measured circuit data</th><th>Units</th></tr>
<tr><td>$R_{d.c.}$</td><td>Not used</td><td>1,125</td><td>1,125</td><td>Ω</td></tr>
<tr><td>$R_{load_2}$</td><td>625</td><td>Not used</td><td>Not used</td><td>Ω</td></tr>
<tr><td>$f$</td><td colspan="2">1</td><td>1.0005</td><td>MHz</td></tr>
<tr><td>$L_{S_2}$</td><td colspan="2">1.73</td><td>1.78</td><td>µH</td></tr>
<tr><td>$Q_{L_{S2}}$</td><td colspan="2">72</td><td>72</td><td>–</td></tr>
<tr><td>$C_2$</td><td colspan="2">14.63</td><td>14.46</td><td>nF</td></tr>
<tr><td>$D$</td><td colspan="2">10</td><td>10</td><td>%</td></tr>
<tr><td>$L_{S_1}$</td><td colspan="2">1.73</td><td>1.76</td><td>µH</td></tr>
<tr><td>$Q_{L_{S1}}$</td><td colspan="2">115</td><td>115</td><td>–</td></tr>
<tr><td>$\eta_{link}$</td><td colspan="2">8.49</td><td>8.49[a]</td><td>%</td></tr>
<tr><td>$C_1$</td><td>14.55</td><td>14.38 + 0.06[b]</td><td>14.71</td><td>nF</td></tr>
<tr><td>$\hat{V}_{prim}$</td><td>22.4</td><td>22.4</td><td>21.9[c]</td><td>V</td></tr>
<tr><td>$THD\ V_{prim}$</td><td>0</td><td>2.0</td><td>1.16[d]</td><td>%</td></tr>
<tr><td>$\hat{I}_{prim}$</td><td>2.06</td><td>2.07</td><td>1.98[a]</td><td>A</td></tr>
<tr><td>$THD\ I_{prim}$</td><td>0</td><td>0.8</td><td>Not measured</td><td>%</td></tr>
<tr><td>$v_{S_{max}}$</td><td>43.9</td><td>44.0</td><td>43.8</td><td>V</td></tr>
<tr><td>$i_{S_{max}}$</td><td>619</td><td>675</td><td>Not measured</td><td>mA</td></tr>
<tr><td>$I^*_{cc}$</td><td>11.75</td><td>11.2</td><td>15.3</td><td>mA</td></tr>
<tr><td>$V_{cc}$</td><td>21.4</td><td>21.4</td><td>22.0</td><td>V</td></tr>
<tr><td>$\eta_{driver}$</td><td>93.6</td><td>95.3</td><td>Not measured</td><td>%</td></tr>
<tr><td>$\hat{V}_{sec}$</td><td>5</td><td>4.98</td><td>5.0</td><td>V</td></tr>
<tr><td>$\eta_{rectif}$</td><td>90</td><td>90.3</td><td>Not measured</td><td>%</td></tr>
<tr><td>$V_{d.c.}$</td><td>4.5</td><td>4.49</td><td>4.50</td><td>V</td></tr>
<tr><td>$P_{R_{d.c.}}$</td><td>18</td><td>17.9</td><td>18.0[a]</td><td>mW</td></tr>
<tr><td>$\eta$</td><td>7.15</td><td>7.28</td><td>5.35[a]</td><td>%</td></tr>
</table>

[a]*These values are not measured but computed from the other data.*
[b]*$C_1$ includes the 60 pF parasitic source-drain capacitors of two BS170 transistors.*
[c]*$\hat{V}_{prim}$ is estimated as $v_{S_{max}}/2$.*
[d]*This value is computed from spectrum analyser readings of the first nine harmonics.*

The driver parameters and components are listed in Table 5.3 together with the simulation results and prototype measurements. The first-order simplified calculations are again in good agreement with the Spice simulations and measurements.

*TABLE 5.3.* *THE CLASS-E DRIVEN INDUCTIVE LINK FOR THE IMPACT IMPLANTS. THE FIRST ORDER CALCULATIONS ARE FOR $V_{Ssat}$ = 0 V, $R_{ON}$ = 2.5 Ω, $t_F = t_R$ = 4 ns, $L_S$ = 10 nH, WHICH ARE THE TYPICAL VALUES FOR THE BS170 MOSFET TRANSISTOR.*

<table>
<tr><th></th><th>1st order calculations</th><th>Spice simul. with BS170</th><th>Measured circuit data</th><th>Units</th></tr>
<tr><td>$R_{d.c.}$</td><td>Not used</td><td>1,125</td><td>1,125</td><td>Ω</td></tr>
<tr><td>$R_{load_2}$</td><td>625</td><td>625</td><td>Not used</td><td>Ω</td></tr>
<tr><td>$f$</td><td colspan="2">1</td><td>1.0007</td><td>MHz</td></tr>
<tr><td>$L_{S_2}$</td><td colspan="2">1.73</td><td>1.78</td><td>μH</td></tr>
<tr><td>$Q_{L_{S2}}$</td><td colspan="2">72</td><td>72</td><td>–</td></tr>
<tr><td>$C_2$</td><td colspan="2">14.63</td><td>14.46</td><td>nF</td></tr>
<tr><td>$D$</td><td colspan="2">12.5</td><td>12.5</td><td>%</td></tr>
<tr><td>$L_{S_1}$</td><td colspan="2">54.94</td><td>54.17</td><td>μH</td></tr>
<tr><td>$Q_{L_{S1}}$</td><td colspan="2">235</td><td>235</td><td>–</td></tr>
<tr><td>$f_0$</td><td colspan="2">2.583</td><td>2.583</td><td>MHz</td></tr>
<tr><td>$\eta_{link}$</td><td colspan="2">14.96[a]</td><td>14.96[a]</td><td>%</td></tr>
<tr><td>$C_{1par}$</td><td>4.72</td><td>4.50 + 0.03[b]</td><td>4.50</td><td>nF</td></tr>
<tr><td>$C_{1ser}$</td><td>426</td><td>427</td><td>441.7</td><td>pF</td></tr>
<tr><td>$min.\ L_{choke}$</td><td>0.995</td><td>1</td><td>1.03</td><td>mH</td></tr>
<tr><td>$\hat{V}_{prim}$</td><td>126</td><td>126</td><td>125[c]</td><td>V</td></tr>
<tr><td>$THD\ V_{prim}$</td><td>0</td><td>0.53</td><td>not measured</td><td>%</td></tr>
<tr><td>$\hat{I}_{prim}$</td><td>365</td><td>361</td><td>367[a]</td><td>mA</td></tr>
<tr><td>$THD\ I_{prim}$</td><td>0</td><td>0.28</td><td>Not measured</td><td>%</td></tr>
<tr><td>$v_{S_{max}}$</td><td>19.4</td><td>20.1</td><td>19.2</td><td>V</td></tr>
<tr><td>$i_{S_{max}}$</td><td>223</td><td>205</td><td>Not measured</td><td>mA</td></tr>
<tr><td>$I_{cc}$</td><td>14.62</td><td>14.14</td><td>16.0</td><td>mA</td></tr>
<tr><td>$V_{cc}^*$</td><td>9.56</td><td>9.60</td><td>9.6</td><td>V</td></tr>
<tr><td>$\eta_{driver}$</td><td>95.6</td><td>96.5</td><td>Not measured</td><td>%</td></tr>
<tr><td>$\hat{V}_{sec}$</td><td>5</td><td>5.0</td><td>5</td><td>V</td></tr>
<tr><td>$\eta_{rectif}$</td><td>90</td><td>89.9</td><td>Not measured</td><td>%</td></tr>
<tr><td>$V_{d.c.}$</td><td>4.5</td><td>4.5</td><td>4.5</td><td>V</td></tr>
<tr><td>$P_{R_{d.c.}}$</td><td>18</td><td>18[a]</td><td>18[a]</td><td>mW</td></tr>
<tr><td>$\eta$</td><td>12.9</td><td>13[a]</td><td>11.7[a]</td><td>%</td></tr>
</table>

[a] *These values are not measured but computed from the other data.*

[b] *$C_{1par}$ includes the 30 pF parasitic source-drain capacitors of a BS170 transistor.*

[c] *$\hat{V}_{prim}$ is estimated as $v_{S_{max}}/2$.*

## 5.4 CONCLUSIONS

This chapter described the procedure to optimise weakly coupled inductive links. The disordered tangle of relations between the link and driver parameters was unravelled to a quadripartite design flow:

- The magnetic link optimisation
- Optimisation of the link secondary
- Optimisation of the link primary, including the driver, and
- Verification of the link circuit by SPICE simulations

The interaction between the design kernels was centralised around two parameters. The coupling factor $k$ was taken as the gateway between the magnetic and the electronic design. The primary coil inductance $L_{s1}$ separated the secondary from the primary optimisation. The interactions between the other link parameters were organised such that they act as much as possible inside the design kernels.

The analysis also demonstrated the supremacy of class-E drivers in case of weak coupling. Some caution should be exercised when using them for well-coupled links where they require a regulation mechanism to keep their load circuit optimally tuned.

This fifth chapter concludes our generalised link theory. Automatic tuning of link parameters is now discussed in the next and final chapter.

## 5.5 REFERENCES

**Abramowitz M.** and I.A. Stegun (Eds.), *Handbook of mathematical functions,* Dover, New York, 1965.

**Bramanti M.**, "Circular planar coil as electromagnetic movement detector of inside organs in biosystems: the fundamental design relationships," *Med. Biol. Eng.*, vol. 20, pp. 202–206, 1982.

***Donaldson N. de N.**, "Effect of the metallic seal of a hermetic enclosure on the induction of power to an implant," *Med. Biol. Eng. Comput.*, vol. 30, pp. 63–68, 1992.

***Flack F.C.**, E.D. James and D.M. Schlapp, "Mutual inductance of air-cored coils: effect on design of radio-frequency coupled implants," *Med. Biol. Eng. Comput.*, vol. 9, pp. 79–85, 1971.

**Forster I.C.**, "Preliminary development of a radiotelemetry system for biological applications," *Med. Biol. Eng. Comput.*, vol. 24, pp. 281–291, 1986.

***Geselowitz D.B.**, Q.T.N. Hoang and R.P. Gaumond, "The effects of metals on an transcutaneous energy transmission system," *IEEE Trans. Biomed. Eng.*, vol. BME-39, pp. 928–934, 1992.

***Grover F.W.**, *Inductance calculations,* Van Nostrand, New York, 1946.

**Harnwell G.P.**, *Principles of electricity and electromagnetism,* pp. 329–331, McGraw-Hill, New York, 1949.

**Hart L.W.**, W.H. Ko, J.H. Meyer Jr., D.P. Vasholz and R.I. Joseph, "A noninvasive electromagnetic conductivity sensor for biomedical applications," *IEEE Trans. Biomed. Eng.*, vol. BME-35, pp. 1011–1022, 1988.

**Heck C.**, "Magnetic parameters of materials," ch. 3 in: *Magnetic Materials and their Applications*, Trans. from German by S.S. Hill, Butterworth Co. (Publishers) Ltd., London, ISBN 0-408-70399-7, pp. 16–100, 1974.

***Hoang Q.T.N.**, B.F. Hete and R.P. Gaumond, "Comments on 'Radio-frequency coils in implantable devices: misalignment analysis and design procedure'," *IEEE Trans. Biomed. Eng.*, vol. BME-40, pp. 715, 1993.

**Hochmair E.S.**, "System optimization for improved accuracy in transcutaneous signal and power transmission," *IEEE Trans. Biomed. Eng.*, vol. BME-31, pp. 177–186, 1984.

**Jacob R.**, "Aerial design in inductive loop telemetry," *IEEE Trans. Biomed. Eng.*, vol. BME-23, pp. 118–123, 1976.

***Ko W.H.**, S.P. Liang and C.D.F. Fung, "Design of radio-frequency powered coils for implant instruments," *Med. Biol. Eng. Comput.*, vol. 15, pp. 634–640, 1977.

**Kwok M.C.** and M.G. Pepper, "Noninvasive detection of ventricular wall by electromagnetical coupling. Part 1 theory: the changes in the reflected impedance of a coil over a semi-infinite medium with properties ranging from lossy dielectric to a conductor," *Med. Biol. Eng. Comput.*, vol. 29, pp. 136–140, 1991.

**Matsuki H.**, M. Shiiki, K. Nadehara, T. Yamamoto and K. Murakami, "Transcutaneous energy transmission using amorphous magnetic fiber for an artificial heart system," *Int. J. Appl. Electromagn. Mater. 1*, pp. 201–205, 1990.

**Mohammed O.A.** and W.P. Batina, "Computer aided design tool for implantable telemetry systems," *Proc. IEEE 7th Annu. Conf. EMBS*, pp. 1155–1159, 1985.

**Mohammed O.A.** and W.P. Batina, "A method for pacemaker's coil design using a numerical model," *Proc. IEEE 8th Annu. Conf. EMBS*, pp. 817–820, 1986.

**Mohammed O.A.**, "Evaluation of implantable telemetry system performance using CCETS software," *Proc. IEEE 10th Annu. Conf. EMBS*, New Orleans, USA, pp. 1239–1240, 1988.

**Pepper M.G.**, D.J.E. Taylor and M.C. Kwok, "Noninvasive detection of ventricular wall by electromagnetical coupling. Part 2 Experimental: cardiokyomography," *Med. Biol. Eng. Comput.*, vol. 29, pp. 141–148, 1991.

**Schuder J.C.**, J.H. Gold, H. Stoeckle and J.A. Holland, "The relationship between the electric field in a semi-infinite conductive region and the power input to a circular coil on or above the surface," *Med. Biol. Eng.*, vol. 14, pp. 227–233, 1976.

***Soma M.**, C.G. Douglas and R.L. White, "Radio-frequency coils in implantable devices: misalignment analysis and design procedure," *IEEE Trans. Biomed. Eng.*, vol. BME-34, pp. 276–282, 1987.

***Terman F.E.**, *Radio Engineers' Handbook*, McGraw-Hill, New York/London, pp. 47–109, 1943.

**Troyk P.R.** and M.A.K. Schwan, "Modelling of weakly-coupled inductive links," in: *Biotelemetry XIII (Proc. 13th Int. Symposium on Biotelemetry)*, Eds. C. Cristalli, C.J. Amlaner Jr. and M.R. Neuman, Williamsburg, VI, pp. 63–68, 1996.

**Van Schuylenbergh K.** and R. Puers, 'A computer assisted methodology for inductive link design for implant applications," in: *Biotelemetry XII (Proc. 12th Int. Symposium on Biotelemetry)*, Ancona, Eds. P. Mancini and S. Fioretti, Editrice Universitaria Litografia Felici, Pisa, Italy, pp. 392–400, 1993.

***Zierhofer C.M.** and E.S. Hochmair, "Geometric approach for coupling enhancement of magnetically coupled coils," *IEEE Trans. Biomed. Eng.*, vol. BME-43, pp. 708–714, 1996.

# 6 Automatic link tuning

The methods described so far, optimise the link performance for one specific coupling value and link output voltage. Many applications have a varying coil coupling or power drain. Their links must be optimised to provide adequate power in worst-case conditions of minimal coupling and maximal secondary consumption. Actual conditions are, by definition of the notion worst case, more favourable leading to a powering field that is too strong and not efficiently used. An example of a self-tuning set-up is described now that automatically detects inefficient field use and reduces the driver power accordingly (VAN SCHUYLENBERGH and PUERS, 1996). This reduces the power considerably and allows for smaller batteries in portable units. The energetic impact of driving the link unnecessary hard is easily understood considering that the link output voltage doubles when the coupling doubles for some reason. The latter doesn't take much because of the non-linear relation between coil spacing and coupling. Cutting the driver output voltage by a factor of two brings the link output back to nominal, resulting in a power saving of a factor of roughly four.

The autotuning driver described in this chapter was originally developed for a clinical telemetry implant for operation by non-technical users. User-friendliness was therefore found important and optimal link performance had to be guaranteed without technical assistance or equipment. As the inductive link relied on high-Q resonant tank circuits, the coil driver had to be precisely tuned to the resonance frequency of the individual patient's circuit. These frequencies differed from patient to patient due to component tolerances. Finally, the driver amp had to accommodate variable coil inductances for two reasons. Holding the link primary close to a metal object or mechanically deforming it changed its inductance. We found a flexible primary coil more comfortable to the patient. It could easily be taped to the patient's body and didn't obstruct body movements.



K. Van Schuylenbergh and R. Puers, *Inductive Powering: Basic Theory and Application to Biomedical Systems*, Analog Circuits and Signal Processing, 169–177. 

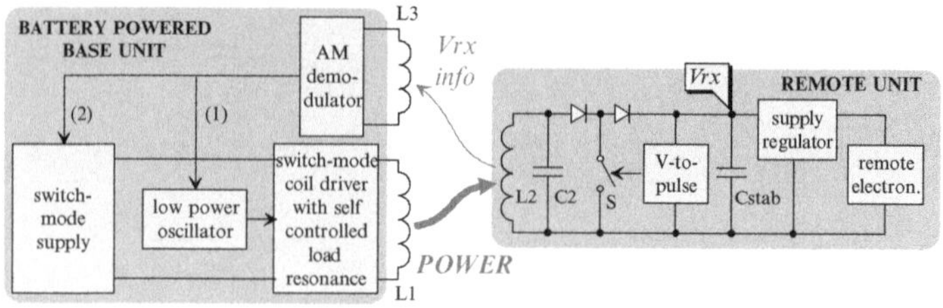


Figure 6.1. *Block diagram of the self-tuning inductive powering system. The two control loops are based on the feedback of the amplitude of the signal $V_{rx}$ received at the remote site.*

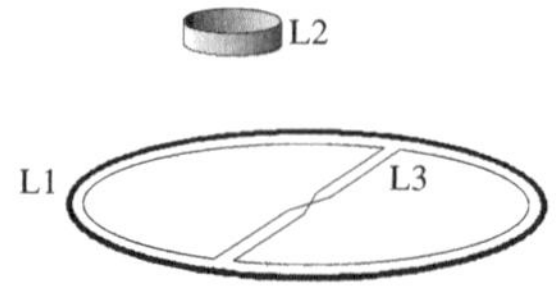


Figure 6.2. *The coil configuration with the dedicated figure-of-eight shape of $L_3$ to minimise its coupling to $L_1$.*

The self-tuning set-up presented here, is designed to deliver a regulated output of 4 $V_{d.c.}$ at 4 mA peak for coupling factors between 0.7% and 2%. It incorporates multiple automatic tuning loops to handle coil movements, inductance variations and component tolerances, and to continuously adjust the power output. A visual and an acoustic indicator give the operator feedback on the quality of the established link. This allows positioning the base coil in search for the best coupling; a useful feature when the exact location of the remote unit is unknown.

## 6.1 AUTOMATIC SEARCH OF THE TRANSFER FREQUENCY AND AUTO-REGULATION OF THE DRIVER POWER

The system, represented in Figure 6.1, enables automatic tuning of the transfer frequency and of the driver output. Both control loops essentially rely on the feedback of the d.c. input $V_{rx}$ of the remote supply regulator. This information is fed back via the inductive link with the backscattering technique described in Figure 1.15. The remote module transfers its data back to the base unit, by modulating its power consumption (DONALDSON, 1986; KIMURA *et al.*, 1986; NEUKOMM *et al.*, 1988; NEUKOMM and KÜNDIG, 1990; TANG *et al.*, 1995). This influences the field strength around the secondary coil and is hence picked up the primary coil because of the magnetic coupling.

In order to prepare the $V_{rx}$ information for transmission, a pulse frequency encoder converts the $V_{rx}$ voltage into a pulse signal with a low duty cycle. At each pulse, a switch $S$ briefly shorts $L_2$, thus modulates the magnetic powering field, and induces a signal across the powering coil. However, this signal is far too small for simple and reliable recovery, because of the weak coupling. We therefore installed a third coil $L_3$ to sense the field modulation.

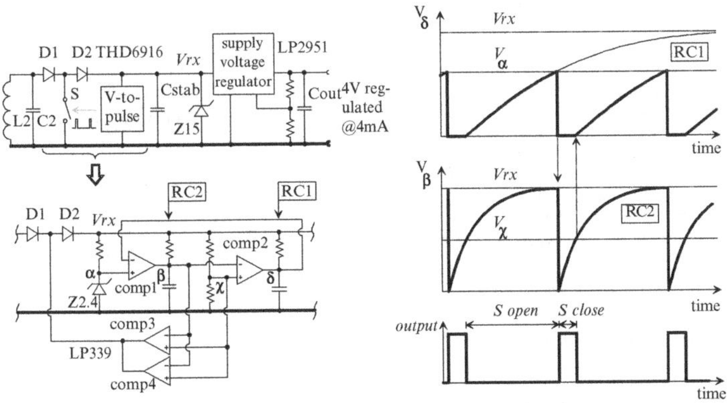


Figure 6.3. *Electronic diagram of the power receiver. The time behaviour of the signals in the voltage-to-pulse encoder is shown on the right side.*

Figure 6.4. *Custom thick-film hybrid implementation of the power-receiver of Figure 6.3. The hybrid has a diameter of 18 mm.*

The coil triplet $L_1$, $L_2$ and $L_3$ is implemented as shown on Figure 6.2. This configuration, with the figure-of-eight shaped $L_3$, minimises the coupling between the sensing coil $L_3$ and the powering coil $L_1$, to increase the modulation depth of the received data signal (DONALDSON, 1979, 1983). The retrieved signal is then amplitude demodulated and pulse-position decoded to provide the $V_{rx}$ information to the other driver circuits.

The link optimisation takes place in two stages. First, an initialisation sweeps the driver frequency until a maximal voltage $V_{rx}$ is found (arrow *(1)* in Figure 6.1). From then on, during normal operation, the $V_{rx}$ information on the coil $L_3$ controls the driver power to maintain a pre-set $V_{rx}$ value at the regulator input (arrow *(2)* in Figure 6.1).

Figure 6.3 zooms in on the remote part of the link circuitry. The a.c. powering signal is picked up by the tank circuit $L_2$-$C_2$, rectified by the Schottky diodes $D_1$ and $D_2$ and stabilised by the capacitor $C_{stab}$. An LP2951 adjustable micro-power regulator keeps the output voltage constant regardless coupling or load variations. The Zener diode $Z_{15}$ protects the supply regulator from an input overvoltage in an unexpected case of exceptionally good coil coupling.

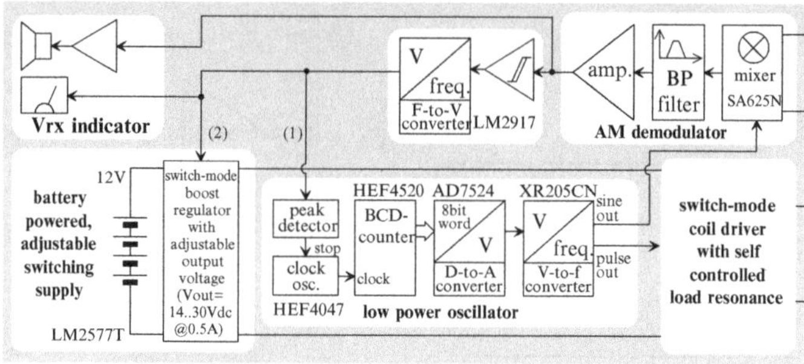


*Figure 6.5.* *Detailed block diagram of the coil driver unit.*

The core of the $V_{rx}$-feedback circuit is the LP339 micro-power quad comparator. A voltage-controlled pulse oscillator built around the comparators $comp_1$ and $comp_2$, converts the $V_{rx}$ voltage into a low duty cycle pulse signal. The voltage $V_{rx}$ is charging the branch $RC_1$ and when the Zener voltage $V_\alpha$ is reached, the capacitor is discharged and a pulse is produced.

The pulse duration is set by comparing the charging of a second RC-branch $RC_2$ with the voltage $V_\chi$ that is half $V_{rx}$. The switch $S$ that shorts the secondary coil $L_2$ consists of the open collector outputs of the two remaining comparators *comp3* and *comp4* in parallel.

This circuit is implemented on a thick-film hybrid (Figure 6.4).

Figure 6.5 reveals the coil driver in further detail. The switch-mode coil driver is triggered by a low-power pulse oscillator. The oscillator frequency is swept during the initialisation stage and then latched. The pulse oscillator (XR205CN) is therefore controlled with a voltage derived from a counter (HEF4520) and a digital-to-analogue (DAC) converter (AD7524). A free-running counter produces a sweeping voltage at the DAC output and hence a sweeping oscillator frequency. Stopping the counter locks the oscillator on the frequency corresponding to the value stored in the counter latch.

In the normal operation mode, the driver output power is continuously adjusted to maintain the input of the remote supply regulator just high enough for a constant and well-regulated output. The coil driver is therefore fed by a high-efficiency voltage-controlled boost regulator (LM2577-Adj.). Its control voltage is derived from the $V_{rx}$ pulses received by coil $L_3$. The amplitude demodulation of the $V_{rx}$ signal on coil $L_3$, is performed by a mixer (SA625N), amplifiers and a band-pass filter. A Schmitt trigger cleans the pulses for the frequency-to-voltage converter (LM2917) that produces a voltage representing the $V_{rx}$ value inside the remote power receiver. This voltage finally controls the supply of the coil driver. A slow control response is deliberately chosen to stabilise the servo loop from the coil driver to the implant and back.

The peak detector is only active during the frequency calibration, where it locks the sweeping pulse generator when the maximal $V_{rx}$ is reached. A panel meter indicates also the output of the frequency-to-voltage converter to give the operator an idea of the quality of the established link.

*A SUMMARY OF ERROR SOURCES IN THE POWERING SET-UP AND HOW THEY ARE CANCELLED.* *TABLE 6.1.*

| Supply error sources | Freq. range | Filtered by: |
|---|---|---|
| Slow coupling variations | d.c. .. 1 Hz | Base unit |
| Rapid coupling variations (e.g. patient movements) | 1–100 Hz | Remote supply regulator |
| Slow to moderate load variations | 10 Hz–5 kHz | $C_{out}$ + supply regulator |
| Rapid load variations + r.f. supply noise | >5 kHz | Decoupling C's in remote circuit |
| Regulator input ripple due to $V_{rx}$ feedback | 1.5–5 kHz | $C_{stab}$ |
| Regulator input ripple due to imperfect rectification | 1 MHz | $C_{stab}$ |
| Regulator output ripple injected by powering field | 1 MHz | $C_{out}$ |

The $V_{rx}$ pulses are in the audible frequency range of 2–5 kHz. An audio amplifier and a loudspeaker are therefore added to give the operator also some acoustic feedback on the link quality while he or she is positioning the primary coil during the initialisation.

Table 6.1. summarises the different error sources in the powering set-up and how they are cancelled out.

## 6.2 SWITCH-MODE COIL DRIVER WITH LOAD-RESONANCE CONTROL

Automatic driver tuning is not very common in the area of inductive powering and publications are scarce. MILLER *et al.* (1993) included a feedback loop in a saturating-class-C driver that kept the switch current and voltage in phase by adjusting the driver frequency. It was designed for high-coupling applications where pole splitting caused driver detuning. TROYK and SCHWAN (1992) developed a class-E driver to power miniature implants. They automatically regulated the transfer frequency and the duty cycle to maintain driver tuning regardless the inductance variations caused by nearby metal parts.

These autotuning techniques basically protect the driver switch when external factors, like pole splitting or inductance variations, imperil its integrity. However, they are not applicable here as they alter the frequency to maintain the driver tuning. The fact that the frequency variations compromise link efficiency and gain is not considered in their work. This is not acceptable for the set-up discussed here and a different regulation concept was therefore conceived. It was built around a saturating-class-C amplifier because there is only one parameter to tune for low-loss operation.

Two aspects in our application call for a third servo loop in the driver circuit to automatically and continuously adjust its tuning: first, the link frequency has to be adjustable to that of the individual patient, and second, the driver needs to accommodate varying inductances since the primary is implemented as a deformable ring in order not to hinder the patient's mobility.

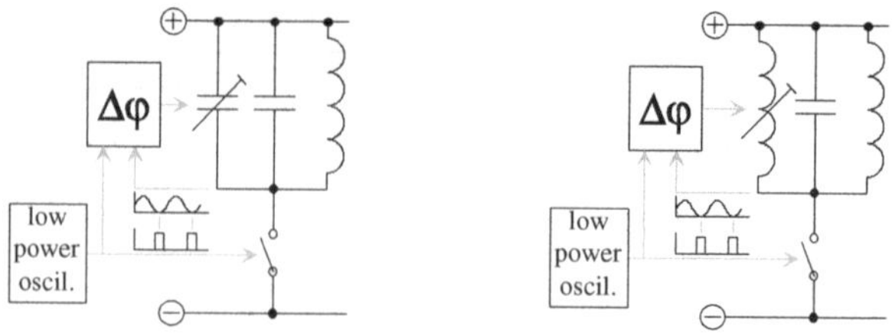


*Figure 6.6.* *Alternatives for the auto-tuning of the load resonance of a switch-mode coil driver.*

*Figure 6.7.* *A transductor.*
*The transductor for the circuit in Figure 6.8 consists of two EF25/7.5 E-cores made of N30 MgZn ferrite. The d.c. coil has 1,500 windings and the a.c. coil two times three windings. This results in an inductance range that starts at 45 μH for a zero control current in the d.c. coil and that saturates at 3.2 μH for 7 mA. Parasitic resonances and core losses make that only the 20–3.2 μH range is actually useful around 1 MHz.*

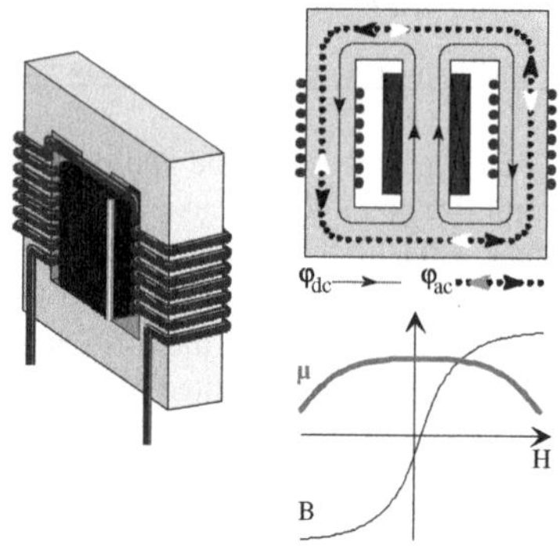


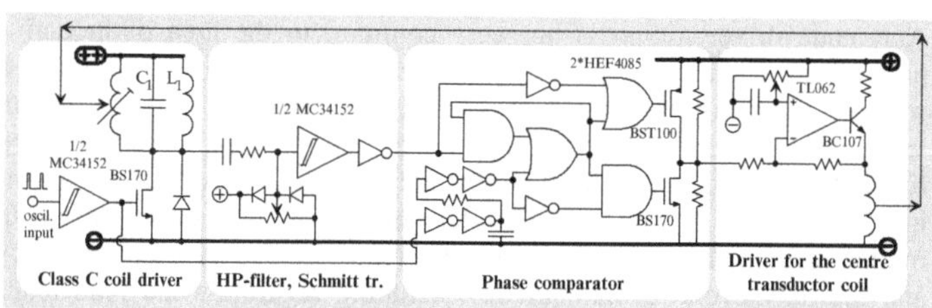


*Figure 6.8.* *Electronic diagram of the switch-mode coil driver with automatic controlled load resonance.*

There are two candidate components for driver tuning: the primary tank capacitance and the tank inductance (Figure 6.6).

Capacitance tuning with a varicap diode or a Miller capacitor in a variable gain stage was ruled out because of the high voltages and currents in the tank $L_1$-$C_1$. These easily range up to 50$V_{pp}$ and several Ampères!

That left us with inductance tuning by means of a transductor as the only viable solution (Figure 6.7). A transductor is a current-controlled a.c. inductance assembled on a double ferrite E-core. A constant current flowing in the centre coil determines the level of core saturation and hence the permeability of the core material. Increasing the centre coil current saturates the core and lowers the a.c. inductance. The a.c. coil is wound in two equal parts on the outer legs to avoid magnetic coupling to the d.c. coil.

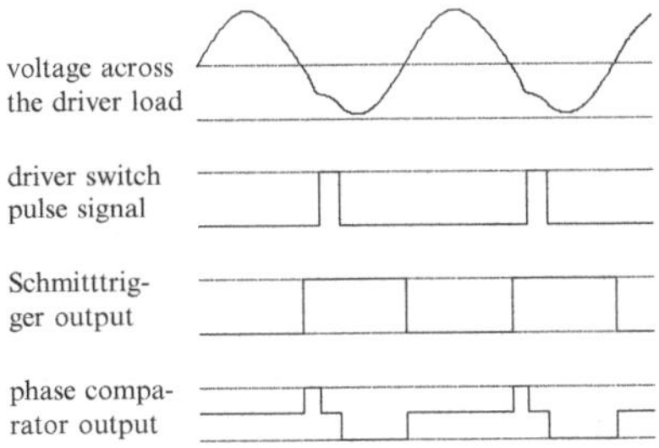


Figure 6.9. *Signal timing in the circuit of Figure 6.8, for an open-loop situation with the coil driver lagging behind the load response.*

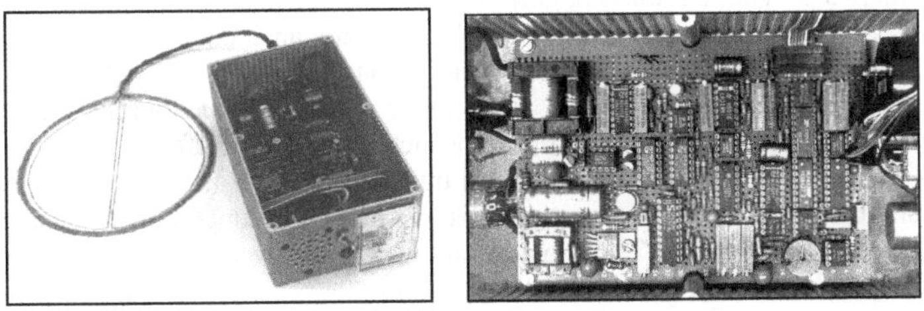

Figure 6.10. *Breadboard prototype of the primary coil driver unit. Note the transductor in the upper left corner of the circuit board.*

Detuning of the driver load is detected as a phase difference between the signal across the switch and the pulse signal that drives the switch. A phase feedback loop adjusts the transductor a.c. inductance until both signals are in phase. Figure 6.8 shows the electronic diagram of the coil driver with auto-controlled load resonance. The voltage across the load $L_1$-$C_1$ is converted into a block signal by the high-pass filter and Schmitt trigger. Its phase is then compared with that of the switch input pulse signal. A small delay is added to this pulse signal to compensate for the delay of the high-pass filter and the Schmitt trigger. The output of the phase comparator is a bi-phase block signal with a positive d.c. component when the coil driver is running ahead of the load response and a negative d.c. component if the driver is lagging behind (Figure 6.9). A driver amplifier feeds the phase comparator output into the centre coil of the transductor. The transductor centre coil current and the saturation level of the transductor core follow the d.c. component only, because of the high coil inductance.

## 6.3 PROTOTYPE RESULTS AND DISCUSSION

We built a 1 MHz prototype of this self-tuning system to power a wireless strain monitor on an experimental nail-plate femoral implant. It is a battery-powered unit with a rechargeable 12 V, 2 Ah cell that lasts 3–8 h depending on the driver output power (Figure 6.10). The primary coil is implemented as a flexible ring with a diameter of 160 mm that can be easily taped onto the patient's leg. The secondary coil has a diameter of 20 mm and length of 6 mm. Litze wire minimises the coil losses.

TABLE 6.2. PERFORMANCE OF THE 1 MHZ IMPLEMENTATION OF THE INDUCTIVE POWERING SYSTEM. THE MEASUREMENTS WERE CARRIED OUT WHILE POWERING A TELEMETER TUNED AT 1,059 kHz.

| Parameter | Minimal driver outp. | Maximal driver outp. | Unit |
|---|---|---|---|
| Transfer frequency | 1,059 | 1,059 | kHz |
| System current drain from 12 V battery | 147 | 550 | mA |
| Coil driver supply voltage (after LM2577) | 13.9 | 30.5 | V |
| Maximal distance $L_1$-$L_2$ for a regul. 4 V link outp. | 32 | 70 | mm |
| Maximal distance $L_1$-$L_3$ for a reliable $V_{rx}$ feedback | 70 | 70 | mm |

The primary coil driver is based on the saturating-class-C example in Chapter 5. That circuit is slightly modified to include a transductor. We used two times three transductor windings for an inductance range of 3.2–20 μH. The inductance of the link primary in parallel with the transductor is then 1.22–1.8 μH. A wider range may be obtained with a lower transductor inductance, but the transductor then bypasses more of the primary coil current. There is thus a trade-off between the control range and the amount of tank current that is lost in the transductor. The capacitor C1 is increased to 28 nF to accommodate the lower inductance of the transductor in parallel with the link primary.

The link output is a regulated 4 Vd.c. at a maximal current drain of 4 mA. The sweep range goes from 850 to 1,150 kHz. The driver servo circuit excluding the actual switch-mode coil driver, consumes 50–54 mA. Table 6.2. further summarises the system performance. The measurements are carried out while powering telemeter tuned at 1,059 kHz. Note that the coupling distances are confined by the limited capabilities of the saturating-class-C circuit. The 70 mm distance already hits the long-term voltage and dissipation limits of the BS170 switch. A class-E topology could therefore perform better in this application. Running it in a saturating-class-C mode is possibly a good compromise since it requires adjusting only one parameter for low-loss operation rather than the two that keep a class E amp in tune.

Although the transductor worked beautifully in the servo circuit, it has its drawbacks. It doesn't allow that much of tuning range and it bypasses part of the link current. It also takes a fair amount of control current (5–10 mA) and its quality factor was limited to 25–35 with the available ferrites.

The self-oscillating coil drivers, presented in Section 2.3.4, may therefore offer an interesting alternative at high link frequencies where the driver capacitors are within the range of adjustable air capacitors. Air capacitors have excellent quality factors, wide tuning range, and high current and voltage ratings. Tuneable tank capacitors also don't bypass the primary link current. Since we don't need fast servo control, a motor-driven adjustable capacitor in a self-oscillating driver may be an elegant solution to obtain a variable link frequency or compensate primary deformations. The motor could be a piezo micromotor[1] that has micrometer accuracy and only takes moderate power during actuation and no power at all to maintain position.

[1] Ultrasonic piezo motors are widely used in auto-focus camera lenses. Linear piezo micromotors are commercialised by PI (Physik Instrumente) GmbH & Co. (www.pi.ws), PiezoMotor AB (www.piezomotor.se), New Scale Technologies Inc. (www.newscaletech.com) and many others. PI (Physik Instrumente) GmbH & Co. also markets rotary piezo motors.

# 6.4 REFERENCES

**Donaldson N. de N.**, "Morphognostic coils: a technique for transmitting several near-field radio signals through the same space," *Med. Biol. Eng. Comput.*, vol. 17, pp. 271–274, 1979.

**Donaldson N. de N.**, "Orthogonal-coils receiver: a configuration for improving the position tolerance of coupled morphognostic coils," *Med. Biol. Eng. Comput.*, vol. 21, pp. 224–226, 1983.

***Donaldson N. de N.**, "Passive signalling via inductive coupling," *Med. Biol. Eng. Comput.*, vol. 24, pp. 223–224, 1986.

**Kimura M.**, T. Sugiura, Y. Harada, T. Hasewaga and K. Yoshimura "Addition to an RF-coupled phrenic nerve stimulator implant to provide outward transmission of body temperature," *Med. Biol. Eng. Comput.*, vol. 24, pp. 659–661, 1986.

**Miller J.A.**, G. Bélanger and T. Mussivand, "Development of an autotuned transcutaneous energy transfer system," *ASAIO J.*, vol. 39, pp. M706–M710, 1993.

**Neukomm P.A.** and H. Kündig, "Passive wireless actuator control and sensor signal transmission," *Sens. Actuat.*, A21, nos. 1–3, pp. 258–262, 1990.

**Neukomm P.A.**, H. Kündig, H. Baggenstos and K. Zerobim, "Passive Telemetry by absorption modulation – A new principle for long term transabdominal monitoring of pressure and EMG of the uterus of cows," in: *Biotelemetry X (Proc. 10th Int. Symposium on Biotelemetry)*, Ed. C.J. Amlaner Jr., University of Arkansas Press, Fayetteville, AR, pp. 487–496, 1988.

**Tang Z.**, B. Smith, J.H. Schild and P.H. Peckham, "Data transmission from an implantable biotelemeter by load-shift keying using circuit configuration modulator," *IEEE Trans. Biomed. Eng.*, vol. BME-42, pp. 524–528, 1995.

**Troyk P.R.** and M.A.K. Schwan, "Closed-loop class E transcutaneous power and data link for MicroImplants," *IEEE Trans. Biomed. Eng.*, vol. BME-39, pp. 589–599, 1992.

**Van Schuylenbergh K.** and R. Puers, "Self-tuning inductive powering for implantable telemetric monitoring systems," *Sens. Actuat.*, A52, pp. 1–7, 1996.

# A1 Vector formulae

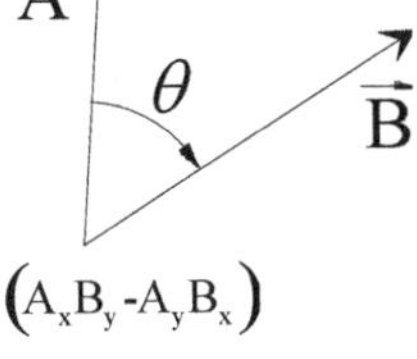


Scalar product of vectors:

$$\vec{A}.\vec{B} = A_x B_x + A_y B_y + A_z B_z = AB\cos\theta$$

Vector product of vectors:

$$\vec{A} \times \vec{B} = \vec{C} = \vec{u}_x\left(A_y B_z - A_z B_y\right) + \vec{u}_y\left(A_z B_x - A_x B_z\right) + \vec{u}_z\left(A_x B_y - A_y B_x\right)$$

$\vec{C}$ is perpendicular to $\vec{A}$ and $\vec{B}$ and $\left|\vec{C}\right| = AB\sin\theta$

$\vec{u}_x, \vec{u}_x$ and $\vec{u}_x$ are the three unit vectors along the x-, y- and z-axes.

The triple scalar product of vectors: $\vec{A}.\vec{B} \times \vec{C} = \vec{B}.\vec{C} \times \vec{A} = \vec{C}.\vec{A} \times \vec{B}$

The triple vector product of vectors: $\vec{A} \times \left(\vec{B} \times \vec{C}\right) = \left(\vec{A}.\vec{C}\right)\vec{B} - \left(\vec{A}.\vec{B}\right)\vec{C}$

The gradient of a scalar field p: $\text{grad}\, p = \vec{\nabla} p = \vec{u}_x \frac{\partial p}{\partial x} + \vec{u}_y \frac{\partial p}{\partial y} + \vec{u}_z \frac{\partial p}{\partial z}$

The divergence of a vector field $\vec{F}$: $\operatorname{div}\vec{F} = \vec{\nabla}.\vec{F} = \dfrac{\partial F_x}{\partial x} + \dfrac{\partial F_y}{\partial y} + \dfrac{\partial F_z}{\partial z}$

The Laplacian of a scalar field p: $\nabla^2 p = \vec{\nabla}.\vec{\nabla}p = \dfrac{\partial^2 p}{\partial x^2} + \dfrac{\partial^2 p}{\partial y^2} + \dfrac{\partial^2 p}{\partial z^2}$

The Laplacian of a vector field $\vec{F}$:

$$\nabla^2\vec{F} = \vec{\nabla}\vec{\nabla}.\vec{F} = \vec{u}_x\nabla^2 F_x + \vec{u}_y\nabla^2 F_y + \vec{u}_z\nabla^2 F_z$$

The curl vector of a vector field $\vec{F}$:

$$\operatorname{curl}\vec{F} = \vec{\nabla}\times\vec{F} = \left[\frac{\partial F_z}{\partial y} - \frac{\partial F_y}{\partial z}\right]\vec{u}_x + \left[\frac{\partial F_x}{\partial z} - \frac{\partial F_z}{\partial x}\right]\vec{u}_y + \left[\frac{\partial F_y}{\partial x} - \frac{\partial F_x}{\partial y}\right]\vec{u}_z$$

Gauss' Theorem: $\displaystyle\oiint_{\substack{\text{closed}\\ \text{surface A}}} \vec{F}.d\vec{a} = \iiint_{\substack{\text{volume V}\\ \text{enclosed by A}}} \operatorname{div}\vec{F}\,dv$

Stokes' Theorem: $\displaystyle\oint_{\substack{\text{closed}\\ \text{line L}}} \vec{F}.d\vec{l} = \iint_{\substack{\text{surface A}\\ \text{enclosed by L}}} \operatorname{curl}\vec{F}.d\vec{a}$

# A2 Coil models and measurements

This short story on coils, their parasitics and how to measure them will sound fairly familiar to anyone who already tried to squeeze the maximum out of a coil design. The knowledge of how and where coil parasitics occur, and how to interpret the data produced by measurement equipment, allows the novel designer to avoid common pitfalls and improve coil performance.

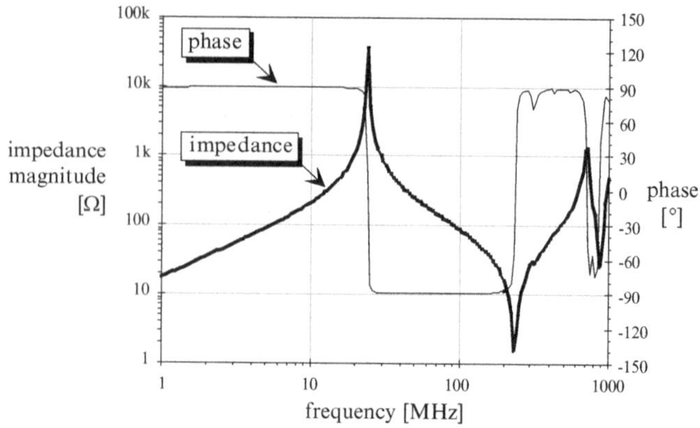


*Figure A2.1.* *"Real world" coils show resonance phenomena because of parasitic capacitances. A measured impedance-versus-frequency plot is shown here, for an air-cored, cylindrical coil with a diameter of 11 mm and a length of 22 mm. It has 25 windings of 0.7 mm enamelled solid copper wire. The coil impedance is measured on an HP8753C network analyser with an HP85047A S-parameter test set. The first resonance is obviously a parallel resonance and is caused by the coil inductance and the parasitic interwinding capacitance. The second, series resonance comes from the inductance of the connection wires and the coil that turned capacitive because of the parallel resonance.*

*Figure A2.2.* *The coil interwinding capacitance.*

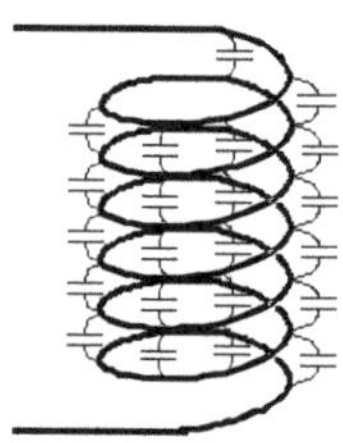

# A2.1 COIL MODELS

"Real world" coils are not ideal. They have parasitic coil losses and parasitic resonances. The coil losses come from the Ohmic dissipation in the wire resistance, the magnetic hysteresis of the coil core and surrounding materials, electromagnetic radiation and the eddy currents induced in the surrounding materials. These losses are best modelled by a series resistor because the Ohmic losses are directly related to the coil current, and because all other losses are function of the magnetic field strength that is also proportional to coil current.

The parasitic resonances originate from parasitic coil capacitors. Each part of the coil winding has a parasitic capacitance to its neighbours. This forms a lumped inter-winding capacitor, causing the coil to resonate at its ***self-resonance frequency*** (Figures A2.1 and A2.2). This interwinding capacitor appears in parallel to the coil inductance. Above the resonance frequency, the coil loses its inductive characteristic because the current then flows through the parasitic interwinding capacitors instead of the actual windings. As a result, no magnetic field is induced.

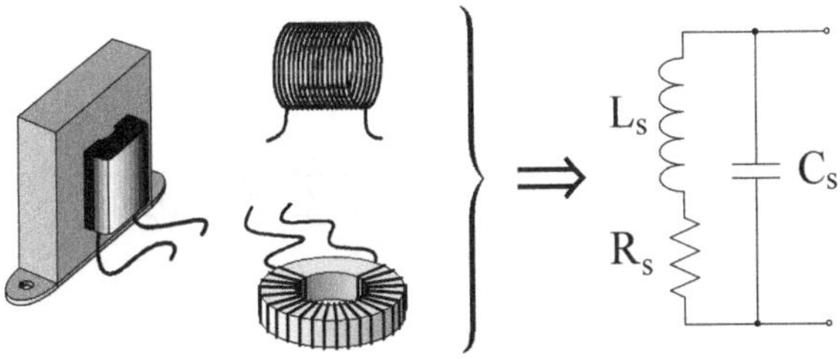


Figure A2.3. *"Real" coils are well modelled for frequencies below their first self-resonance, by an ideal inductance, a series loss resistor $R_s$ and a parallel capacitor $C_s$. The inductor and capacitor values are pretty constant as they largely relate to the coil geometry. The resistor is frequency and amplitude dependent because it models the Ohmic losses in the coil windings, the radiation and the magnetic and eddy-current losses in the coil core and surrounding materials all of which are frequency and amplitude dependent.*

## A2.1.1 SERIES L–R–C MODEL

The lumped series resistance and the lumped parallel capacitor are not practical for circuit calculations. Measurements however indicated that the lumped coil model can be accurately represented by its discrete equivalent of an ideal coil $L_S$, a series loss resistor $R_S$ and a parallel capacitor $C_S$ (Figure A2.3) **for frequencies below and around their first resonance frequency**. Inductor behaviour is more complex past this resonance frequency and irrelevant to inductive link design.

### A2.1.1.1 The model inductor

The model inductance $L_S$ is determined by the geometry of the coil and the surrounding structures (in case these contain substances with a higher permeability than vacuum). Some simple coil geometries, like long solenoids and perfect toroids, have analytic expressions for their inductance. One has to rely on measurements and empirical expressions for the other geometries (TERMAN, 1943). The coreless circular loop is very common with inductive links. The inductance of such a loop with centreline diameter $D$ and a circular cross section of diameter $d$, is:

$$L_S = \frac{\mu_0 D}{2} \ln\frac{8D}{d} - 2 + \frac{\mu_r}{4} .n^2$$ Equation A2.1.

where $n$ equals the number of turns (this is expression (29) from (TERMAN, 1943) reworked for SI units and low frequency, thus $\delta = 1/4$). The inductance of the also popular circular pancake coil is (expression (46) in SI units from (TERMAN, 1943)):

$$L_S \approx \frac{a^2}{0.2032a + 0.2794c} .n^2 \quad [\mu H]$$ Equation A2.2.

where $a$ is the average coil radius (in meter) and $c$ the radial winding depth (in meter). This formula is accurate to within 5 % for $c > 0.2a$. The inductance of a short single-layer solenoid of radius $a$ and length $b$ is (expression (53) in (TERMAN, 1943) in SI units and for $c = 0$ and $b = d$),

$$L_S \approx \mu_0 a \frac{b^2}{128a^2} - 0.5 + 1 + \frac{b^2}{24a^2} \ln\frac{8a}{b} .n^2$$ Equation A2.3.

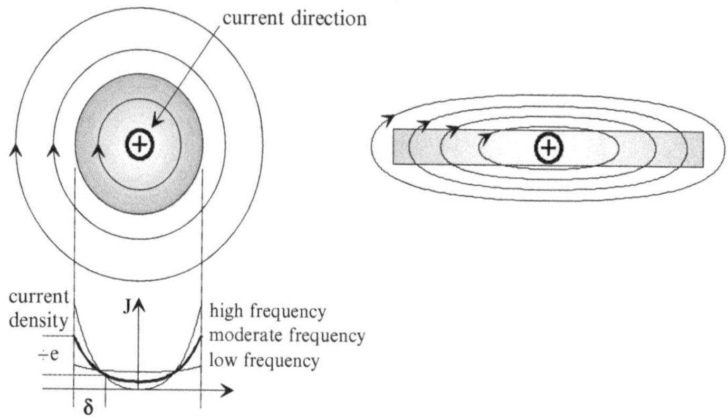


*Figure A2.4.* *Flux lines and current distribution in a plain wire and a thin strip. This shows how skin effect causes the current to concentrate at the outer edges as far away from the conductor centre as possible (TERMAN et al., 1955). The current density is indicated by the shading density.*

These are all low-frequency inductances. The wire permeability is assumed to be $\mu_0$ and the influence of the wire insulation was neglected. Coil inductances decrease slightly with the frequency because skin and proximity effect change the current distribution. The influence is however, negligible in the frequency ranges used for inductive powering as there is no gross current change. The redistribution occurs only inside the cross section of the coil wire and the overall current flow remains unchanged. On the other hand, their influence on the winding resistance is important at higher frequencies because it reduces the active wire cross-section.

The words "skin" and "proximity effect" express two identical effects with a different cause. ***Skin effect*** is caused by the internal a.c. magnetic field in a current-conducting wire. This field pushes the current charges to the outer layer near the surface of the conductor. Most of the current then flows where it is encircled by the smallest number of flux lines, i.e. the outer conductor surface (Figure A2.4). The ***skin depth*** $\delta$ is that distance below the surface of a conductor where the current has diminished to $1/e$ of its value at the surface. It decreases with the square of the frequency $f$:

*Equation A2.4.* $$\delta = \frac{1}{\sqrt{\pi \sigma \mu f}}$$ where $\sigma$ is the medium conductivity and $\mu$ its permeability.

This current-redistribution effect is called ***proximity effect*** when it is caused by the magnetic fields of currents in nearby conductors (e.g. an adjacent winding). This effect adds to the skin effect and makes the resistance increase even more prevalent. The relation between frequency and skin depth entirely depends of the given geometry and cannot be simply expressed like for skin effect.

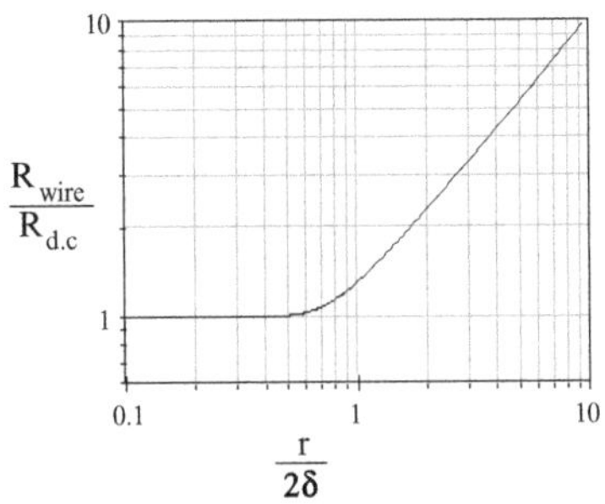


| | $\rho$ at 20°C | $D$ | $\delta$ [µm] 100 kHz | 1 MHz | 10 MHz |
|---|---|---|---|---|---|
| Cu | 1.72 | 66 | 209 | 66 | 21 |
| Au | 2.2 | 75 | 237 | 75 | 24 |
| Ag | 1.63 | 64 | 202 | 64 | 20 |
| Al | 2.67 | 82 | 259 | 82 | 26 |
| Pt | 10.6 | 164 | 519 | 164 | 52 |
| 316L | 74 | 433 | 1370 | 433 | 137 |

$\rho$ is expressed in $\mu\Omega.cm$, $D$ in $mm\sqrt{Hz}$

$$\delta = \frac{D}{\sqrt{f}} \quad where\ D = \sqrt{\frac{\rho}{\pi\mu}}$$

*The ratio between the a.c. wire resistance and its d.c. value, in function of half the wire radius normalised to the skin depth. The table on the right summarises some numerical data on the specific resistance and the skin depth (based on FINK and CHRISTIANSEN, 1982 and WESTMAN, 1956).* *Figure A2.5.*

### A2.1.1.2 The model resistor

The resistor $R_S$ in the discrete coil model represents the coil losses. These are the wire resistance, the hysteresis core losses, losses by eddy currents in surrounding materials, radiation[1]:

$$R_S = R_{wire} + \frac{P_{hysteresis} + P_{eddy} + P_{radiation}}{I_{R_S}^2}$$ *Equation A2.5.*

The resistance of a plain wire with radius $r$ and length $l$, is given by its d.c. resistance multiplied by a factor that models the skin effect:

$$\begin{cases} R_{wire} = R_{d.c.}\left[1 + \frac{1}{3}\left(\frac{r}{2\delta}\right)^4\right] & for\ \delta > r/2\ (i.e. low\ f) \\ R_{wire} = R_{d.c.}\left[\frac{r}{2\delta} + \frac{1}{4} + \frac{3}{64}\left(\frac{2\delta}{r}\right)\right] & for\ \delta < r/2\ (i.e. high\ f) \end{cases} \quad \text{and } R_{d.c.} = \frac{l}{\sigma\pi r^2}$$ *Equation A2.6.*

Skin effect is only relevant when the skin depth is smaller than the wire diameter (Figure A2.5). The substitution of Equation A2.4. for the skin depth in Equation A2.6. shows that the low-frequency wire resistance is proportional to a constant plus a quadratic frequency term. The high frequency expression is proportional to a constant and a square term. Proximity effect increases the powers of these frequency-dependent terms or adds higher order terms that become dominant at a higher frequency.

The effective a.c. resistance of a conductor can be made to approach the d.c. resistance at low and moderate frequencies, by assembling the conductor from a number of thoroughly interwoven strands of thin enamelled wire connected in parallel at their ends, and insulated throughout the rest of their length (TERMAN, 1955).

[1] If the coil under test is the primary coil of an inductive link, the secondary power consumption appears as a resistor $R_{eq}$ in series with $R_S$.

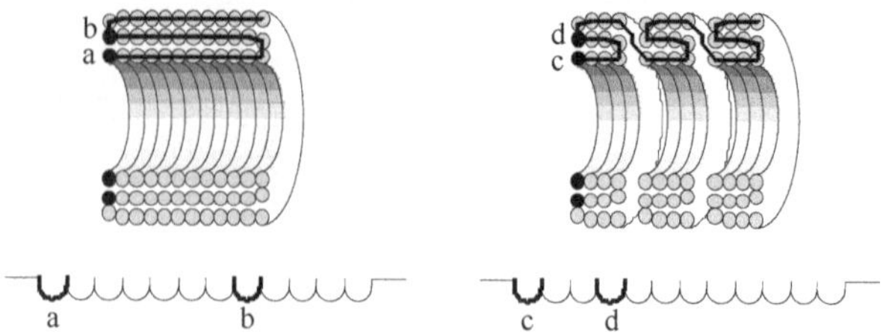


*Figure A2.6. Transversal cross section through a normal multilayer coil and a bank-wound coil. Bank winding (right) avoids that coil turns that are well apart in the electric coil structure are laid out close to each other in the geometric construction.*

Such a stranded cable is called a ***Litze*** conductor. If the stranding is properly done, each wire links, on the average, with the same number of flux lines as every other wire, and the current divides evenly among the strands. If at the same time each strand is of small diameter, it will have relatively little skin effect over its cross section. Practical Litze conductors are very effective at frequencies below about 1 MHz. As the frequency becomes higher, the benefits disappear because the capacitance between the strands allows the current to hop across the strand insulation.

The hysteresis in the B–H curve of magnetic materials is responsible for the magnetic losses that warm up coil cores. The area enclosed by the B-H curve is a measure for the magnetic energy lost during one cycle. The hysteresis losses are proportional to the frequency and to the magnetic flux density to the power 1.6:

*Equation A2.7.*

$$P_{hysteresis} = \eta f \hat{B}^{1.6}$$

$\eta$ is a material constant and $\hat{B}$ the magnitude of the magnetic flux density in the material.

The magnetic energy lost by eddy currents that are induced in a conductor with conductivity $\sigma$, volume $V$ and thickness $d$, perpendicular on the B-field, is given by

*Equation A2.8.*

$$P_{eddy} = \frac{\sigma \omega^2 \hat{B}^2 d^2 V}{24}$$

$\hat{B}$ is the magnetic flux density in the conductor, $\omega$ is the radian frequency.

The radiated power is set by the actual shape of the coil and its dimensions. It is only relevant at coil dimensions comparable to the wavelength.

#### A2.1.1.3 The model capacitor

The capacitor $C_S$ in the discrete model is preferably derived from a measured value of the phase resonance $\omega_0$ of the coil:

*Equation A2.9.*

$$\omega_0 = \frac{1}{\sqrt{L_S C_S}}$$

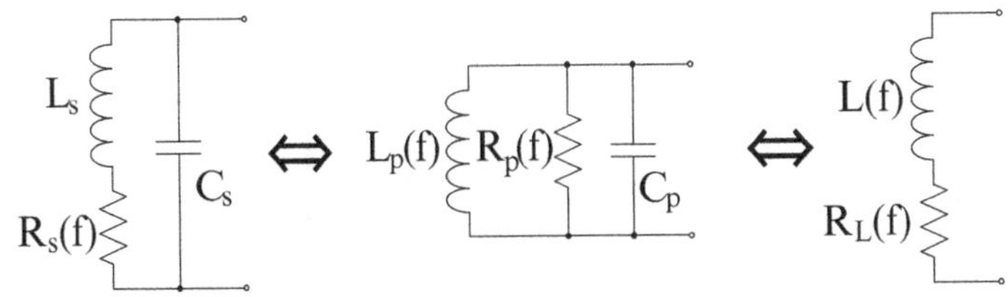


*The series L–R–C model of Figure A2.3 is most naturally related to the "real life" lumped component. The coil inductance and its parasitic capacitance can be regarded as constants in the frequency bands used for inductive powering. It has an equivalent parallel L–R–C model and an equivalent L–R model. The L–R model is only valid for frequencies below the first self-resonance frequency, as it cannot model the capacitive behaviour above the first resonance.* *Figure A2.7.*

$C_S$ cannot simply be predicted from the coil geometry (winding area and separation) because the interwinding voltage is not constant along the coil. Everything depends on how the coil is constructed and the wire layers are laid out. The parts near the coil terminals exhibit a larger voltage swing than the central windings. The apparent capacitance of such outer turns is larger if they are placed close to each other in a multilayer coil structure, as more charge is displaced (Figure A2.6). Multilayer constructions are therefore not advisable if a high self-resonance frequency is desired. Opt for bank winding if a single layer cannot accept enough windings.

## A2.1.2 PARALLEL L–R–C MODEL

The series L–C–R model relates most closely to the actual coil parasitics. But, for some circuit calculations, it is more practical to use a parallel model with the loss resistor in parallel to the inductor (Figure A2.7). The relation between both models is given by

$$L_P = L_S + \frac{R_S^2}{\omega^2 L_S} = L_S\left(1 + \frac{1}{Q_{L_S}^2}\right)$$ *Equation A2.10.*

$$C_P = C_S$$ *Equation A2.11.*

$$R_P = R_S + \frac{\omega^2 L_S^2}{R_S} = R_S\left(1 + Q_{L_S}^2\right)$$ *Equation A2.12.*

## A2.1.3 SERIES L–R MODEL

LRC meters and vector network analysers measure either impedance amplitudes and phase angles, or real and imaginary impedances. A positive imaginary part is then displayed as an equivalent inductance and a negative part as an equivalent capacitance. Coils are thus implicitly considered as an inductor $L$ in series with a resistor $R_L$ and without capacitor. The measured components $L$ and $R_L$ cannot simply be set equal to the inductor and resistor in the series or parallel coil model. Measured $L$ and $R_L$ always vary with the frequency because of the coil resonance, even in case of constant $L_S$, $R_S$ and $C_S$ (Figure A2.8).

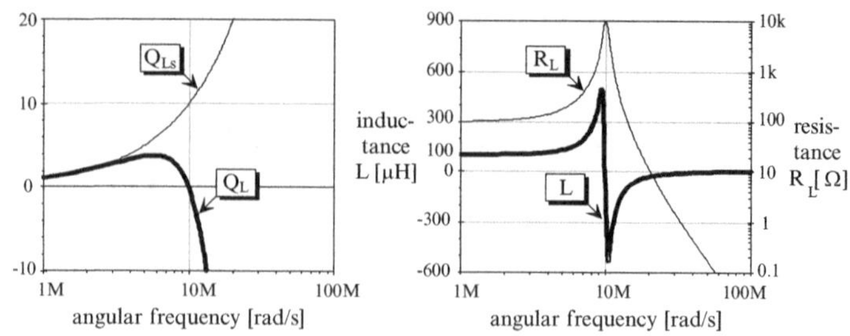


*Figure A2.8.* *The $L$–$R_L$ series model of a series $L_S$–$C_S$–$R_S$ tank with constant $L_S$, $C_S$ and $R_S$ ($L_S = 100\ \mu H$, $C_S = 100\ pF$ and $R_S = 100\ \Omega$). The $Q_{LS} = \omega L_S/R_S$ keeps increasing with the frequency, whereas the $Q_L = \omega L/R_L$ reaches a maximum, then decreases to zero at coil resonance and becomes negative because of the negative apparent inductance L. The $R_L$ increase has nothing to do with skin or proximity effect. The resonance effect amplifies the current in the $L_S$–$R_S$–$C_S$ loop and increases the dissipation in $R_S$. This augmented dissipation shows up as an increased $R_L$ in the $L$–$R_L$ coil model.*

Measured $L$ and $R_L$ data are converted into $L_S$ and $R_S$ data with the following expressions. $C_S$ is best derived from a $\omega_0$ measurement.

*Equation A2.13.*

$$R_S \approx R_L\left[1-\left(\frac{\omega}{\omega_0}\right)^2\right]^2$$

*Equation A2.14.*

$$L_S \approx L\left[1-\left(\frac{\omega}{\omega_0}\right)^2\right]$$

These approximate expressions are valid for $\left|(\omega_0/\omega)^2-1\right| > 1/Q_{L_S}$ [(2)]. They proved more useful than their exact counterparts that are numerically unstable near the coil resonance. Away from this resonance, there is no significant precision difference between both.

The coil quality factor $Q_{Ls}$ ($= \omega L_S/R_S$) is found from $Q_L$ data with this exact expression:

*Equation A2.15.*

$$Q_{L_S} = \frac{Q_L \pm \sqrt{Q_L^2 + 4\left(\frac{\omega}{\omega_0}\right)^2\left[1-\left(\frac{\omega}{\omega_0}\right)^2\right]}}{2\left[1-\left(\frac{\omega}{\omega_0}\right)^2\right]} \qquad \begin{array}{l} +\ for\ \omega < \omega_0 \\ -\ for\ \omega > \omega_0 \end{array}$$

---

2 This condition corresponds to $|\Delta\omega| \equiv |\omega_0 - \omega| > \dfrac{\omega_0}{2Q_{LS}}$. The 3 dB frequency band [-$B_{3dB}$, $B_{3dB}$] around $\omega_0$ where the impedance of the series model remains higher than $Z_{max}/2$ is given by $B_{3dB} = \dfrac{\omega_0\sqrt{3}}{2Q_{LS}}$. The above condition for $|\Delta\omega|$ thus means that $|\Delta\omega|$ should exceed some 3 dB bandwidths of the impedance peak.

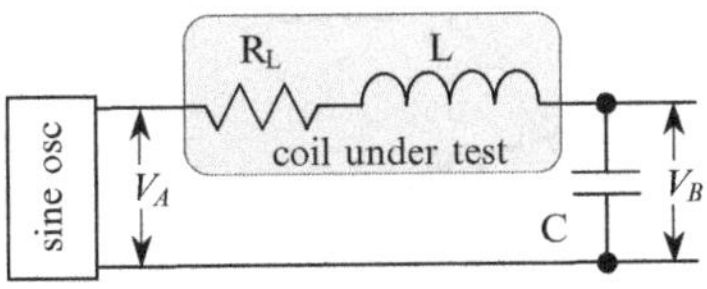


Figure A2.9. *The measurement of the coil quality factor $Q_L$ as a ratio of the voltages $V_B$ and $V_A$. The source frequency is adjusted until the coil L resonates with C. $V_B$ is then maximal and lags exactly 90° behind $V_A$. The coil quality factor $Q_L$ is then found as $V_B/V_A$.*

The frequency $\omega_0$ is the phase resonance frequency of the coil impedance. The quality factor $Q_L$ of the series-LR model is defined as $Q_L \equiv \omega L/R_L$. It becomes negative, together with the inductance $L$, above the coil resonance.

## A2.2 COIL MEASUREMENTS

### A2.2.1 COIL INDUCTANCE

The measurement of the inductance $L$ is rarely problematic as it dominates the impedance $Z_L = R_L + j\omega L$. As the inductance $L_S$ of the series-LRC model is assumed to be constant, it can be set equal to the low frequency value of $L$ (cf. Equation A2.14. at low $\omega$).

### A2.2.2 COIL QUALITY FACTOR

Note that vector network analysers are not very well suited for measuring coil quality factors. Such analysers derive the device impedance magnitude and phase angle from scattering effects in a 50 Ω set-up, and then compute the real and imaginary impedance components. As our coil impedance is mostly reactive, small phase-angle errors cause large errors on the real impedance component. Moreover, the impedance measurement itself becomes less and less accurate the further it deviates from 50 Ω. A long warm-up (several hours!) and careful calibration of the measurement equipment to cancel out probe impedances, make a significant difference, but all-in-all, measurements of coil resistance or quality factor remain unreliable with vector network analysers.

Coil quality factors $Q_L$ are best measured with either an RF-IV impedance analyser (preferably with calibration), a dedicated Q-meter, or an oscilloscope and the following method. The quality factor $Q_L$ is found as a ratio of two voltages $V_A$ and $V_B$ with the set-up depicted in Figure A2.9. Connect a high-Q capacitor $C$ to the coil under test and adjust the source frequency until a maximal $V_B$ and a phase angle of 90° is observed. At this point, the capacitor admittance $j\omega C$ completely cancels the coil admittance $1/(j\omega L)$. The coil quality factor $Q_L$ is then found as $V_B/V_A$. Note that the capacitor $C$ includes the probe capacitance. Its exact value is of no importance but its losses should be small compared to the coil losses. Repeat this procedure with different capacitors $C$ to obtain $Q_L$ data at multiple frequencies.

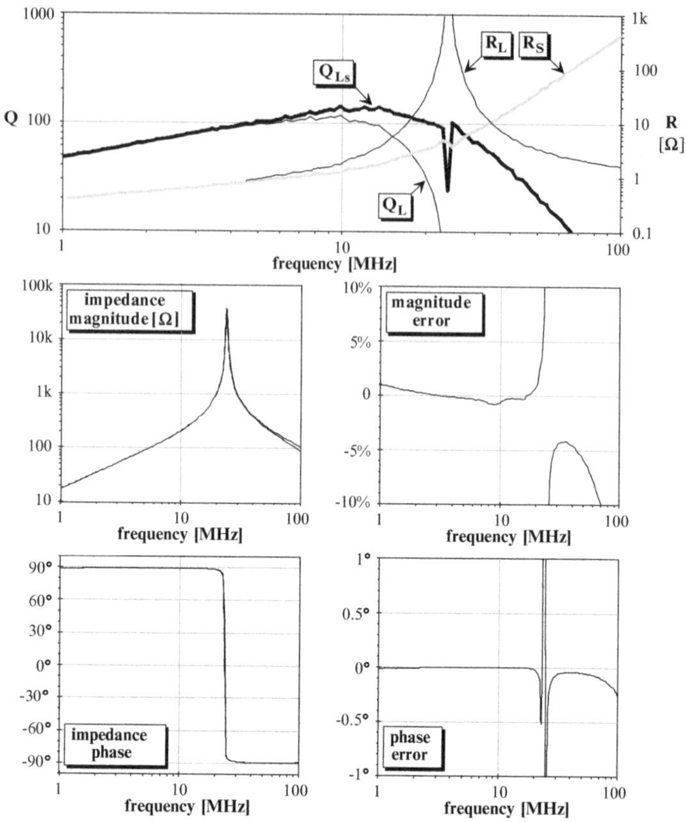


*Figure A2.10.* *The measured data of the coil of Figure A2.1 have been converted into a series-LRC model ($L_S$ = 2.74 μH, $f_0$ = 24.096 MHz, $C_S$ = 15.9 pF). The $R_S$ and $Q_{L_S}$ on the top diagram were computed with Equations A2.13. and A2.15. respectively.*
*The impedance of the resulting LRC model was then verified against the measured impedance data. Very good agreement was found, except in the small region where $|(\omega_0/\omega)^2 - 1| < 1/Q_{LS}$ and the approximate formulae are not valid, and above 50 MHz where the second coil resonance comes into play.*

DONALDSON and PERKINS (1983) mentioned a similar method to measure the $R_P$ of the parallel R–L–C coil model (Figure A2.11). A resistor $R$ and a high-Q capacitor $C$ are added to the coil. The source frequency is adjusted to the resonance of $L_P$ with $C_P$ and $C$. The voltage $V_B$ is then maximal and in phase with the source voltage $V_A$. The coil resistor $R_P$ is then found as $R_P = R\frac{V_A}{V_A - V_B}$. The capacitor $C$ includes again the probe capacitance. Its losses should be small compared to the coil losses.

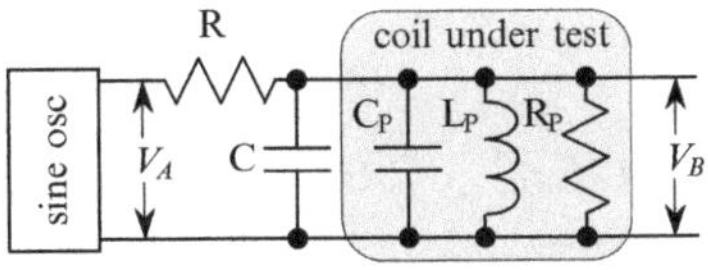


Figure A2.11. *The measurement of the resistor $R_P$ of the parallel-LRC model. The source frequency is adjusted until the coil L resonates with C. $V_B$ is then maximal and exactly in phase with $V_A$. The coil resistor $R_P$ is then found as*

$$R_P = R\frac{V_A}{V_A - V_B}.$$

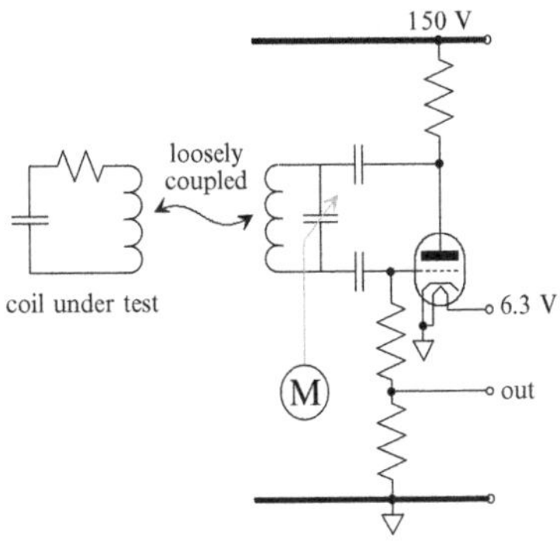


Figure A2.12. *A scanning grid-dip oscillator (from COLLINS, 1967). Sweeping the frequency is controlled by a servo or motor that varies the parallel capacitor of the oscillator tank. The grid current sensed at the output terminal reaches a maximum at the resonance of the coil under test when it absorbs most electromagnetic energy.*

## A2.2.3 COIL SELF-RESONANCE FREQUENCY

Probing a coil to measure its resonance frequency is often doubtful as the coil resonance is changed by the probe capacitance. It is possible to use an impedance or network analyser where the probe capacitance has been cancelled out by careful calibration. But the accuracy of such procedure becomes problematic for smaller coils with small parasitic capacitances compared to the probe capacitance. The contact less method employed by the grid-dip technique is therefore preferred. A grid-dip meter is basically a sweep oscillator with its oscillator coil loosely coupled to the coil under test (Figure A2.12). At its resonance frequency, the coil under test absorbs a maximal amount of electromagnetic power. This is reflected in the meter coil and sensed as a peak in the oscillator-tube grid current. A tube version is shown here to explain the name of the technique, but solid-state versions are available.

## A2.2.4 COIL COUPLING

The coupling of a coil set is measured in a two-step procedure (Figure A2.13, VAN PAEMEL, 1990). The primary coil is first connected to a sinusoidal voltage source, the frequency is swept and the secondary coil voltage is measured. (The capacitances of the source, its cable and the primary coil are represented by the capacitor $C_{1A}$. The secondary coil and the probe capacitors are combined in capacitor $C_{2A}$.) The voltage transfer $V_{2A}/V_{1A}$ is flat between the zero where the primary leakage impedance becomes larger than the source resistor, and the coil resonance peaks. The voltage $v_1$ is then equal to $v_{1A}$ and all capacitors can be neglected. Suppose that the probe resistor $R_{probe}$ is much larger than the leakage impedance of the secondary coil plus its loss resistor $R_{S2}$.

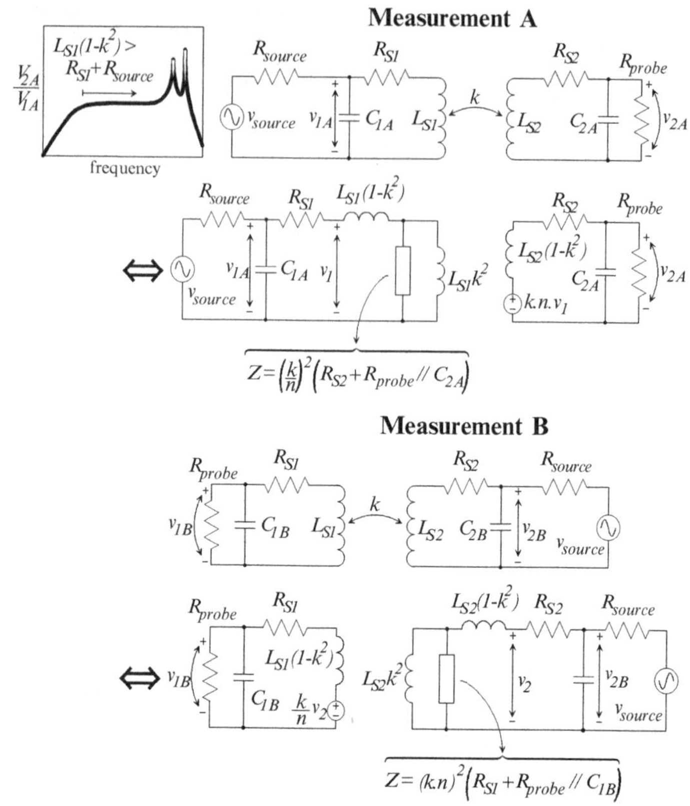


Figure A2.13. *Measuring the coil coupling in two steps.*

Hence,

$$A_A = \frac{V_{2A}}{V_{1A}} = k.n \text{ provided } R_{probe} >> j\omega L_{S_2}\left(1-k^2\right)+R_{S_2}$$

Change now the source and the probe and measure the voltage at the primary while driving the secondary coil. The transformer ratio is *1/n* instead of *n*. The voltage gain in the flat section of the transfer function is

$$A_B \equiv \frac{V_{1B}}{V_{2B}} = \frac{k}{n} \text{ provided } R_{probe} >> j\omega L_{S_1}\left(1-k^2\right)+R_{S_1}$$

The coupling factor and the transformer ratio are then found as

$$k = \sqrt{A_A.A_B} \text{ and } n = \sqrt{\frac{A_A}{A_B}}$$

Determining the coupling of the same coils in a different position requires only one $A_A$ measurement. Their transformer ratio $n$ remains unchanged.

Note that the method fails with very small coils as the flat section in the transfer function then disappears.

# A2.3 REFERENCES

**Collins C.C.**, "Miniature passive pressure transensor for implantation in the eye," *IEEE Trans. Biomed. Eng.*, vol. BME-14, pp. 74–83, 1967.

**Donaldson N. de N.** and T.A. Perkins, "Analysis of resonant coupled coils in the radio frequency transcutaneous links," *Med. Biol. Eng. Comput.*, vol. 21, pp. 612–627, 1983.

**Fink D.G.** and D. Christiansen, *Electronics Engineers' Handbook*, McGraw-Hill, New York, pp. 6–4, 6–6, 1982.

***Terman F.E.**, *Radio Engineers' Handbook*, McGraw-Hill, New York/London, pp. 47–109, 1943.

**Terman F.E.**, R.A. Helliwell, J.M. Pettit, D.A. Watkins and W.R. Rambo, *Electronic and Radio Engineering*, McGraw-Hill, New York/Toronto/London, pp. 21–24, 1955.

**Van Paemel M.**, *Ontwerp van een prothese voor elektrische gehoorstimulatie* (*Design of a prosthesis for electrical auditory stimulation*), Ph.D. dissertation (in Dutch), Katholieke Universiteit Leuven, Leuven, Belgium, 1990.

**Westman H.P.** (Ed.), *Reference Data for Radio Engineers, 4th edn.* (International Telephone and Telegraph Corporation, 320 Park Ave., New York 22), American Book-Stratford Press, New York, pp. 45–46, 1956.

# A3 Saturating-class-C amplifiers

This appendix delves into the mathematical details of saturating-class-C amplifiers. A description of how these stages operate, can be found in Chapter 3. First-order expressions are derived here for the driver efficiency, and for the currents and voltages in the switch transistor and the load coil. By first order, we mean that all circuit parasitics are set to zero for the computation of the circuit signals. The contribution of one specific parasitic on the driver efficiency is then calculated considering the first-order simplified circuit signals, together with a small but non-zero value for the parasitic of interest. First-order expressions thus assume that the parasitics are too small to distort the circuit signals significantly. This assumption cannot be held when driver efficiencies are below 90%). This appendix revises and extends the signal expressions for a saturating-class-C driver by VAN PAEMEL (1990) to higher frequencies where the switch current fall time $t_F$ comes into play, and to bipolar switches that show a saturation voltage $V_{S_{sat}}$ during conduction.

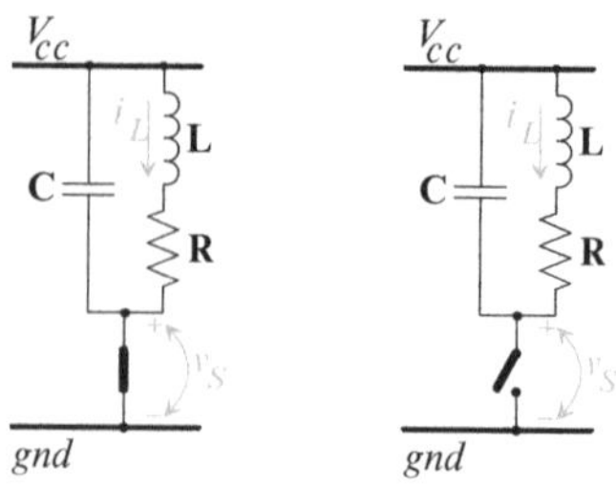


*Figure A3.1. First-order simplifications of the switch-on (left) and the switch-off (right) situations.*

# A3.1 THE STEADY-STATE SWITCH VOLTAGE AND COIL CURRENT

The steady-state components of the switch voltage $v_S(t)$ and coil current $i_L(t)$ are calculated in this paragraph, without any assumption on the timing characteristics of the switch driver signal. The timing requirements for the circuit to run in a saturating-class-C mode or in class E (with one coil and one capacitor) are derived in sections A3.3 and A3.4, respectively.

The calculations are simplified to a ***first order*** approximation to obtain manageable expressions. This ***means that $R_{ON}$, $V_{S_{sat}}$ and $t_F$ are assumed to be zero for the computation of the circuit signals $v_S$ and $i_L$. The contribution of e.g. $R_{ON}$ on the collector efficiency is then calculated considering the very same simplified circuit signals with $V_{S_{sat}}$ and $t_F$ equal to zero, together with a small but non-zero value for $R_{ON}$.***

## A3.1.1 SWITCH ON

The first-order simplification of the switch-on situation is depicted in Figure A3.1 (left) and corresponds to the circuit equation:

$$V_{cc} = L.\frac{di_{L_{ON}}(t)}{dt} + R.i_{L_{ON}}(t)$$

This results in a zero switch voltage $v_{S_{ON}}(t)$ and an exponentially rising coil current $i_{L_{ON}}(t)$ with a time constant $\alpha$ (Figure A3.2.). $I_{ON}$ is the current at the moment of switch on.

*Equation A3.1.*

*Equation A3.2.*

$$\begin{cases} i_{L_{ON}}(t) = i_S(t) = \frac{V_{cc}}{R} + \left(I_{ON} - \frac{V_{cc}}{R}\right).\exp\left(-\frac{t}{\alpha}\right) \\ v_{S_{ON}}(t) = 0 \end{cases}$$

where $\alpha = L/R$ and $I_{ON}$ stands for the coil current at switch on.

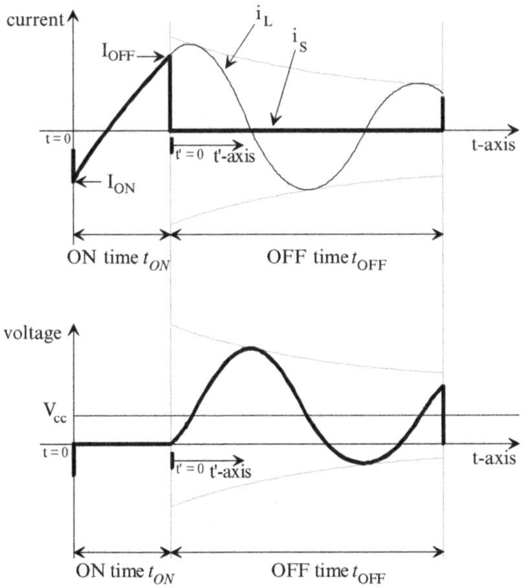


*Figure A3.2.* *Signals in the circuits of Figure A3.1 during one arbitrary, not even a steady-state, on–off cycle (the bold line on the top diagram, indicates the current through the switch, the hairline denotes the coil current).*

## A3.1.2 SWITCH OFF

### A3.1.2.1 The coil current $i_{L_{OFF}}$

The opening of the switch changes the circuit equation into (cf. Figure A3.1):

$$v_C(t') = L.\frac{di_{L_{OFF}}(t')}{dt'} + R.i_{L_{OFF}}(t')$$

Note that the time variable $t'$ equals the variable $t$ used in previous paragraph A3.1.1, except that the origin $t' = 0$ is shifted to the moment of switching off ($t = t' + t_{ON}$).

The capacitor voltage is expressed in terms of the coil current $i_{L_{OFF}}(t')$:

$$\Leftrightarrow -\frac{1}{C}\int_0^{t'} i_{L_{OFF}}(t)dt + \frac{q(t'=0)}{C} = L.\frac{di_{L_{OFF}}(t')}{dt'} + R.i_{L_{OFF}}(t')$$

$$\Leftrightarrow L.\frac{d^2 i_{L_{OFF}}(t')}{dt'^2} + R.\frac{di_{L_{OFF}}(t')}{dt'} + \frac{i_{L_{OFF}}(t')}{C} = 0$$

The capacitor $C$ restrains discontinuous voltage jumps during the switch turn off:

$$v_{C_{ON}}(t = t_{ON}) = v_{C_{OFF}}(t' = 0)$$

$$\Leftrightarrow V_{cc} = L.\frac{di_{L_{OFF}}(t'=0)}{dt'} + R.i_{L_{OFF}}(t'=0)$$

The presence of the coil $L$ makes that the current $i_L$ cannot change instantaneously. Hence,

$$i_{L_{OFF}}(t'=0)=i_{L_{ON}}(t=t_{ON})=I_{OFF}$$

The solution of the differential equation for $i_{L_{OFF}}(t')$ with the boundary condition for $di_{L_{OFF}}(t'=0)/dt'$ and $i_{L_{OFF}}(t'=0)$ is a damped oscillation. It decays with a time constant $\alpha'$ which is twice the time constant $\alpha$ during switch-on (Figure A3.2.).

*Equation A3.3.*

$$i_{L_{OFF}}(t')=I_{OFF}\cdot\frac{\cos(\omega_{tank}t'-\theta)}{\cos\theta}\exp\left(-\frac{t'}{\alpha'}\right)$$

where $I_{OFF}$ is the coil current at the moment of turning off the switch

and
$$\begin{cases}\alpha'=\dfrac{2L}{R}=2\alpha\\ \omega_{tank}=\sqrt{\dfrac{1}{LC}-\dfrac{R^2}{4L^2}}\\ x=\dfrac{I_{OFF}\,\omega_{tank}\,L}{V_{cc}-\dfrac{RI_{OFF}}{2}}\\ \theta=\arctan\left(\dfrac{1}{x}\right)=\arccos\left(\dfrac{x}{\sqrt{1+x^2}}\right)\\ I_{OFF}=i_{L_{ON}}(t=t_{ON})=\dfrac{V_{cc}}{R}+\left(I_{ON}-\dfrac{V_{cc}}{R}\right)\exp\left(-\dfrac{t_{ON}}{\alpha}\right)\end{cases}$$

The symbol $\omega_{tank}$ is the damped-oscillation frequency of the L–R–C tank. An oscillating response requires an underdamped L–R–C tank. Therefore, $R$ should not exceed the limit:

*Equation A3.4.*

$$R<\sqrt{4L/C}$$

### A3.1.2.2 The switch voltage $v_{S_{OFF}}$

The expression for $v_{S_{OFF}}$ is calculated from

$$v_{S_{OFF}}(t')=V_{cc}-v_C(t')=Vcc-\int_0^{t'}i_{L_{OFF}}(t)dt+\text{constant, with } v_{S_{OFF}}(0)=0$$

$$\Leftrightarrow v_{S_{OFF}}(t')=V_{cc}-v_C(t')=Vcc-L\frac{dI_{L_{OFF}}(t')}{dt'}-R.I_{L_{OFF}}(t').$$

*Equation A3.5.*

$$v_{S_{OFF}}(t')=V_{cc}\cdot\left[1+\frac{\sin(\omega_{tank}t'-\tau)}{\sin\tau}\exp\left(-\frac{t'}{\alpha'}\right)\right]$$

$$\text{with } \tau=\theta+\arctan\left(\frac{1}{\omega_{tank}\alpha'}\right)$$

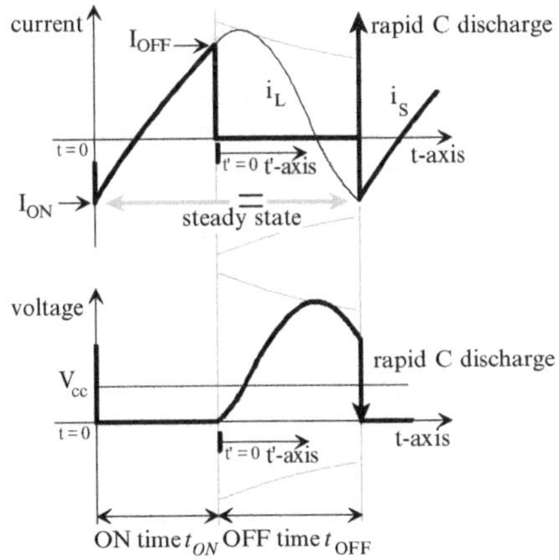


*Signals in the circuits of Figure A3.1 during one arbitrary steady-state, but non-class C on–off cycle. The occurrence of a current peak for non-zero switch-on voltages is discussed in paragraph A3.6.* *Figure A3.3.*

### A3.1.3 STEADY-STATE CIRCUIT BEHAVIOUR[1]

Steady-state circuit behaviour requires that the coil current $i_L$ at the end of an on–off cycle is equal to the initial coil current $I_{ON}$ (Figure A3.3):

$$i_{L_{OFF}}(t' = t_{OFF}) = I_{ON}$$

$$\Leftrightarrow \begin{cases} I_{OFF}\left[\cos\left(\omega_{tank}t_{OFF}\right)+\dfrac{1}{x}\sin\left(\omega_{tank}t_{OFF}\right)\right]\exp\left(-\dfrac{t_{OFF}}{\alpha'}\right) = I_{ON} \\ \text{with } x = \dfrac{I_{OFF}\,\omega_{tank}L}{Vcc-\dfrac{RI_{OFF}}{2}} \end{cases}$$

The combination of this equation and the expression for $I_{OFF} = i_{LON}(t = t_{ON})$ (cf. Equation A3.3.) forms a system of three equations in the three unknowns $I_{ON}$, $I_{OFF}$ and $x$ with solution:

$$\begin{cases} I_{OFF} = \dfrac{V_{cc}}{R}\dfrac{2x}{a+x} \\ I_{ON} = I_{OFF}\dfrac{c+\dfrac{d}{x}}{\exp\left(-\dfrac{t_{ON}}{\alpha}\right)} \\ x = \dfrac{ab+d}{1-b-c} \end{cases} \text{where} \begin{cases} a = \omega_{tank}\alpha' \\ b = \dfrac{1-\exp\left(-\dfrac{t_{ON}}{\alpha}\right)}{2} \\ c = \cos\left(\omega_{tank}t_{OFF}\right).\exp\left(-\dfrac{t_{ON}}{\alpha}\right).\exp\left(-\dfrac{t_{OFF}}{\alpha'}\right) \\ d = \sin\left(\omega_{tank}t_{OFF}\right).\exp\left(-\dfrac{t_{ON}}{\alpha}\right).\exp\left(-\dfrac{t_{OFF}}{\alpha'}\right) \end{cases}$$ *Equation A3.6.*

[1] Some useful expressions for these calculations:

$$a\cos\alpha + b\sin\alpha = \sqrt{a^2+b^2}\cos(\alpha-\theta) = a\frac{\cos(\alpha-\theta)}{\cos\theta} \text{ where } \begin{cases} \vartheta \equiv \arctan\left(\dfrac{b}{a}\right) \\ \cos\theta = \dfrac{a}{\sqrt{a^2+b^2}} \end{cases}$$

## A3.2 NORMALISED EXPRESSIONS

Normalisation to dimensionless parameters often simplifies expressions, reduces the number of parameters and enhances the insight into the influence of a given parameter on a signal shape.

The voltages, currents and impedances are now normalised according to

$$v_N \equiv \frac{v}{V_{cc}},\ i_N \equiv \frac{\omega L}{V_{cc}} i \text{ and } Z_N \equiv \frac{Z}{\omega L}$$

The parameter $\omega$ stands for the pulsation if one repeats the on–off-switching pattern over and over again. Hence:

$$\omega \equiv 2\pi f = \frac{2\pi}{T} = \frac{2\pi}{t_{ON} + t_{OFF}}$$

Normalisation of the $i_L$ and $v_S$ expressions found in the previous paragraph, reduces the parameters to the following three independent and normalised variables:

- the **normalised pulsation** $\boldsymbol{W}$ of the tank ringing during switch off:

$$W \equiv \frac{\omega_{tank}}{\omega}$$

- the **normalised resistance** $\boldsymbol{S}$ of the L–R–C tank: $S \equiv \dfrac{R}{\omega L}$
- the **duty cycle** $\boldsymbol{D}$ of the switch: $D \equiv \dfrac{t_{ON}}{t_{ON} + t_{OFF}} = \dfrac{t_{ON}}{T} = \dfrac{\omega\, t_{ON}}{2\pi}$

Both time variables $t$ and $t'$ are combined into a single normalised time $y$:

$$y \equiv \omega\, t = \omega(t' + t_{ON})$$

### A3.2.1 SWITCH ON ($0 \le y \le 2\pi D$)

Normalising Equation A3.1. and Equation A3.2. produces:

*Equation A3.7.*
*Equation A3.8.*

$$\begin{cases} i_{L_N}(y) = \dfrac{1}{S} + \left(I_{ON_N} - \dfrac{1}{S}\right).\exp(-yS) \\ v_{S_N}(y) = 0 \end{cases}$$

### A3.2.2 SWITCH OFF ($2\pi D \le y \le 2\pi$)

The normalised versions of Equation A3.3. and Equation A3.4. are:

*Equation A3.9.*
*Equation A3.10.*

$$\begin{cases} i_{L_N}(y) = I_{OFF_N}.\dfrac{\cos\left[W(y-2\pi D)-\theta\right]}{\cos\theta}\exp\left[-(y-2\pi D)\dfrac{S}{2}\right] \\ v_{S_N}(y) = 1 + \dfrac{\sin\left[W(y-2\pi D)-\tau\right]}{\sin\tau}\exp\left[-(y-2\pi D)\dfrac{S}{2}\right] \end{cases}$$

$$\text{with}\quad \theta = \arctan\left(\frac{1}{x}\right) \quad\text{and}\quad \tau = \theta + \arctan\left(\frac{S}{2W}\right)$$

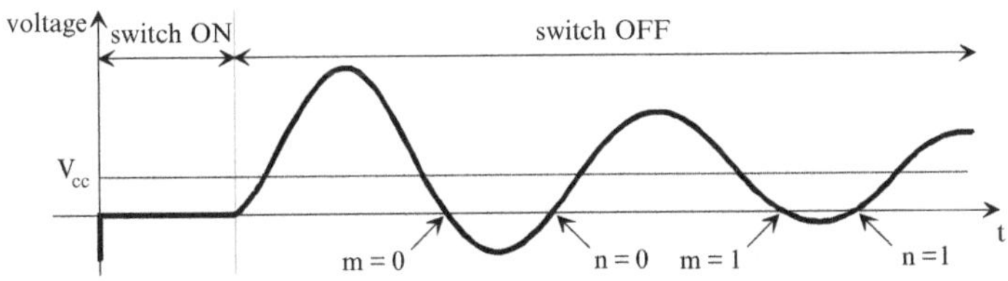


Figure A3.4. *A diagram of the switch voltage in the circuit of Figure A3.1, indicating the potential switch-on moments for saturating-class-C operation, according to Equation A3.12. and Equation A3.13.*

### A3.2.3 STEADY-STATE CONDITIONS

$$\begin{cases} I_{OFF_N} = \dfrac{1}{S}\dfrac{2x}{a+x} \\ I_{ON_N} = \dfrac{I_{OFF_N}\left(c+\dfrac{d}{x}\right)}{\exp(-2\pi DS)} \\ x = \dfrac{ab+d}{1-b-c} \end{cases} \text{with} \begin{cases} a = \dfrac{2W}{S} \\ b = \dfrac{1-\exp(-2\pi DS)}{2} \\ c = \cos\left[2\pi(1-D)W\right].\exp\left[-\pi(1+D)S\right] \\ d = \sin\left[2\pi(1-D)W\right].\exp\left[-\pi(1+D)S\right] \end{cases} \quad \text{Equation A3.11.}$$

## A3.3 SATURATING-CLASS-C OPERATION

All previous expressions for $i_L$ and $v_S$ describe the steady state signals for an arbitrary repetition of on–off cycles. Nothing is assumed about a switch timing that corresponds to efficient operation. The saturating-class-C mode inquires that the switch is turned on, when the voltage $v_S$ across its terminals has returned to zero:

$$v_{S_N}(y = 2\pi) = 0$$

$$\Leftrightarrow \sin[2\pi W(1-D) - \tau] = -\sin\tau.\exp[\pi(1-D)S]$$

This means, strictly mathematical speaking, that the frequency and the duty cycle of the applied switch signal must conform either to

$$2\pi W(1-D) - \tau = m.2\pi + \arcsin(-p) \quad \text{Equation A3.12.}$$

or to

$$2\pi W(1-D) - \tau = n.2\pi + \pi + \arcsin(-p) \quad \text{Equation A3.13.}$$

The parameter $p$ stands for $\sin\tau.\exp[\pi(1-D)S]$ and $m$ and $n$ are arbitrary integers. The meaning of both expressions is shown in Figure A3.4.

The switch mode described by Equation A3.12. with $m$ equal to 0, is the classic saturating-class-C mode. An $m$-value of 1 corresponds to a saturating-class-C frequency doubler (because the ringing frequency of the L–R–C tank is about twice the switching frequency). Higher $m$-values feature saturating-class-C frequency multipliers with a multiplication factor $(m-1)$.

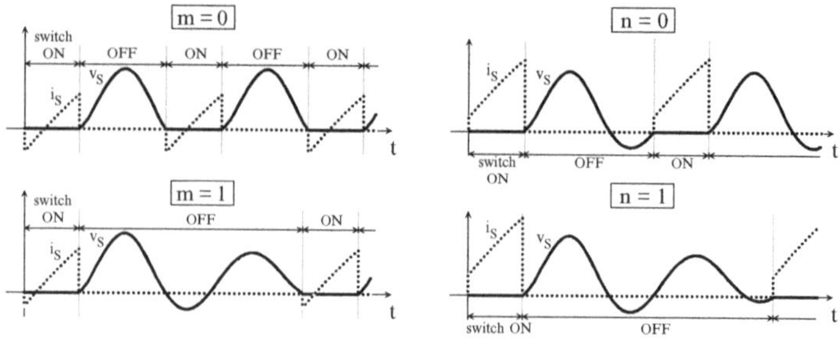


*Figure A3.5.* *Switch current $i_S$ and switch voltage $v_S$ for different m and n values. The left traces (Equation A3.12.) describe proper saturated-class-C behaviour with current cycling back-and-forth between the L–R–C tank and the power supply. The right traces (Equation A3.13.) only draw a positive supply current and don't offer a high driver efficiency.*

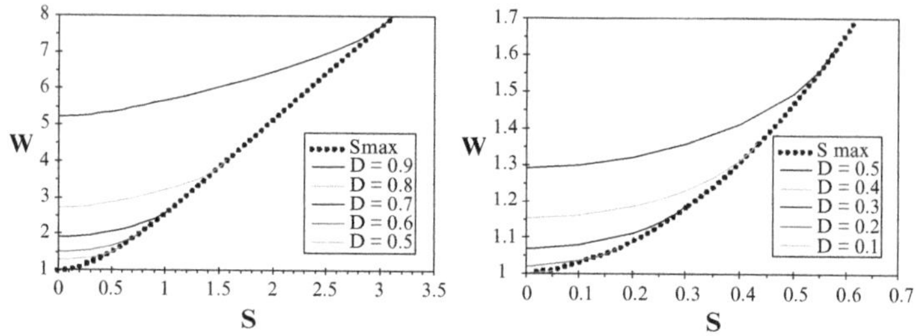


*Figure A3.6.* *W for classic saturating-class-C operation (m = 0), in function of S for a range of duty-cycle values D.*

Equation A3.13. can be ignored as none of the switching modes it describes feature the low power dissipation that results from a back-and-forth energy exchange between the L–R–C tank and the supply. Figure A3.5. right shows that this doesn't occur in this mode; the supply current, which equals the switch current $i_S$, doesn't oscillate around zero but is always positive.

Finally, the normalised switching frequency *W* for saturating-class-C operation is derived from Equation A3.12.:

*Equation A3.14.*

$$W = \frac{m}{1-D} + \frac{\tau - \arcsin(p)}{2\pi(1-D)}$$

where $p = \sin\tau \,.\, \exp[\pi(1-D)S]$

$m$ = integer number

$m = 0$ for classic saturating class C, $m = 1$ for a frequency doubler, higher $m$ values for frequency multipliers with a frequency-multiplication factor equal to $(m + 1)$

This is an iterative expression: the *W* value is required to calculate $\tau$ which is needed to calculate *W*. Its numerical computation is discussed in paragraph A3.5. The results are graphically represented in Figure A3.6. Note the *S*-maximum. Higher *S* values increase the tank damping, thereby reducing the ringing amplitude. Above $S_{max}$, the switch voltage $v_S$ doesn't reach zero and efficient saturating-class-C operation becomes impossible (e.g. $S = 0.3$ in Figure A3.7.).

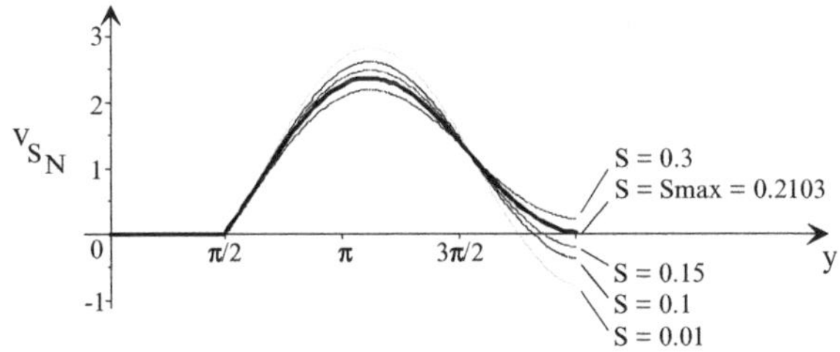


*The influence of an increasing load S on the switch signal (for D = 25% and W = 1.1). An increasing load S reduces the ringing amplitude and retards the appropriate switch-on moment for saturating-class-C operation.* Figure A3.7.

*At a certain $S_{max}$, it becomes impossible to switch on at zero $v_S$ voltage ($S_{max}$ is 0.2103 for this example). The damping in the L–R–C tank is too high to create enough undershoot to reach the y-axis.*

*Switching with an S value equal to $S_{max}$ corresponds to the class-E switching mode.*

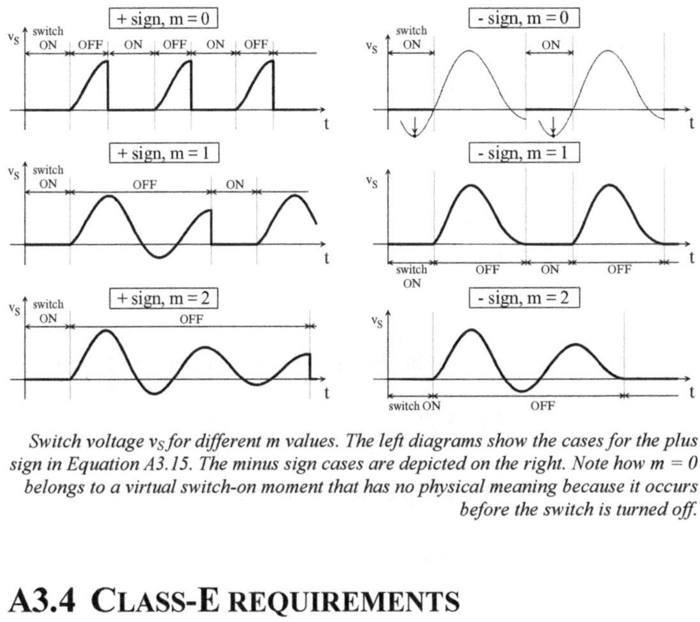


*Switch voltage $v_S$ for different m values. The left diagrams show the cases for the plus sign in Equation A3.15. The minus sign cases are depicted on the right. Note how m = 0 belongs to a virtual switch-on moment that has no physical meaning because it occurs before the switch is turned off.* Figure A3.8.

# A3.4 CLASS-E REQUIREMENTS

The definition of the class-E mode states that the switch voltage has a zero value and slope, at the moment of switch-on. This can be achieved with a saturating-class-C circuit topology, by running at $S = S_{max}$:

$$\begin{cases} v_S(t' = t_{OFF}) = 0 \\ \dfrac{dv_S(t' = t_{OFF})}{dt'} = 0 \end{cases}$$

The latter expression is equivalent to

$$\frac{dv_C(t' = t_{OFF})}{dt'} = 0 \qquad \Leftrightarrow i_C(t' = t_{OFF}) = 0$$

$$\Leftrightarrow i_L(t' = t_{OFF}) = 0$$

$$\Leftrightarrow i_{L_N}\left(t' = t_{OFF}\right) = 0$$

$$\overset{\text{Eqn.A3.9}}{\Leftrightarrow} \cos\left[2\pi W\left(1 - D\right) - \theta\right] = 0$$

*Equation A3.15.*

$$\Leftrightarrow 2\pi W\left(1 - D\right) - \theta = \pm\frac{\pi}{2} + 2\pi.m$$

where $m$ is an integer number (0, 1, 2, 3, ...)

The minus sign in the above expression corresponds to the zero slope at a local minimum in the $v_S$ curve (Figure A3.8.). The plus sign belongs to a local maximum in this curve and should thus be omitted because this switch mode never meets the saturating-class-C condition of zero voltage at switch-on. Hence, to have the circuit running in a class-E mode, $W$ should meet both Equation A3.14. for a zero switch voltage at turnon and Equation A3.16. for a zero slope at turnon.

*Equation A3.16.*

$$W = \frac{3 + 4m}{4\left(1 - D\right)} + \frac{\theta}{2\pi\left(1 - D\right)} = W_{\max}$$

where $m$ is an integer number
$m = 0$ for normal class E operation, $m = 1$ for a frequency doubler,
higher $m$ values for frequency multipliers with
a frequency multiplication factor equal to $(m + 1)$

This is again an iterative expression because the value for $W$ is needed to compute $\theta$. This requirement for operation in class-E mode can also be expressed in terms of the parameter $p$:

$$\text{saturating class C} \quad \Leftrightarrow v_{S_N}\left(y = 2\pi\right) = 0$$

$$\Leftrightarrow \sin\left[2\pi W\left(1 - D\right) - \tau\right] = -p$$

$$\Leftrightarrow \sin\left[2\pi W\left(1 - D\right) - \theta - \arctan\left(\frac{1}{a}\right)\right] = -p$$

$$\text{class E} \quad \overset{\text{Eqn. A3.16}}{\Leftrightarrow} \quad 2\pi W\left(1 - D\right) - \theta = \frac{3\pi}{2} + m.2\pi$$

The substitution of this last expression in the one above yields:

$$\Leftrightarrow \sin\left[\frac{3\pi}{2} + m.2\pi - \arctan\left(\frac{1}{a}\right)\right] = -p$$

$$\Leftrightarrow p = \frac{1}{\sqrt{1 + \tan^2\left[\arctan\left(\frac{1}{a}\right)\right]}} = \frac{a}{\sqrt{a^2 + 1}} = p_{\max}$$

Thus, as an alternative to the class-E requirements stated above, $W$ should meet Equation A3.14. for a zero switch voltage at turnon and $p$ should equal $p_{max}$ according to Equation A3.17. for a zero slope at turnon.

*Equation A3.17.*

$$p = p_{\max} = \frac{a}{\sqrt{a^2 + 1}}$$

This is not an iterative formula and is therefore easier to handle than the iterative Equation A3.16.

## A3.5 THE NUMERICAL *W* COMPUTATION

Non-linear equations can be numerically solved in many different ways. One of the most popular simple methods was first published by Isaac Newton in 1685 and improved in 1690 by Raphson to its current version. The method calculates numerical values for the zeros of a given function $f(x)$ by iterative approximation. A new (and hopefully better) approximation $x_{k+1}$ is computed by evaluating the function $f(x)$ and its derivative in the previous approximation $x_k$:

$$x_{k+1} = x_k - \frac{f(x_k)}{\frac{df(x_k)}{dx}}$$

The successive approximations converge to a zero if there exists a closed interval $[a, b]$ where the function $f(x)$ has only one zero and where the function $f(x)$ has a continuous second derivative. The function should also be either pure concave or pure convex over the entire interval $[a, b]$. Finally, the tangents to the function $f(x)$ in each point of the interval $[a, b]$, should cross the $x$-axis within the interval.

A major drawback of this method is that it requires an explicit expression for the first derivative of the function $f(x)$. This seriously impedes the computation of $W$ because the parameters are heavily interrelated by many different links.

The Whittacker method is a solution to this problem. Whittacker's approximations are computed by:

$$x_{k+1} = x_k - \frac{f(x_k)}{r} \quad \textit{where r is a constant real number}$$

The method converges in the same conditions as Newton-Raphson's, if

$$0 < \frac{df(x)/dx}{r} < 2$$

This method converges substantially slower than the previous one, but the closer $r$ approximates $f(x)$'s first derivative, the better the convergence.

The $W$ for saturating-class-C operation as described by Equation A3.14., is the zero of $\frac{m}{1-D} + \frac{\tau(W) - \arcsin(p(W))}{2\pi(1-D)} - W = 0$

Hence, the iterative expression to find $W$, according to the Whittacker method, becomes:

$$W_{k+1} = W_k + \left[\frac{m}{1-D} + \frac{\tau(W_k) - \arcsin(p(W_k))}{2\pi(1-D)} - W_k\right] \cdot \frac{D}{2}$$

The operation point for class-E mode is found by starting in a saturating-class-C point and iteratively increasing $S$ until $p$ becomes equal to $p_{max}$.

Table A3.1. was computed this way. The same results were obtained with a similar numerical method in (KAZIMIERCZUK, 1983b) where only class-E operation was studied and no saturating class C.

*TABLE A3.1. NUMERICAL SOLUTIONS FOR W FOR SATURATING-CLASS-C (EQUATION A3.14.) AND CLASS-E OPERATION (EQUATION A3.14. AND EQUATION A3.16., SUBTABLE BOTTOM RIGHT).*

| | S | | | | | | | |
|---|---|---|---|---|---|---|---|---|
| **D** | 0.001 | 0.01 | 0.1 | 0.2 | 0.3 | 0.4 | 0.5 | 0.6 |
| 0.05 | 1.00041 | 1.00091 | – | – | – | – | – | – |
| 0.1 | 1.00314 | 1.00339 | – | – | – | – | – | – |
| 0.2 | 1.02274 | 1.02287 | 1.03655 | – | – | – | – | – |
| 0.3 | 1.06960 | 1.06969 | 1.07912 | 1.11072 | – | – | – | – |
| 0.4 | 1.15368 | 1.15375 | 1.16145 | 1.18575 | 1.23072 | 1.31306 | – | – |
| 0.5 | 1.29155 | 1.29162 | 1.29848 | 1.31962 | 1.35619 | 1.41152 | 1.49592 | – |
| 0.6 | 1.51762 | 1.51768 | 1.52415 | 1.54376 | 1.57660 | 1.62322 | 1.68529 | 1.76735 |
| 0.7 | 1.91414 | 1.91421 | 1.92051 | 1.93946 | 1.97053 | 2.01317 | 2.06695 | 2.13187 |
| 0.8 | 2.73032 | 2.73038 | 2.73668 | 2.75546 | 2.78582 | 2.82659 | 2.87645 | 2.93418 |
| 0.9 | 5.21553 | 5.21559 | 5.22198 | 5.24093 | 5.27125 | 5.31133 | 5.35938 | 5.41369 |

| | S | | | | | | | |
|---|---|---|---|---|---|---|---|---|
| **D** | 0.7 | 0.8 | 0.9 | 1 | 1.25 | 1.5 | 1.75 | 2 |
| 0.05 | – | – | – | – | – | – | – | – |
| 0.1 | – | – | – | – | – | – | – | – |
| 0.2 | – | – | – | – | – | – | – | – |
| 0.3 | – | – | – | – | – | – | – | – |
| 0.4 | – | – | – | – | – | – | – | – |
| 0.5 | – | – | – | – | – | – | – | – |
| 0.6 | 1.88481 | – | – | – | – | – | – | – |
| 0.7 | 2.20885 | 2.30062 | 2.41450 | 2.57744 | – | – | – | – |
| 0.8 | 2.99885 | 3.06986 | 3.14705 | 3.23075 | 3.47732 | 3.84527 | – | – |
| 0.9 | 5.47281 | 5.53560 | 5.60122 | 5.66912 | 5.84688 | 6.03515 | 6.23520 | 6.45019 |

| | S | | | |
|---|---|---|---|---|
| **D** | 2.25 | 2.5 | 2.75 | 3 |
| 0.05 | – | – | – | – |
| 0.1 | – | – | – | – |
| 0.2 | – | – | – | – |
| 0.3 | – | – | – | – |
| 0.4 | – | – | – | – |
| 0.5 | – | – | – | – |
| 0.6 | – | – | – | – |
| 0.7 | – | – | – | – |
| 0.8 | – | – | – | – |
| 0.9 | 6.68553 | 6.95063 | 7.26511 | 7.68870 |

| | CLASS E | |
|---|---|---|
| **D** | **Smax** | **Wmax** |
| 0.05 | 0.01008 | 1.000919 |
| 0.1 | 0.05039 | 1.009933 |
| 0.2 | 0.17025 | 1.069447 |
| 0.3 | 0.2995 | 1.185227 |
| 0.4 | 0.4312 | 1.361024 |
| 0.5 | 0.5772 | 1.619615 |
| 0.6 | 0.7614 | 2.015671 |
| 0.7 | 1.0367 | 2.682191 |
| 0.8 | 1.5611 | 4.020074 |
| 0.9 | 3.1155 | 8.006054 |

## A3.6 DRIVER EFFICIENCY $\eta_{driver}$:

The ***driver efficiency*** is defined as ***the ratio between the average power dissipated in the resistive load R and the average power delivered by the supply source $V_{cc}$***. It can be computed looking at the energy balance during one on–off cycle (Figure A3.9.). The first-order simplification means that the loss terms are added one by one to the lossless circuit ($R_{ON} = t_F = V_{S_{sat}} = 0$), and that their influence on the circuit signals $v_S$, $i_S$ and $i_L$ is neglected.

$$\eta_{driver} = \frac{\Delta E_R}{V_{cc}\int_0^{t_{ON}} i_S\,dt + R_{ON}\int_0^{t_{ON}} i_S^2\,dt + V_{S_{sat}}\int_0^{t_{ON}} |i_S|\,dt + \int_0^{t_F} i'_S v'_S\,dt' + \int_0^{t_R} i''_S v''_S\,dt''}$$

The signals $i'_S$ and $v'_S$ are the switch current and voltage during the current rise time $t_R$, $i''_S$ and $v''_S$ correspond to current fall time $t_F$. Note the t'- and t''-axes in Figure A3.9 to simplify the computation of the last two integrals. Dividing each energy term by the period $T$, results in an expression with power terms instead of energy terms:

$$\eta_{driver} = \frac{P_R}{P_{V_{cc}} + P_{R_{ON}} + P_{V_{Ssat}} + P_{t_F} + P_{t_R}}$$

$$\Leftrightarrow \eta_{driver} = \frac{p_R}{1 + p_{R_{ON}} + p_{V_{Ssat}} + p_{t_F} + p_{t_R}}$$

*Equation A3.18.*

$$\text{where} \begin{cases} p_R \equiv \dfrac{P_R}{P_{V_{cc}}} = \dfrac{\Delta E_R}{V_{cc}\int_0^{t_{ON}} i_S\,dt} \\[2ex] p_{R_{ON}} \equiv \dfrac{P_{R_{ON}}}{P_{V_{cc}}} = \dfrac{R_{ON}\int_0^{t_{ON}} i_S^2\,dt}{V_{cc}\int_0^{t_{ON}} i_S\,dt} \\[2ex] p_{V_{Ssat}} \equiv \dfrac{P_{V_{Ssat}}}{P_{V_{cc}}} = \dfrac{V_{Ssat}\int_0^{t_{ON}} |i_S|\,dt}{V_{cc}\int_0^{t_{ON}} i_S\,dt} \\[2ex] p_{t_R} \equiv \dfrac{P_{t_R}}{P_{V_{cc}}} = \dfrac{\int_0^{t_F} i'_S v'_S\,dt'}{V_{cc}\int_0^{t_{ON}} i_S\,dt} \\[2ex] p_{t_F} \equiv \dfrac{P_{t_F}}{P_{V_{cc}}} = \dfrac{\int_0^{t_R} i''_S v''_S\,dt''}{V_{cc}\int_0^{t_{ON}} i_S\,dt} \end{cases}$$

The useful part of this power dissipation goes into the load $R$:

$$P_R = \frac{R}{T}\left[\int_0^{t_{ON}} i_{L_{ON}}^2(t)\,dt + \int_0^{t_{OFF}} i_{L_{OFF}}^2(t')\,dt'\right]$$

This expression is, however, far too complex to calculate. Therefore, $P_R$ is calculated via the energy balance of the first order simplified circuit taken from $t = 0$ until $t = t_{ON} + t_{OFF}$ (just before the next switch-on). The power $P_{V_{cc}}$, supplied by $V_{cc}$ is spread over two contributions:

- The dissipation in the load $R$.
- The modified amount of energy stored in the capacitor $C$ (in case of improperly tuned switching). Note that the amount of energy in the tank coil $L$ is not changed because the circuit is in the steady-state.

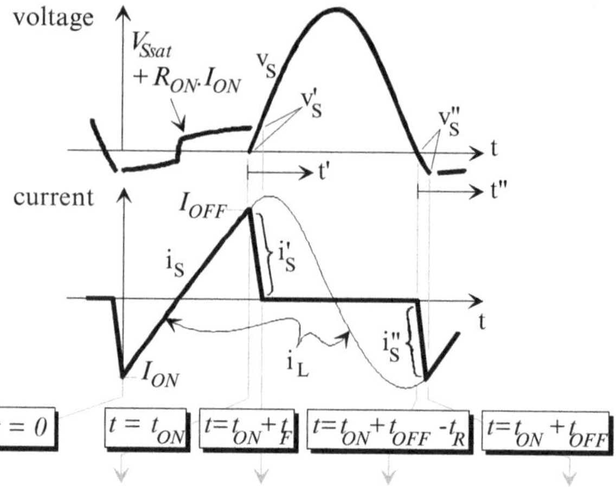


| | $t=0$ … $t=t_{ON}$ | $t=t_{ON}$ … $t=t_{ON}+t_F$ | $t=t_{ON}+t_F$ … $t=t_{ON}+t_{OFF}-t_R$ | $t=t_{ON}+t_{OFF}-t_R$ … $t=t_{ON}+t_{OFF}$ |
|---|---|---|---|---|
| $\Delta E_{V_{cc}}$ | $-V_{cc}\int_0^{t_{ON}} i_S\,dt$<br>$-R_{ON}\int_0^{t_{ON}} i_S^2\,dt$<br>$-V_{S_{sat}}\int_0^{t_{ON}} i_S\,dt$ | $-\int_0^{t_F} i'_S v'_S\,dt'$ | | $-\int_0^{t_R} i''_S v''_S\,dt''$ |
| $\Delta E_{switch}$ | $R_{ON}\int_0^{t_{ON}} i_S^2\,dt$<br>$+V_{S_{sat}}\int_0^{t_{ON}} i_S\,dt$ | $\int_0^{t_F} i'_S v'_S\,dt'$ | | $\int_0^{t_R} i''_S v''_S\,dt''$ |
| $\Delta E_R$ | $R\int_0^{t_{ON}} i_L^2\,dt$ | $R\int_{t_{ON}}^{t_{ON}+t_{OFF}} i_L^2\,dt$ | | |
| $\Delta E_L$ | $\frac{L}{2}\left(I_{OFF}^2 - I_{ON}^2\right)$ | $-\frac{L}{2}\left(I_{OFF}^2 - I_{ON}^2\right)$ | | |

*Figure A3.9.* *Energy balance during one cycle, assuming perfect switch timing. A plus sign denotes a net power consumption, a minus sign means that energy was delivered. The shaded terms correspond to the energy balance of the first-order simplified circuit ($R_{ON} = t_F = V_{S_{sat}} = 0$). The individual loss factors are considered here one by one and corresponding traces are therefore not connected on the above drawing.*

Hence,[2]

$$P_R = P_{V_{cc}} - \frac{\Delta E_C}{T} = P_{V_{cc}}$$

$$\Rightarrow P_R = P_{V_{cc}} = \frac{V_{cc}^2}{R}\left(D - \frac{I_{OFF_N} - I_{ON_N}}{2\pi}\right) \text{ and } p_R = 1$$

*Equation A3.19.*

The relative power terms in Equation A3.18. compare thus the amplifier losses to the amplifier useful output power.

$P_{R_{ON}}$ is the power dissipated in $R_{ON}$, during switch-on. The ratio $p_{R_{ON}}$ is found as follows[3]:

$$p_{R_{ON}} \equiv \frac{P_{R_{ON}}}{P_{V_{cc}}} = \frac{R_{ON}\int_0^{t_{ON}} i_S^2\,dt}{V_{cc}\int_0^{t_{ON}} i_S\,dt} = \frac{R_{ON}}{\omega L}\frac{\int_0^{2\pi D} i_{S_N}^2\,dy}{\int_0^{2\pi D} i_{S_N}\,dy}$$

$$= \frac{R_{ON}}{\omega L}\left[\frac{1}{S} - \frac{I_{OFF_N}^2 - I_{ON_N}^2}{4\pi D - 2\left(I_{OFF_N} - I_{ON_N}\right)}\right]$$

---

2 Improperly tuned on-switching (i.e. at a switch voltage different from zero), is not accepted here because of the abrupt charge of the tank capacitor C via the transistor (or discharge, depending on the switch-voltage polarity at switch-on). Even when the average dissipation is low with small mistunings, the high transient peaks involved with the abrupt capacitor charge or discharge may be harmful for the switch transistor.

3 The averages of the normalised switch current, the squared normalised switch current and the absolute value of the switch current, are useful expressions for these calculations:

$$\overline{i_{S_N}} = \frac{1}{2\pi}\int_0^{2\pi D} i_{S_N}(y)dy = \frac{1}{S}\left(D - \frac{I_{OFF_N} - I_{ON_N}}{2\pi}\right)$$

$$\overline{i_{S_N}^2} = \frac{1}{2\pi}\int_0^{2\pi D} i_{S_N}^2(y)\,dy = \frac{1}{S^2}\left[D - \frac{I_{OFF_N} - I_{ON_N}}{2\pi}\left(1 + \frac{S}{2}\left(I_{OFF_N} + I_{ON_N}\right)\right)\right]$$

$$\overline{|i_{S_N}|} = \frac{1}{2\pi}\int_0^{2\pi D} |i_{S_N}(y)|\,dy = \frac{1}{2\pi}\left[-\int_0^{y_0} i_{S_N}(y)dy + \int_{y_0}^{2\pi D} i_{S_N}(y)dy\right] \text{ where } y_0 = \frac{\ln\left(1 - SI_{ON_N}\right)}{S}$$

$$= \frac{1}{S}\left[-\frac{\ln\left(1 - I_{ON_N}S\right)}{\pi S} + D - \frac{I_{OFF_N} + I_{ON_N}}{2\pi}\right]$$

$$\Rightarrow \begin{cases} \dfrac{\overline{i_{S_N}^2}}{\overline{i_{S_N}}} = \dfrac{\int_0^{2\pi D} i_{S_N}^2(y)dy}{\int_0^{2\pi D} i_{S_N}(y)dy} = \dfrac{1}{S} - \dfrac{I_{OFF_N}^2 - I_{ON_N}^2}{4\pi D - 2\left(I_{OFF_N} - I_{ON_N}\right)} \\[2ex] \dfrac{\overline{|i_{S_N}|}}{\overline{i_{S_N}}} = \dfrac{\int_0^{2\pi D} |i_{S_N}(y)|\,dy}{\int_0^{2\pi D} i_{S_N}(y)dy} = 1 - \dfrac{2}{S}\dfrac{SI_{ON_N} + \ln\left(1 - SI_{ON_N}\right)}{2\pi D - \left(I_{OFF_N} - I_{ON_N}\right)} \end{cases}$$

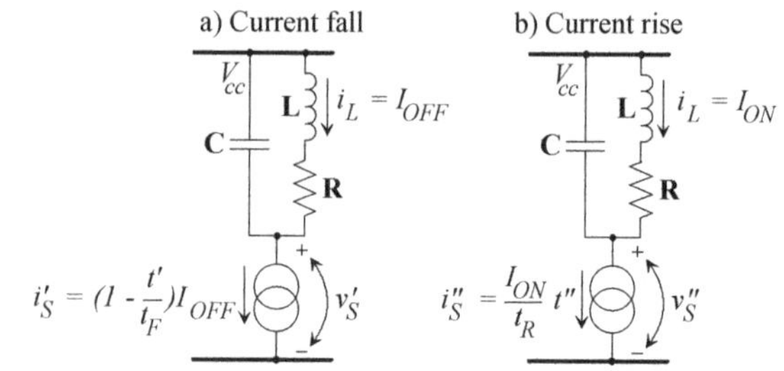


*Figure A3.10. Approximate model of a saturating-class-C during the switch fall (a) and rise (b) time.*

*Equation A3.20.*

$$p_{R_{ON}} = \frac{R_{ON}}{R} - \frac{R_{ON}}{2\omega L}\left(\frac{I^2_{OFF_N} - I^2_{ON_N}}{2\pi D - I_{OFF_N} + I_{ON_N}}\right)$$

$P_{V_{S_{sat}}}$ is the power dissipated in $V_{S_{sat}}$, during switch-on[3]:

$$p_{V_{S_{sat}}} \equiv \frac{P_{V_{S_{sat}}}}{P_{V_{cc}}} = \frac{V_{S_{sat}} \int_0^{t_{ON}} |i_S|\, dt}{V_{cc} \int_0^{t_{ON}} i_S\, dt} = \frac{V_{S_{sat}}}{V_{cc}} \frac{\int_0^{2\pi D} |i_{S_N}|\, dy}{\int_0^{2\pi D} i_{S_N}\, dy}$$

*Equation A3.21.*

$$p_{V_{S_{sat}}} = \frac{V_{S_{sat}}}{V_{cc}}\left[1 - \frac{2}{S}\frac{S I_{ON_N} + \ln\left(1 - S I_{ON_N}\right)}{2\pi D - I_{OFF_N} + I_{ON_N}}\right]$$

The power dissipation during the current fall time is estimated by assuming a linear current decrease from the value $I_{OFF}$ to zero. The coil current $i_L$ is further assumed to be constant (= $I_{OFF}$) in this short interval. The linear switch current and the constant coil current are charging the tank capacitor to produce a parabolic switch voltage (Figure A3.10a):

$$\begin{cases} i'_S(t') = I_{OFF}\left(1 - \dfrac{t'}{t_F}\right) \\ v'_S(t') = V_{cc} - v_C(t') = V_{cc} - \left(\displaystyle\int_0^{t'} \frac{i'_S(\vartheta) - i_L}{C} d\vartheta + v_C(t' = 0)\right) = \dfrac{I_{OFF}}{C}\dfrac{t'^2}{2t_F} \end{cases}$$

$$P_{t_F} = \frac{1}{T}\int_0^{t_F} v'_S i'_S\, dt' = \frac{I^2_{OFF}}{2TCt_F}\int_0^{t_F}\left(t'^2 - \frac{t'^3}{t_F}\right) dt' = \frac{I^2_{OFF} t_F^2}{24TC} = \frac{V_{cc}^2 I^2_{OFF_N} t_F^2}{48\pi\omega CL^2}$$

*Equation A3.22.*

$$p_{t_F} = \frac{I^2_{OFF_N} t_F^2 R}{24\omega CL^2\left(2\pi D - I_{OFF_N} + I_{ON_N}\right)}$$

RAAB and SOKAL (1978) used the same method to estimate the influence of current fall times on the class-E efficiency. This approximation is, however, only valid for current fall times much smaller than the period. A more precise calculation of the impact of a substantial current fall time on class-E efficiency was carried out by KAZIMIERCZUK (1983b). His method readily modifies to saturated class C mode and current rise times.

The power dissipation during the current rise time is approximated in the very same way: the switch current is built-up linearly and the coil current is constant (= $I_{ON}$) (Figure A3.10b).

$$\begin{cases} i''_S(t'') = I_{ON}\dfrac{t''}{t_R} \\ v''_S(t'') = V_{cc} - v_C(t'') = V_{cc} - \left(\displaystyle\int_0^{t''}\frac{i''_S(\vartheta) - i_L}{C}d\vartheta + v_C(t''=0)\right) \\ \quad = \dfrac{I_{ON}}{2C}\left(\dfrac{t''^2}{t_R} - 2t'' + t_R\right) \end{cases}$$

$$P_{t_R} = \frac{1}{T}\int_0^{t_R} v''_S i''_S\, dt'' = \frac{I_{ON}^2}{2TCt_R}\int_0^{t_R}\left(t''t_R - 2t''^2 + \frac{t'^3}{t_R}\right)dt'' = \frac{I_{ON}^2 t_R^2}{24TC}$$
$$= \frac{V_{cc}^2 I_{ON_N}^2 t_R^2}{48\pi\omega CL^2}$$

$$p_{t_R} = \frac{I_{ON_N}^2 t_R^2 R}{24\omega CL^2\left(2\pi D - I_{OFF_N} + I_{ON_N}\right)} \qquad \text{Equation A3.23.}$$

The saturating-class-C circuit is fed by a power supply $V_{cc}$ with low output impedance. The combination of supply value and the duty cycle directly determine the circuit voltages. The circuit currents are set by the coil voltage and the impedance of the loaded coil and not by the switch parasitics. The $I_{cc}$ value found for in the first-order simplified circuit does not take the driver losses into account. The "real life" $I_{cc}^*$ is thus somewhat larger than $I_{cc}$ to compensate for the losses that were not included in the first-order model:

$$V_{cc}I_{cc}^* \equiv V_{cc}\int_0^{t_{ON}} i_S\, dt = T\left(P_R + P_{R_{ON}} + P_{V_{S_{sat}}} + P_{t_F} + P_{t_R}\right)$$

$$\Leftrightarrow I_{cc}^* = I_{cc}\left(1 + p_{R_{ON}} + p_{V_{S_{sat}}} + p_{t_F} + p_{t_R}\right) = \frac{I_{cc}}{\eta_{driver}} \qquad \text{Equation A3.24.}$$

# A3.7 HIGH $Q_{tank}$ EXPRESSIONS

Primary coil drivers typically feature high tank quality factors, especially at low coil coupling. Assuming an infinite tank Q significantly simplifies the signal expressions.

It makes, however, no sense to use high-Q assumptions for the class-E operation mode. It is just because of the tank damping that the switch voltage can reach zero with a zero slope.

## A3.7.1 SWITCH ON

The first-order switch-on situation is described by the net equation:

$$V_{cc} = L.\frac{di_{L_{ON}}(t)}{dt}$$

This results in a zero switch voltage $v_{S_{ON}}(t)$ and a linearly rising coil current $i_{L_{ON}}(t)$:

*Equation A3.25.*

*Equation A3.2.*

$$\begin{cases} i_{L_{ON}}(t) = i_S(t) = I_{ON} + \dfrac{V_{cc}}{L}t \\ v_{S_{ON}}(t) = 0 \end{cases}$$

where $I_{ON}$ is the coil current at switch on.

## A3.7.2 SWITCH OFF

### A3.7.2.1 The coil current $i_{L_{OFF}}$

The net equation now becomes:

$$v_C(t') = L.\frac{di_{L_{OFF}}(t')}{dt'}$$

The capacitor voltage is expressed in terms of the coil current $i_{L_{OFF}}(t')$:

$$\Leftrightarrow -\frac{1}{C}\int_0^{t'} i_{L_{OFF}}(t)dt + \frac{q(t'=0)}{C} = L.\frac{di_{L_{OFF}}(t')}{dt'}$$

$$\Leftrightarrow L.\frac{d^2 i_{L\,OFF}(t')}{dt'^2} + \frac{i_{L_{OFF}}(t')}{C} = 0$$

The capacitor $C$ inhibits sudden voltage jumps during the switch turn off:

$$v_{C_{ON}}(t = t_{ON}) = v_{C_{OFF}}(t' = 0)$$

$$\Leftrightarrow V_{cc} = L.\frac{di_{L_{OFF}}(t'=0)}{dt'}$$

The coil $L$ makes current $i_L$ impossible. Hence,

$$i_{L_{OFF}}(t' = 0) = i_{L_{ON}}(t = t_{ON}) = I_{OFF}$$

The solution of the differential equation for $i_{L_{OFF}}(t')$ with the boundary conditions $di_{L\,OFF}(t'=0)/dt'$ and $i_{L_{OFF}}(t'=0)$, is a non-damped oscillation.

*Equation A3.26.*

$$i_{L_{OFF}}(t') = I_{OFF}.\frac{\cos(\omega_{tank}t' - \theta)}{\cos\theta}$$

where $I_{OFF}$ is the coil current when turning off the switch

and
$$\begin{cases} \omega_{tank} \equiv \sqrt{\dfrac{1}{LC}} \\ x = \dfrac{I_{OFF}}{V_{cc}\,\omega_{tank}\,C} \\ \theta = \arctan\left(\dfrac{1}{x}\right) \\ I_{OFF} = i_{L_{ON}}(t = t_{ON}) = I_{ON} + \dfrac{V_{cc}}{L}t_{ON} \end{cases}$$

#### A3.7.2.2 The switch voltage $v_{S_{OFF}}$

The expression for $v_{S_{OFF}}$ is calculated from

$$v_{S_{OFF}}(t') = V_{cc} - v_C(t') = Vcc - \int_0^{t'} i_{L_{OFF}}(t)dt + \text{constant, with } v_{S_{OFF}}(0) = 0$$

which is equivalent to $v_{S_{OFF}}(t') = V_{cc} - v_C(t') = Vcc - L\dfrac{dI_{L_{OFF}}(t')}{dt'}$.

$$\boxed{v_{S_{OFF}}(t') = V_{cc}.\left[1 + \frac{\sin(\omega_{tank}t' - \theta)}{\sin\theta}\right]}$$ *Equation A3.27.*

### A3.7.3 STEADY-STATE CIRCUIT BEHAVIOUR

Steady-state circuit behaviour requires that the coil current $i_L$ at the end of an on–off cycle is equal to the initial coil current $I_{ON}$:

$$i_{L_{OFF}}(t' = t_{OFF}) = I_{ON}$$ *Equation A3.28.*

It's not required to expand this condition to explicit expressions for $I_{ON}$ and $I_{OFF}$, like in paragraph A3.1.3.[4] It is simpler to use Equation A3.28. as such, when deriving the conditions for saturating-class-C behaviour.

### A3.7.4 SATURATING-CLASS-C OPERATION

The saturating-class-C mode states that the switch is turned on, when the voltage $v_S$ across its terminals has returned to zero:

$$v_{S_{OFF}}(t' = t_{OFF}) = 0$$

$$\overset{\text{Eqn.A3.20}}{\Leftrightarrow} \quad \sin(\omega_{tank}t_{OFF} - \theta) = -\sin\theta$$

$$\Leftrightarrow 2\sin\frac{\omega_{tank}t_{OFF}}{2}\cos\frac{\omega_{tank}t_{OFF} - 2\theta}{2} = 0$$

$$\Leftrightarrow \frac{\omega_{tank}t_{OFF} - 2\theta}{2} = n.\pi + \frac{\pi}{2}$$

$$\Leftrightarrow \omega_{tank}t_{OFF} = n.2\pi + \pi + 2\theta$$ *Equation A3.29.*

---

[4] The explicit expressions for $I_{ON}$ and $I_{OFF}$ are found as follows.

$$\text{Equation A3.21} \overset{\text{Eqn.A3.19}}{\Leftrightarrow} \begin{cases} I_{OFF}\left[\cos(\omega_{tank}t_{OFF}) + \dfrac{1}{x}\sin(\omega_{tank}t_{OFF})\right] = I_{ON} \\ \text{where } x = \dfrac{I_{OFF}}{V_{cc}\omega_{tank}C} \end{cases}$$

This equation and the expression for $I_{OFF} = i_{L_{ON}}(t = t_{ON})$ from Equation A3.26., form a system of three equations in the three unknowns $I_{ON}$, $I_{OFF}$ and $x$. Substitute $x$:

$$\Leftrightarrow \begin{cases} I_{OFF}\cos(\omega_{tank}t_{OFF}) + V_{cc}\omega_{tank}C\sin(\omega_{tank}t_{OFF}) = I_{ON} \\ I_{OFF} = I_{ON} + \dfrac{V_{cc}}{L}t_{ON} \end{cases}$$

$$\Leftrightarrow \begin{cases} I_{ON} = V_{cc}\sqrt{\dfrac{C}{L}}\dfrac{\omega_{tank}t_{ON}\cos(\omega_{tank}t_{OFF}) + \sin(\omega_{tank}t_{OFF})}{1 - \cos(\omega_{tank}t_{OFF})} \\ I_{OFF} = V_{cc}\sqrt{\dfrac{C}{L}}\dfrac{\omega_{tank}t_{ON} + \sin(\omega_{tank}t_{OFF})}{1 - \cos(\omega_{tank}t_{OFF})} \end{cases}$$

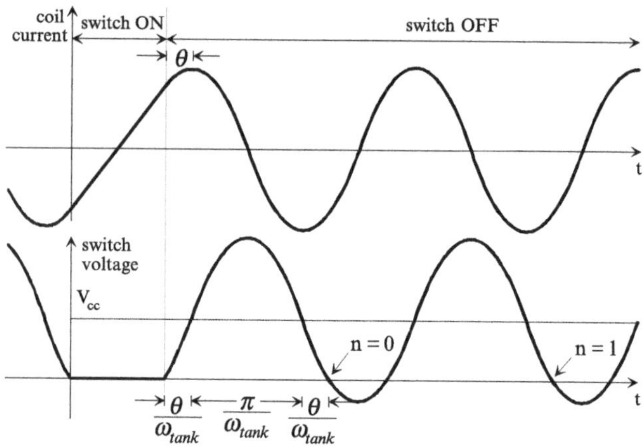


*Figure A3.11. Saturating-class-C operation with the infinite $Q_{tank}$ assumption.*

The mode with $n = 0$ is the classic saturating class C, the other modes are frequency multipliers with a multiplication factor $n$ (Figure A3.11).

Substitute this value for $\omega_{tank}\, t_{OFF}$ and Equation A3.28. for steady-state behaviour, into Equation A3.26. to proof that $I_{ON} = -I_{OFF}$:

$$I_{OFF} = i_{L_{OFF}}\left(t' = t_{OFF}\right) = I_{OFF}\frac{\cos\left(\omega_{tank}t_{ON} - \theta\right)}{\cos\theta}$$
$$= I_{ON}\frac{\cos\left(n.2\pi + \pi + \theta\right)}{\cos\theta} = -I_{ON}$$

Combine this with Equation A3.25. for $t = t_{ON}$ to find $I_{ON}$:

Equation A3.30.
$$\begin{cases} I_{ON} = -\dfrac{V_{cc}}{2L}t_{ON} \\ I_{OFF} = -I_{ON} \end{cases}$$

We still need to find what combination of operation frequency $\omega$ and duty cycle $D$ meets Equation A3.29. The same definitions for the duty cycle $D$ and the normalised frequency $W$ of paragraph A3.2 are used.

$$\text{Equation A3.29.} \Leftrightarrow W2\pi(1-D) = n.2\pi + \pi + 2\arctan\left(\frac{1}{x}\right)$$

$$\text{where } x = \frac{I_{OFF}}{V_{cc}\,\omega_{tank}\,C}$$

Substitute $I_{OFF}$ from Equation A3.30. to find:

Equation A3.31.

Equation A3.32.
$$\Leftrightarrow \begin{cases} x = \pi D W \\ W = \dfrac{1}{1-D}\left[n + \dfrac{1}{2} + \dfrac{1}{\pi}\arctan\left(\dfrac{1}{\pi D W}\right)\right] \end{cases}$$

where $n = 0$ corresponds to the classic saturating class C, and higher $n$-values describe frequency multipliers with a multiplication factor $(n+1)$.

*NUMERICAL SOLUTIONS FOR W FOR SATURATING-CLASS-C OPERATION, ASSUMING AN INFINITE TANK QUALITY FACTOR $Q_{tank}$.* *TABLE A3.2.*

| | W | | |
|---|---|---|---|
| **D** | **n = 0** | **n = 1** | **n = 2** |
| 0 | 1 | 2 | 3 |
| 0.05 | 1.000406 | 2.003122 | 3.009915 |
| 0.1 | 1.003136 | 2.022063 | 3.062390 |
| 0.2 | 1.022742 | 2.130302 | 3.303631 |
| 0.3 | 1.069600 | 2.336680 | 3.698471 |
| 0.4 | 1.153675 | 2.654516 | 4.264537 |
| 0.5 | 1.291547 | 3.127828 | 5.079376 |
| 0.6 | 1.517614 | 3.858725 | 6.316678 |
| 0.7 | 1.914141 | 5.094457 | 8.390779 |
| 0.8 | 2.730314 | 7.583429 | 12.55044 |
| 0.9 | 5.215524 | 15.07467 | 25.04495 |

The last expression is a non-linear equation in $W$ without explicit solution. Numerical solutions are found by iteration (Table A3.2).

## A3.7.5 DRIVER OUTPUT POWER AND EFFICIENCY

The driver efficiency is calculated like in paragraph A3.6 by integration of the expressions for the circuit currents. However, the expression for $P_{V_{cc}}$ and the idea to compute the driver output power $P_R$ via the supply power, are meaningless with the infinite-$Q_{tank}$ circuit model. The infinite-$Q_{tank}$ assumption sets the circuit dissipation and $P_{V_{cc}}$ to a theoretical zero. $P_R$ is to be calculated from direct integration of $i_L$. Hence,

$$P_R = \frac{R}{T}\left[\int_0^{t_{ON}} i_{L_{ON}}^2(t)dt + \int_0^{t_{OFF}} i_{L_{OFF}}^2(t')dt'\right] = \frac{RV_{cc}^2}{6\omega^2 L^2}\left[\pi^2 D^2(3-D) + \frac{3}{W^2}\right]$$ *Equation A3.33.*

The other terms in the expression of the driver efficiency are easily found by integrating the circuit currents.

$$\eta_{driver} = \frac{p_R}{1 + p_{R_{ON}} + p_{V_{S_{sat}}} + p_{t_F} + p_{t_R}}$$ *Equation A3.34.*

where
$$\begin{cases} p_R = 1 \\ p_{R_{ON}} = A\dfrac{R_{ON}}{R} 2\pi D \\ p_{V_{S_{sat}}} = A\dfrac{V_{S_{sat}}}{V_{cc}}\dfrac{3\omega L}{R} \\ p_{t_F} = A\dfrac{t_F^2\omega}{8CR} \\ p_{t_R} = A\dfrac{t_R^2\omega}{8CR} \end{cases}$$
and
$$A \equiv \frac{\pi D^2 W^2}{3 + (3-D)\pi^2 D^2 W^2}$$

The "real" supply current $I_{cc}^*$ is again larger than $I_{cc}$ to compensate for the driver losses:

*Equation A3.24.*
$$I_{cc}^* = \frac{I_{cc}}{\eta_{driver}} = \frac{P_R}{V_{cc}\eta_{driver}}$$

The ***driver output power capability p*** is defined as ***the ratio between the output power $P_R$ and the product of the maximum switch voltage with the maximum switch current***. This parameter allows comparing the output capabilities of different driver types for a transistor with given maximum voltage and current ratings.

*Equation A3.35.*
$$p \equiv \frac{P_R}{v_{S_{max}} i_{S_{max}}} = \frac{P_R}{V_{cc}\left(1+\dfrac{1}{\sin\theta}\right) I_{OFF}} = \frac{W^2\pi^2 D^2 (3-D)+3}{6Q_{tank} W^2 \pi D\left(1+\sqrt{1+W^2\pi^2 D^2}\right)}$$

## A3.7.6 First harmonic component

The most important signal component for a saturating-class-C amplifier in an inductive link is the first harmonic of the voltage across the coil and the resistor. This voltage equals $V_{cc} - v_S$:

$$\begin{cases} v_{L_{ON}}(t) = V_{cc} \\ v_{L_{OFF}}(t') = -V_{cc}\dfrac{\sin(\omega_{tank}t' - \theta)}{\sin\theta} \end{cases}$$

The Fourier development of this voltage is given by:

$$v_L(t') = V_{L_{d.c.}} + \hat{V}_{L_\omega}\cos(\omega t' - \psi_\omega) + \hat{V}_{L_{2\omega}}\cos(2\omega t' - \psi_{2\omega}) + \ldots$$

where $\hat{V}_{L_\omega} = \sqrt{a_\omega^2 + b_\omega^2}$ and

$$\begin{cases} a_\omega = \dfrac{2}{T}\displaystyle\int_0^T v_L(t')\cos(t')dt' = \dfrac{\omega}{\pi}\left[\int_0^{t_{OFF}} v_{L_{OFF}}(t')\cos(t')dt' + \int_{t_{OFF}}^T v_{L_{ON}}(t')\cos(t')dt'\right] \\ b_\omega = \dfrac{2}{T}\displaystyle\int_0^T v_L(t')\sin(t')dt' = \dfrac{\omega}{\pi}\left[\int_0^{t_{OFF}} v_{L_{OFF}}(t')\sin(t')dt' + \int_{t_{OFF}}^T v_{L_{ON}}(t')\sin(t')dt'\right] \end{cases}$$

The voltage $v_L$ is expressed on a t'-axis that starts with an off-cycle, to obtain the simplest function description for $v_L$.

$$\begin{cases} a_\omega = \dfrac{2}{T}\displaystyle\int_0^T v_L(t')\cos(t')dt' = \dfrac{\omega}{\pi}\left[\int_0^{t_{OFF}} v_{L_{OFF}}(t')\cos(t')dt' + \int_{t_{OFF}}^T v_{L_{ON}}(t')\cos(t')dt'\right] \\ b_\omega = \dfrac{2}{T}\displaystyle\int_0^T v_L(t')\sin(t')dt' = \dfrac{\omega}{\pi}\left[\int_0^{t_{OFF}} v_{L_{OFF}}(t')\sin(t')dt' + \int_{t_{OFF}}^T v_{L_{ON}}(t')\sin(t')dt'\right] \end{cases}$$

The expansion of these integrals with the substitution of the condition for $t_{OFF}$ to operate in saturating class C, $t_{OFF} = (n.2\pi + \pi + 2\theta)/\omega_{tank}$ (from Equation A3.29.), produces:

$$\Leftrightarrow \begin{cases} a_\omega = \dfrac{-V_{cc}\left(\omega_{tank}\omega\cos A\cos\theta + \omega_{tank}\omega\cos\theta + \omega_{tank}^2\sin A\sin\theta\right)}{\pi\sin\theta\left(\omega_{tank}^2 - \omega^2\right)} \\ b_\omega = \dfrac{-V_{cc}\left(\omega_{tank}\omega\sin A\cos\theta + \omega_{tank}^2\sin\theta - \omega_{tank}^2\cos A\sin\theta\right)}{\pi\sin\theta\left(\omega_{tank}^2 - \omega^2\right)} \end{cases}$$

$$\text{where } A = \frac{\omega}{\omega_{tank}}(\pi + n.2\pi + \theta)$$

$W$ and $\theta$ were previously defined as $W = \dfrac{\omega_{tank}}{\omega}$ and $\tan(\theta) = \dfrac{1}{x}$:

$$\Leftrightarrow \begin{cases} a_\omega = \dfrac{-V_{cc}}{\pi\left(W^2-1\right)}\left[Wx\cos\left(\dfrac{\pi+n.2\pi+2\theta}{W}\right) + Wx + W^2\sin\left(\dfrac{\pi+n.2\pi+2\theta}{W}\right)\right] \\ b_\omega = \dfrac{-V_{cc}}{\pi\left(W^2-1\right)}\left[Wx\sin\left(\dfrac{\pi+n.2\pi+2\theta}{W}\right) + W^2 - W^2\cos\left(\dfrac{\pi+n.2\pi+2\theta}{W}\right)\right] \end{cases}$$

The operation in the saturating-class-C mode allows to replace $W$ by $\dfrac{n.2\pi+\pi+2\theta}{2\pi(1-D)}$ (Equation A3.32.) and $x$ by $\pi DW$ (Equation A3.32.):

$$\hat{V}_{L_\omega} = \frac{2V_{cc}W^2\left|\pi D\cos(\pi D) - \sin(\pi D)\right|}{\pi\left(W^2-1\right)} \qquad \textit{Equation A3.36.}$$

# A3.8 REFERENCES


***Kazimierczuk M.K.**, "Effects of the collector current fall time on the class E tuned power amplifier," *IEEE J. Solid-State Circuits*, vol. SC-18, pp. 181–193, 1983a.

***Kazimierczuk M.K.**, "Exact analysis of class E tuned power amplifier with only one inductor and one capacitor in load network," *IEEE J. Solid-State Circuits*, vol. SC-18, pp. 214–221, 1983b.

***Raab F.H.** and N.O. Sokal, "Transistor power losses in the class E tuned power amplifier," *IEEE J. Solid-State Circuits*, vol. SC-13, pp. 912–914, 1978.

***Sokal N.O.**, "Class E high-efficiency switching-mode tuned power amplifier with only one inductor and one capacitor in load network – approximate analysis," *IEEE J. Solid-State Circuits*, vol. SC-16, pp. 380–384, 1981.

**Van Paemel M.**, *Ontwerp van een prothese voor elektrische gehoorstimulatie (Design of a prosthesis for electrical auditory stimulation)*, Ph.D. dissertation (in Dutch), Katholieke Universiteit Leuven, Leuven, Belgium, 1990.

# INDEX

CPSIA information can be obtained at www.ICGtesting.com
Printed in the USA
LVOW01*2344010915

452394LV00009B/155/P

9 789048 124114